# On
# Scientific
# Thinking

# On Scientific Thinking

edited by

RYAN D. TWENEY

MICHAEL E. DOHERTY

CLIFFORD R. MYNATT

**Columbia University Press**
**New York**
**1981**

Columbia University Press
New York          Guildford, Surrey

Copyright © 1981 by Columbia University Press
All Rights Reserved
Printed in the United States of America
Printed on permanent and durable acid-free paper.


Library of Congress Cataloging in Publication Data
Main entry under title:

On scientific thinking.

    Bibliography:
    Includes index.
    1. Science—Philosophy.   2. Science—Methodology.
I. Tweney, Ryan D., 1943–   II. Doherty, Michael E., 1935–
III. Mynatt, Clifford R., 1945– [DNLM:   1. Science.
2. Philosophy. Q 175 058]
Q175.0478        501        80-28373
ISBN 0-231-04814-9
ISBN 0-231-04815-7 (pbk.)

We take great pleasure
in dedicating this book to our wives,
Kathy, Dot, and Barbee

# Contents

# Contents

# Preface

THE PURPOSES OF this book have been set out in the introduction: a few words here will indicate how the book came to be, and express our thanks to those whose efforts have influenced it. Our work in the psychology of science has been a collaboration in the truest sense. Each of us brought to the program a long-standing curiosity about science, and some prior research of at least tangential relevance. However, we did not actively begin to investigate science as a problem for psychology until after extensive discussions, first between two of us (Michael Doherty and Clifford Mynatt), and then among the three of us. These discussions, in 1975 and 1976, led to our first research efforts, and to new problems and new research.

When the idea of doing a book first arose, we certainly had no sense of having completed our inquiries, nor even of being able to see the end of our efforts. Nonetheless, it was clear that we had gained, in addition to certain fascinating results, a perspective on *possible* research which was far wider than we three could explore alone. That, ultimately, is why this book is a cross between an anthology and a monograph, rather than a true example of either. We felt that there was no better way to illustrate for others the potential that we saw than to present rich and fertile statements by writers and thinkers of great ability and stature, yet we felt a need to place these statements in the context of possible future research.

As it is, we have written about one-third of this book. If we prove to have been too wordy or too didactic in tone, we hope that that will not stand in the way of appreciation of the anthologized authors. Our hope is that this work will lead to a greater sense of the coherence of the psychology of science.

We must express gratitude to many people who have aided us in this endeavor, in one fashion or another. We have profited greatly from discussions with our colleagues and students, and especially with Michael Bradie, Howard Gruber, Barbee Mynatt, and Kirk Smith. Michael Bradie and Steve Yachanin read and commented upon the entire manuscript, and helped us avoid many errors and misrepresentations. Ron Westrum provided valuable comments on the Introduction.

Many individual scientists have helped us by discussing their work with us. We are grateful to all, and especially to Lew Fulcher, Gary Heberlein, Stan and Iris Ovshinsky, Roger Thibault, and Charles Titus who spent many hours sometimes at great inconvenience, to assist us.

The research we have conducted was supported in part by grants from the Faculty Research Center of Bowling Green State University and in part by funds from the Department of Psychology. Our department chairman, Donald V. DeRosa has been extremely supportive of our work, and we gratefully acknowledge his assistance. Trisha Tatam cheerfully and rapidly typed seemingly endless drafts. We owe a great deal to our editor, Dr. Vicki P. Raeburn of Columbia University Press, for her encouragement and advice.

Finally, very special thanks are due to Peter C. Wason, of University College, London, who has provided advice, encouragement, and many specific suggestions that have been invaluable to us.

Bowling Green, Ohio<br>September 20, 1980

# General

# Introduction

A DOMAIN OF INQUIRY can be defined formally, as when biology is defined as the science of life. A less succinct but more informative characterization is provided by pointing to specific studies that exemplify the domain in question. This book attempts the second sort of definition. By directing the reader's attention to a set of readings, some old and perhaps familiar, some new and perhaps unfamiliar, we hope to make the point that a particular domain of inquiry, the psychology of scientific thinking, is coming into existence. Because that domain promises to be a fruitful arena for research, we feel that this book may be of value to potential researchers in the same way that collections of biological specimens are of value to biologists in the formation of taxonomic systems: as a starting point.

Since the inception of the Scientific Revolution some four centuries ago, science has grown from an avocation of a small number of people to a major historical force in the production of social and political change. Of that revolution, Butterfield has said:

> It outshines everything since the rise of Christianity and reduces the Renaissance and Reformation to the rank of mere episodes. . . . It changed the character of men's habitual mental operations even in the conduct of the nonmaterial sciences, while transforming the whole diagram of the physical universe and the very texture of human life itself. (1957:7–8).

Expressions of this sort are so frequent that it is surprising that the investigation of scientists' activities using the methodological techniques of science itself is quite recent. It is most surprising to us that psychologists have rarely turned their attention to this problem. Just as the act of artistic creation is regarded by many as ineffable and unknowable, so also has scientific inquiry been regarded as mysterious, and left unexplored. This situation is especially curious since the activities of scientists have recently been the subject of extensive analyses by other social scientists (especially sociologists and historians), by philosophers (for whom science constitutes a topic of major importance), and by scientists themselves (who have been nearly as prolific as politicians and statesmen in the production of memoirs). But psychologists have, until recently, ignored the psychological issues involved in scientific thinking.

To be sure, isolated studies of scientists have been conducted by psychologists. Wertheimer's (1945) analysis of Einstein's thinking process is a well-known example, as is Roe's (1953) study of the personality characteristics of scientists. Recently, a number of investigators have called for concerted investigation of science, using the methods of psychology (Mitroff 1974a; Mahoney 1976; Fisch 1977; and others). Little of this work has focused upon what we regard as the central question: What are the cognitive mechanisms that underlie scientific thinking? Interesting questions can be raised from a noncognitive point of view, of course, questions concerning, say, the motivation of scientists, or the relation of scientific inquiry to ethical values, but none of these, in our view, is likely to be answerable until we know more about the cognitive activity that constitutes the heart of scientific activity. While some lip service has been paid to this goal, there has been no systematic effort to understand the reasoning processes of scientists and no effort to relate such processes to generalizations about science made in other disciplines, such as the sociology of science, which have sometimes assumed that every scientist is a psychological "black box," accepting data as input and producing theories as output.

There is no a priori reason to think scientific thinking is intrinsically unknowable, in spite of the complexity of its various manifestations. John Locke (1690), for example, argued that careful analysis of the nature of empirical knowledge was a necessary prerequisite for further scientific advance. He spoke of himself as an "underlabouror, . . . clearing ground a little, and removing some of the rubbish that lies in the way to knowledge" (1690, "The Epistle to the Reader"). Locke sought to achieve this goal through analysis of what we would today regard as the psychological sources of knowledge. René Descartes (1637) pursued a similar goal through consideration of the rational sources of knowledge. While neither of these efforts would be considered scientific today, it is worth noting that both of these figures tried to understand how science was possible by an examination of the psychological processes of sensation, perception, and reasoning which result in knowledge.

Similar efforts were made in later centuries but almost always in a spirit of philosophical inquiry directed toward human knowledge in general, rather than toward scientific inquiry as such. Philosophical analysis has usually regarded scientific knowledge as a finished product, apart from the thought processes that gave rise to that knowledge. Yet scientific thinking is a natural phenomenon, part of the natural universe, and an activity of the greatest significance for the adaptation of our species to the planet. John Tyndall, in 1874, claimed for science the right to investigate any domain to which its methods applied, no matter what the consequences for other sorts of inquiry. More than a century after that claim, science itself is largely exempt from the same scrutiny.

What are the psychological processes that underlie scientific inquiry? Many testable questions flow from this larger one, once science is conceptualized as itself an object of scientific study: How is evidence related by scientific thinkers to hypotheses? How are hypotheses formulated? What constitutes the scientist's criterion for the relevance of evidence? Under what conditions are hypotheses abandoned? Under what conditions are hypotheses *not* abandoned? What cognitive limitations constrain the activities of the scientist? How does facility in conducting science develop in the individual? Can this facility be taught? Can it be stultified by the educational system? These questions are but a sample of the *empirically testable* issues that arise in the psychology of science. Very few of these questions have had appropriate research directed at them, and none has been answered definitively.

In the last few years, a few publications by psychologists

have appeared on this topic. Maslow, in a book entitled *The Psychology of Science* (1966) sought to identify some of the motivational underpinnings of science (though he was, on the whole, quite critical of science). Mitroff (1974a, 1974b) conducted an extensive interview study of NASA geophysicists. Weimer (1975) made an explicit appeal for study of the psychological processes of science, and Mahoney (1976), in a highly polemical tract, called for more research on the subject and sought to identify some of the reasons why it has not been an active research area. Specific experimental studies designed to be laboratory analogs of "real-world" science have also appeared in recent years. Gerwin and Newsted (1977) tested an explicit model of induction. Mynatt, Doherty, and Tweney (1977, 1978) used a computer-simulated "universe" to study inference in complex problem spaces. Ross and his colleagues (Ross 1977; Ross, Lepper, Strack, and Steinmetz 1977; Nisbett and Ross, 1980) have drawn an analogy between social psychological attributions of causality and scientific inference. All of these studies point to the potential testability of the questions raised by the psychology of science.

This book will consist of a set of reprinted papers, and excerpts from books and papers, dealing with the emergent psychology of science. We have included certain materials that call for programmatic research, much as we are calling for programmatic research now. But our principal goal is to demonstrate to the reader that appropriate scientific questions can be asked about science, that the tools of scientific psychology are relevant to answering those questions, and that (as suggested by the few cases where research has begun) interesting answers are possible. Because we are asking the reader to read many of the selections in a new way, with new questions in mind, we will provide interpretive introductions. These should provide orientation toward the issues raised in the selections.

In the remainder of this introduction, we elaborate the arguments for a scientific approach to the understanding of science; and we review some of the research that points to the possibilities. After considering the relation of the psychology of science to other disciplines, we examine its relation to certain subareas of psychology, paying special attention to studies of thinking and reasoning. In particular, experimental approaches hold promise for understanding the process by which scientific hypotheses are utilized. Such approaches are introduced here and presented in detail later. The experimental work, new as it is, has uncovered specific processes which may be very general and which may have broad implications for the conduct of science. We argue that one can validate, at least partially, the approach of using laboratory analogs to science by briefly considering what this research can tell us when its findings are used to illuminate an actual case history of scientific research—Michael Faraday's experiments on the possible relation between gravity and electricity.

## The Relation

## of Psychology of Science

## to Other Disciplines

It is necessary to make some preliminary comments about the relation between the psychology of science and those other areas which have attempted to provide objective analyses of science. In particular, we need to establish that the questions addressed by the psychology of science are not addressable, or not addressable in the same fashion, by the philosophy of science, by the sociology of science, or by the history of science.

PHILOSOPHY OF SCIENCE
As a distinct subdiscipline of philosophy, philosophy of science is of fairly recent origin. Before the seventeenth-century inquiries of Bacon, Locke, and Descartes, there were few attempts to understand science using the special tools of philosophical analysis (Harre 1967; Losee 1972). Systematic efforts at an understanding of science, as distinct from knowledge in general, began only in the nineteenth century, with the work of such figures as John Stuart Mill (1843) and William Whewell (1860). In the twentieth century, such efforts accelerated. Partly as a result of the development of quantum mechanics, the philosophical analysis of science burgeoned. The work of

the philosophers of the "Vienna circle" (e.g., Carnap 1935) was of special importance. By positing verifiability as the criterion for the meaningfulness of a given proposition, they articulated a method of deciding upon the boundary between science and nonscience, a method which has been influential among scientists in spite of certain philosophical difficulties (Popper 1962; Harre 1967).

As valuable and informative as these analyses have been, they do not provide a descriptive account of science. The philosophical approach to science may show unequivocally the logical invalidity of certain types of inferences made by scientists (Salmon 1967; Popper 1934), which from one point of view can be seen as critical of science. But such analysis has not changed the activity of scientists. The *pragmatic* criterion for the results of scientific inquiry, the fact that such inquiry leads to consistent and useful knowledge, does not rest on formal logic. Ultimately, there is no logical necessity behind what James called the "manufactured world" of science, "about which rational propositions may be formed" (1890: 2:665).

To be sure, philosophers have not generally insisted that scientists *ought* to regard their analyses as prescriptive. Some, for example Feyerabend (1970), have argued that scientists should pursue whatever method seems reasonable in the context of a particular problem, ignoring completely whether or not canons of logical inference are being met. For many philosophers, the philosophical analysis of science is a formal affair, carried out to illuminate the logical relationships among the results of science, the relation of scientific theories to underlying assumptions, and the logical relation between a theory and its supporting data. Thus, much effort has been directed at questions involving the interpretation of relativity theory (e.g., Grunbaum 1963). Such work does not usually pretend to tell physicists how to make further progress.

Some philosophers of science, Popper in particular, have taken the stance that philosophical analyses can result in normative guidelines for science. In some instances (e.g., Eccles 1975, selection 14), scientists have adopted the guidelines and have attested to their utility. Most scientists, however, have not paid particular attention to the claims made by philosophy of science.

In part because of Popper's influence, recent philosophy of science has distinguished between issues of justification (how to justify the results of science) and issues of discovery (how to understand the process by which science comes about). The former need not concern itself with psychological questions. Can there be a *philosophy* of discovery? Some have answered in the affirmative (e.g., Hanson 1958; Simon 1973). It is our belief, however, that many of these attempts raise questions that can only be answered empirically—that is, by a scientific, rather than a philosophical, analysis.

Norms for the conduct of scientific inquiry based only on a logical analysis of the finished products of that inquiry are not likely to be successful. Scientific thinking is not governed by formal rules of logic, and, in fact, is often even illogical. This does not mean that normative judgments are impossible—only that they are unlikely to be of value to the conduct of science unless they are formulated in light of knowledge—scientific knowledge—about the nature of science as it is practiced. For this reason, we feel that the psychology of science may eventually lead to prescriptions for better science, to specific analyses of inquiry processes enabling the design of better inquiry, and perhaps, through analyses of scientists' thinking processes, to better science education. Though it is a distant hope, perhaps insights into scientific thinking will promote a fuller understanding of human thinking in general.

HISTORY OF SCIENCE
While specific attempts to chronicle the emergence of science are found in all periods, ranging from Aristotle's account of the history of psychology in the *De Anima* (c. 350 B.C.), to Whewell's *History of the Inductive Sciences* in 1837, it is only in this century that organized, scholarly efforts have been made to understand science using the techniques of historical analysis (Sarton 1962; Thackery and Merton 1972; Kuhn 1977).

Historians of science are primarily concerned with what actually happened in the conduct of science. Questions about the history of a particular concept (e.g., Olby's 1974 history of DNA in molecular biology) concern the actual development and use of the concept in question. Scientific biographies (e.g., Williams 1965, on Faraday) concern the actual behavior of individual scientists. Accounts of scientific change, and attempts to explain such change (e.g., Kuhn 1962), are concerned with the actual forces that produce change. Kuhn's analyses have been especially influential, having been widely discussed by philosophers (e.g., Lakatos & Musgrave 1970).

Kuhn's work has been of special significance in creating awareness of the importance of the nonlogical aspects of scientific progress. He has argued, for example, that acceptance of Copernicus' heliocentric model was based on questions of aesthetic form rather than on strictly empirical issues (Kuhn 1957). Further, his distinction between normal and revolutionary science (between, that is, science that methodically explores an empirical domain using paradigmatic methods and science that creates new paradigms) has directed attention toward the social and psychological factors involved in science (Kuhn 1962). Kuhn has urged that attention be directed toward such issues as the role of textbook examples of "good science" in the socialization of young scientists (Kuhn 1961) and the relation between conceptual processes and scientific theory construction (Kuhn 1974). Though they emerged first in the history of science, these are clearly concerns to which the psychology of science is relevant.

Ultimately, it appears to us that the history of science will stand in relation to the psychology of science as astronomy once did to physics. It is widely acknowledged that progress in understanding the movements of the sun, moon, and planets was dependent upon progress in understanding the dynamics of moving bodies on earth. Thus, the heliocentric system of Copernicus led to a new approach to physics, since accepting it required the abandonment of the then-prevailing Aristotelian dynamics (Butterfield 1957). The physical experiments of Galileo made it possible to reconstruct dynamics in a way compatible with the new astronomy (Galileo 1632). The final step occurred with Isaac Newton, whose *Principia Mathematica* (1687) demonstrated that the motions of the planets could be rigorously derived from the general laws of motion of physical dynamics.

The psychology of science lacks, as yet, even its Galileo, much less its Newton. Even so, we feel that inquiry into the psychological dynamics of scientists ought to be revealing and useful to the historian of science. On the other side, studies in the history of science ought to be especially useful in the psychology of science. Thus, Michael Faraday kept extensive laboratory diary records of his experiments, diaries which have been published (Faraday 1932–1936). He has thus made available to historians a minutely detailed account of his thinking, an account which has served as the raw material for several studies of methodology in science (Williams 1965; Agassi 1971). Such accounts are of considerable interest to the psychology of science. They do not, however, obviate the need for other sorts of inquiry. The limitations of historical analysis are the same as those of astronomy prior to the modern era. First, there is no way to manipulate the phenomena under study and no way to develop unambiguous laws of the relationship between important variables. Second, lacking a coherent account of the psychological dynamics of science, one cannot formulate theories of the history of science which rest on secure foundations. This problem can be seen with great clarity in the debates over scientific imagination. Does insight stem from the intuition of great geniuses? Does it rely on the accidental combination of elements (e.g., James 1880; Crovitz 1969; Toulmin 1972; Campbell 1974)? Does it depend on the growth and development of a gradually crystallizing metaphor applied over the course of years to a domain of empirical facts (Gruber 1974)? All of these possibilities can be supported by appeals to the proper historical data, just as the heliocentric theory of Copernicus and the geocentric theory of Ptolemy can both be supported by huge amounts of astronomical data (Kuhn 1957). The availability of precise physical laws in the seventeenth century allowed Galileo, Kepler, and finally Newton, to validate the heliocentric cosmology. We believe that the same situation could occur in the history of science, which should be similarly served by the psychology of science.

A similar point has been made by Piaget (1929), and by Kuhn, though without mention of the critical role of psychology:

> Closely related to the sociology of science . . . is a field that, though it scarcely yet exists, is widely described as the science of science. Its goal, in the words of its leading exponent Derek Price, is nothing less than "the theoretic analysis of the structure and behavior of science itself," and its techniques are an eclectic combination of the historian's, the sociologist's, and the econometrician's. No one can yet guess to what extent that

goal is attainable, but any progress toward it will inevitably and immediately enhance the significance both to social scientists and to society of continuing scholarship in the history of science. (Kuhn 1977:122).

Add psychology to the "eclectic combination," and our view is very close to Kuhn's.


SOCIOLOGY OF SCIENCE

The sociology of science is even more recent than the history of science, dating back only some 40 or so years. A principal figure in the movement has been Robert Merton (Storer 1973). Initiating his inquiries with the historical question of the relation of seventeenth-century English science to its social and religious context, Merton later argued that the concepts and methods of sociology could be applied to an understanding of science (Merton 1973). Sociologists view science as essentially similar to other social institutions, and utilize standard tools of sociological inquiry in attempting to understand it. Further, underlying all of their work is the assumption that knowledge has an intrinsically social component, that nothing can be known outside of the socially prescribed context (Mannheim 1936).

Because of the availability of empirical techniques such as citation analysis (Cole and Cole 1972), and the relevance of the discipline to such issues as the formulation of social policy (Spiegel-Rossing and Price 1977), the sociology of science has grown rapidly. By focusing on scientists in social contexts, it has been possible to analyze, for example, the social evaluation of scientific discoveries (Merton 1960; Zuckerman and Merton 1971) and the nature of socialization processes in science (e.g., Zuckerman 1977). Nevertheless, the sociology of science has typically ignored the role of the individual scientist. Instead, there has been an assumed "scientific man," comparable to the "economic man" of classical economics (Whitley 1974). While this approach certainly can lead to valuable insights about the nature of scientific inquiry, it cannot provide knowledge about the cognitive activity which constitutes that inquiry. This point is illustrated by the attempts to understand why simultaneous discoveries are so frequent in science (Merton 1963). Cognitive processes are not explicitly dismissed as irrelevant by sociologists. (Merton, certainly, is a very thorough scholar, well aware of all relevant literature.[1]) Instead, it is simply that the explanatory foundation of a sociological account is not cognitive but social. Some sociologists have focused upon the relation between norms of scientific activity and the process of scientific discovery and communication. For example, Edge and Mulkay (1976) examined the context of the discovery of pulsars by British radio astronomers, and tried to relate subsequent controversy about the discovery to general sociological principles. In contrast to the Mertonian approach, which has relied upon quantified indicators derived from, say, scientific publication, Edge and Mulkay relied more upon qualitative analyses of the content of science and its social context. While such studies, like Merton's, reveal much about science, in the end, the sociologist of science creates anew an idealized scientist-as-social-product and contributes relatively little to our understanding of the actual psychological processes.

## The Relation
## of the Psychology of Science
## to Psychology

Research on the psychology of science has only just begun, but it is possible to detect clear areas in which it can benefit from existing research in other areas of psychology. Research areas directly related to our concern include creativity, problem solving, judgment and decision making, and others. It is clear that the psychology of science must concern itself with the nature of creativity if it is to understand the role of innovation in science. Thus, questions about the origin of imaginative insights in science (Holton 1973, 1978) are ultimately questions about the psychological conditions of creativity (Ghiselin 1952; Getzels and Csikszentmihalyi 1976). Studies of human judgment are relevant because they can provide explicit, quantitative models of some aspects of scientific behavior (Mitroff 1974b). Many studies of problem solving have

been reported by experimental psychologists, and their relevance to the behavior of scientists needs to be carefully established (Bruner, Goodnow, and Austin 1956; Wason and Johnson-Laird 1972; Mynatt, Doherty, and Tweney 1977, 1978;
Tweney et al. 1980).

The work of Piaget occupies a special position with regard
to the psychology of science. Since his earliest research, Piaget's quest for an understanding of how children acquire cognitive capabilities has been rooted in an attempt to understand
the sources of all knowledge (Piaget 1965; Woodward 1979).
He has argued for decades (see, e.g., Piaget 1929) that psychology ought to be the foundation of epistemology, since it
is only through the understanding of how knowledge grows
that we can hope to understand the achievements and the limits of knowledge (Piaget 1970).

Piaget's unique developmental approach to epistemology
has produced a number of important contributions to the understanding of science (see, for example, Inhelder and Piaget
1958; Beth and Piaget 1966; Karmiloff-Smith and Inhelder
1975). It has inspired the work of Gruber, who has formulated
a theory of scientific creativity based in part on Piagetian principles (1974, selection 40). We expect the "clinical method"
that Piaget developed in his pioneering studies of children
will prove to be a valuable tool in the psychology of science
as well. At present, however, few investigators have sought
to use such direct methods to explore scientific thinking.
(Mitroff and Holton, as well as Gruber, are clear exceptions
to this generalization.)

Recently, there has been greater awareness in cognitive psychology of the potential power of verbal self-report as a tool
in the analysis of problem solving. Such reports can not be
taken as veridical indicators of cognitive process in all circumstances, especially when subjects already possess a set
of beliefs attributing causality to their own cognitions (as occurs in attributions of social responsibility, cf. Nisbett and
Wilson 1977). Nevertheless, as Ericsson and Simon (1980)
have argued, verbal self-reports can be used to monitor information which is being used "consciously," i.e., which is in
short-term memory and which is being attended to, and hence
such reports can play a critical role in the evaluation of models
of cognition. Further, except at a very molecular level, the

choice of one strategy over another in a problem-solving context is often accessible via self-report. It is precisely this fact which has made process-tracing possible in the analysis of problem solving (Newell and Simon 1972), and, as shall be argued later, which can be used in the analysis of scientific thinking.

Some areas of psychology are less directly related to the psychology of science because they bear less directly on the principal cognitive questions. Still, studies of the development of scientific interests in childhood (e.g., Roe 1953) may contribute to an understanding of scientific inquiry, as may studies of the motivation of scientists (Fisch 1977) and studies of the relation between particular personality types and styles of inquiry. An example of the latter is Mitroff's attempt to relate differences between experimental and theoretical scientists to differences in personality type (Mitroff and Kilmann 1978; Wilkes, 1979). Finally, the study of social influences on beliefs and attitudes is of relevance to an understanding of the occurrence of bias and of error in science (Rostand 1960; Rosenthal 1966).

What can the psychology of science offer to other areas of psychology? The most immediate application is to cognitive psychology. While experimental investigations of the nature of human thinking have expanded greatly in recent years (see, e.g., Johnson-Laird and Wason 1977), there is still far too little analysis of the type of cognition that occurs in nonlaboratory situations. Allport (1975), Neisser (1976), and others have criticized cognitive psychology for staying too close to laboratory tasks. Newell (1973), among others, has recommended the study of "natural cognition" as a solution to the present difficulties. While ambitious studies of, for example, chess playing, have appeared (Chase and Simon 1973), such studies are infrequent, and generalizations limited. The psychology of science may contribute to filling this gap by providing examples of complex cognitive activity in "open" problem spaces: domains where no algorithm or heuristic can guarantee solution (Newell and Simon 1972).

The question of how science ought to be taught, an issue of concern to educational psychologists, is also a potential field of application for the psychology of science. Some studies of the development of scientific reasoning have been conducted, most notably those of Piaget and Roe cited above. These have not directly addressed questions of application to particular teaching methods. In fact, science education, in the United States at least, is still dominated by a kind of "try-it-and-see" approach which is likely to be less efficient than approaches based on a foundation of psychological knowledge about science. In a pilot investigation, concepts derived from our empirical findings about the use of confirmation and disconfirmation were taught to groups of undergraduate students participating in summer research programs sponsored by the National Science Foundation at Bowling Green State University. The concepts proved to be interesting to the students, and perhaps useful in their research education as well.

Finally, the psychology of science may have implications for science policy. Kevles (1978) has shown that the development of physics in the United States was greatly influenced by conceptions of how research ought to be conducted. Many such issues are certainly sociological in nature rather than psychological (questions of, for example, the optimal size of research groups), but others will, we believe, yield to psychological approaches. For example, should individual scientists be encouraged to seek confirmatory evidence for their own hypotheses, or should they be encouraged to gather evidence relevant to a number of alternative hypotheses? This is a question that should be answerable on the basis of research like that described in part 4.

## Current

## Research

A complete review is, of course, beyond the scope of this introduction. In fact, this entire book can be seen as a presentation of current (and some past) work in this area. Nonetheless, a few broad outlines are possible.

The most interesting work has been based on attempts to determine whether scientists do, in fact, use logical inference techniques of the sort prescribed by philosophers. A long tradition of research in experimental psychology has been directed to a similar issue, but only using nonscientists (see Woodworth 1938; Henle 1962). In fact, process models have been developed for certain sorts of reasoning tasks: syllogisms (e.g., Johnson-Laird and Steedman 1978), linear ordering problems (e.g., B. Mynatt and Smith 1977), certain sorts of puzzle problems (e.g., Jeffries, et al. 1977). However, there have been few attempts to generalize beyond the context of the laboratory task and even fewer to address the case of actual scientific thinking.

The few deliberate attempts to capture the essence of scientific reasoning in psychological research will be presented in detail later. In general, the construction of formal models or theories has not played a significant role in this work to date, though, as Simon (1966, selection 5) makes clear, such applications should be rewarding. Instead, most investigators have concerned themselves with describing the phenomena manifested by scientists during the inference process. Issues involving the validation of hypotheses have been paramount: Is confirmatory evidence sought preferentially? Under what conditions is disconfirmatory evidence useful? What is the effect of considering multiple hypotheses? These issues are addressed in parts 2, 3, and 4. Others have sought to determine the conditions responsible for the generation of hypotheses. This work, stemming in part from earlier research on creativity, has concerned itself with the interrelations among concepts in scientific creativity, and has relied heavily on the use of case-study methods. Relevant discussion and selections will be presented in part 7.

Finally, there have been attempts to relate scientific thinking to personality factors and to individual differences (e.g., Roe 1953; Eiduson 1962; Fisch 1977). Some notable results have been achieved in this domain, for example, Mitroff's (1974b) demonstration that there are consistent personality differences between experimental and theoretical scientists. We have chosen to not to emphasize this tradition, in part because it has been reviewed elsewhere (Fisch 1977) and in part because we believe that the investigation of cognitive variables is, at present, a more promising research area. In the long run, of course, such research must concern itself with the nature of individual cognitive differences. Thus, we must eventually ask ourselves about the sources and consequences of the difference in thought processes between, say, an Einstein and a Heisenberg (see Miller 1978, selection 47). We doubt that explanation of the different research strategies of these two scientists will arise out of personality tests, as such tests are currently conceived.

## The Role

## of Nonexperimental

## Studies

The direct observation of actual scientific activity must play a large role in the psychology of science. Ongoing scientific inquiry can be studied directly using such devices as surveys, interviews, and, whenever possible, descriptions by scientists of their own activity. For example, the psychologist Herbert Crovitz has kept a "diary" of his thought processes over a span of years. Recently, he has inspected the entries to determine, for example, whether or not insights (defined introspectively as "aha!" experiences) were scattered randomly across time. He found that they were not, that insights tended to occur only after intensive thought had been directed toward particular issues (Crovitz and Horn 1977). Such studies are certainly open to numerous objections from a rigorous, experimental design point of view, but, nonetheless, they are essential to the future development of the psychology of science. Such records can constitute a proving ground for ideas derived from more carefully controlled research and, equally important, a source of hypotheses to be evaluated experimentally.

Some of the relevant observational data can be found in the historical record. That record is now being preserved and analyzed, but it is our impression that few historians have raised the central psychological questions. That it is possible to do so is manifestly clear from, for example, Gruber's application of a cognitive approach to the understanding of Darwin's notebooks (1974, selections 21 and 40). Generally, however, historians have focused on personality factors in applying psychological analyses to the history of science, thus missing the cognitive underpinnings of science.

# Is an Experimental

# Psychology of Science

# Possible?

We have come a long way since Dugald Stewart asserted that "The difference between experiment and observation consists merely in the comparative rapidity with which they accomplish their discoveries." (1818:33). It is now generally conceded that experimentation, where possible, permits a degree of control over extraneous variables not attainable in any other fashion. Can experimentation be carried out in the psychology of science? Bartlett hinted that such a thing might be possible:

> To approach the study of experimental thinking in the same general way in which I have attempted to approach that of more formal thinking, I ought to state a few experimental questions and get them investigated step by step in the laboratory by a number of experimental scientists. To attempt this might indeed be exceedingly interesting and worth while, but it would take a tremendous amount of time. (1958:114–15).

In fact, in recent years, a number of cognitive inference tasks have been devised which appear to resemble scientific inference tasks. In our own work, we have tried to create a close laboratory simulation of science, in order to directly study scientific inference. In brief, we have relied upon a computer simulated "universe" consisting of interacting geometric shapes that behave in lawful ways. By asking subjects to determine the laws of the system, we have been able to model some of the characteristics of science in the laboratory. This work is described in detail in part 4. In a similar vein, Elstein, Shulman, and Sprafka (1978) and Fox (1980) have used simulations of medical diagnostic problems to study how physicians solve problems.

# The Problem

# of Generalization

All methodologies possess inherent weaknesses that limit the generalizations that can be drawn from them. Simple observation does not usually permit the isolation of particular factors from the tangle of factors that produce any particular phenomenon. Experimentation can overcome this limit, but sometimes only by sacrificing the naturalness of an event for the artificiality of the laboratory. These problems are especially acute in the psychology of science, where the phenomena of interest occur in a limited population, under special conditions, and in a spontaneous way which is seemingly incompatible with the artificiality of a laboratory simulation. Can these limits be overcome?

We believe that they can, but only if many kinds of methodologies are used in an interactive and mutually supportive fashion. For example, we have found in our simulation of scientific inference that the deliberate attempt to disconfirm one's own hypothesis is a difficult strategy, and sometimes not very effective (see part 4). We have made a preliminary application of this finding from our laboratory studies to a historical case study.

If the attempt to disconfirm hypotheses is not always an optimal procedure in a simulation task, then the same ought to be true in actual scientific research. Even further, if our claim is correct, such comparisons ought to reveal aspects of the historical record of science.

On the other side, of course, it is possible that the historical example will *not* match our descriptions, that the relationships we have described may be unique to a laboratory environment. The analysis of the historical case must be sensitive to this possibility also.

We explored the fit between our description and some of the research programs carried out by the English physicist Michael Faraday (1791–1867), who kept very careful, very thorough records of his experimental investigations, and of his hypotheses, speculations, plans, incidental observations, etc. (Faraday, 1932–1936). Faraday "thought aloud" while devising experiments, and thus a comparison between his diary and our subjects' protocols was especially revealing. Such comparisons were made using both successful research, such as Faraday's discovery of electromagnetic induction, and unsuccessful research, such as his repeated attempts to show a relationship between gravity and electricity. As an example of the latter, between March 19, 1848, and October 19, 1849,

Faraday conducted dozens of experiments in an attempt to determine if motion, gravitational force, and electric currents were related. This effort was, of course, entirely reasonable, given his earlier elucidations of the laws that govern the interrelation of motion, electric currents, and magnetic fields (Williams 1965). Faraday's diary for this period reveals an interlinked sequence of speculations, refinements of those speculations into testable propositions, actual experiments, and interpretations of those experiments. The procedure looks very much like Platt's (1964) "strong inference," in that there is a constant search for alternative explanations for any given phenomenon. This is apparent, for example, in the record of August 25, 1849. Having obtained repeated confirmatory results in one experiment, results which suggested that falling coils manifested induced electric currents, Faraday went on to conduct a laborious series of further experiments which revealed the effects to be artifactual. The same kind of interchange between falsification and confirmation characterized the best subjects in our second study (Mynatt, Doherty, and Tweney 1978, selection 18).

We have found the following points of similarity between the inspected diary accounts and the protocols of our subjects in the simulation studies: 1) neither was able to draw on existing literature, since none existed; 2) both Faraday and the best of our subjects used both falsification and confirmation; and 3) both Faraday and the best of our subjects did not place excessive reliance on any one confirmatory or disconfirmatory result. Faraday and our subjects differed, however, in the following respects: 1) Faraday's work was stretched out over much longer periods of time; 2) Faraday appeared to have been greatly motivated by possible analogies between his earlier work and present problems; and 3) much of Faraday's work involved the design of apparatus and research directed at detecting and/or eliminating artifact from his experimental setups.

The last-mentioned point of dissimilarity, the preponderence of relatively mundane "technical" research carried out by Faraday, was especially striking. Such entries were intermingled with entries concerned with, for example, the status of current hypotheses. To what extent does such workaday activity constitute a kind of breeding ground for hypotheses?

Many accounts of the production of new scientific ideas speak of the unconscious cognitive work carried out during periods when the investigator is engaged in some other activity (see, e.g., the accounts of Hadamard 1945 and Simon 1966, selection 5). Perhaps such incubation is to be sought in the immediate context of scientific research, during periods of relatively undemanding, but relevant, "busywork." The point reminds us of Charles Darwin's famous September 28, 1838, notebook entry, in which he gives for the first time a clear statement of the operation of natural selection (Gruber 1974:455–56). This stunning insight is preceded by entries on mosquitoes and on his chambermaid, and followed by an entry on the curiosity of monkeys! How much of what we need to understand about science is hidden in such contexts? This possibility has led us to examine anew the protocols obtained from our subjects, to see if we can detect similar sorts of intermingling of the generation of hypotheses and of activity of a less exalted sort. Perhaps the repetitive pursuit of confirmatory data can be understood in this context. Perhaps it is psychologically necessary to permit the incubation of those new ideas and new alternatives which can then allow falsification to become a useful approach. The examination of Faraday's diaries has thus confirmed some of our expectations, but it has opened new possibilities for future examination of the laboratory protocols as well.

## The Structure
## of this Book

If the psychology of science is, in fact, about to emerge as a coordinate part of the science of science, alongside the philosophy, history, and sociology of science, then it should be possible to assemble a literature from a variety of fields to illustrate the possibility of, and need for, the new subdiscipline. For this reason, in assembling readings we have cast as wide a net as we were able. Readings have been drawn from many disciplines, across four centuries of Western thought concerning science.

We feel that the range of selections is sufficient to establish the point we wish to make. A number of selections have been drawn from works that are sometimes regarded as dated, or as being of interest only to the historian. We have deliberately chosen such selections, even when more recent statements were available, partly because it is important to demonstrate that the concerns we are addressing are not new. What *is* new is the availability of research techniques drawn from contemporary experimental psychology which, in our view, render many classical concerns testable for the first time. Thus, we believe that the psychology of science can be characterized not solely by its subject matter, but by the testability of many of the the problems that arise in connection with that subject matter.

We have tried, whenever possible, to choose selections that were well written. In all cases we have edited selections to make them as concise as possible, consistent with fidelity to the author's original context. The attentive reader will soon notice the presence of leitmotifs—recurrent themes that echo throughout this book. We hope that the reader will derive as much pleasure and excitement from these as we did when they first became apparent to us. The reader should find themes, continuities, and contrasts which have not been made explicit either by us or by the authors of the selections.

The book is divided into nine thematic parts with an epilogue in which a number of general concerns are addressed. Many of the issues we have mentioned in the general introduction are considered in the epilogue in greater detail, and new issues are raised. Each of the nine parts begins with an introduction in which the problem area is discussed and the particular readings introduced.

Following part 1, which includes a number of influential calls for a psychology of science, the parts are arranged in three groups. The first, consisting of parts 2, 3, and 4, focuses upon the use of hypotheses in science. Part 2 shows that hypothetical reasoning is a fundamental activity in science, while parts 3 and 4 provide detailed considerations of how hypotheses are used. The philosopher Karl Popper (selections 9 and 12) has presented an influential argument that suggests scientists ought to attempt disconfirmation of their hypotheses. This counterintuitive claim is so important that part 3 deals with some of its implications for scientific practice. In part 4, evidence is presented that the deliberate search for disconfirmation does not correspond to the psychological reality of most scientific activity. In part 4, the reader will also encounter the most concrete empirical findings that can so far be claimed by the cognitive psychology of science. These findings suggest, first, that Popper's prescription is unrealistic, and, second, that further research will reveal a richer, more complex relation between hypotheses and scientific activity than has yet been imagined.

Parts 5 and 6 consider the role of mathematics and statistics in science. Much less empirical psychological research has been carried out on these important problem areas; the selections accordingly raise more issues than they resolve. Part 5, on the use of mathematics, contains a rich account of Einstein's formulation of special relativity written by the Gestalt psychologist Max Wertheimer (selection 23). While this selection could have been placed in a number of other parts, we include it in part 5 because it makes clear how the use of mathematics was, for Einstein, a heuristically powerful aid to physical intuition. Part 6, on the use of statistics in science, contains selections that are especially pointed for scientific psychology, and often directly critical of it. Further, the paper by Tversky and Kahneman (selection 31) demonstrates the fruitfulness of the empirical investigation of scientists themselves.

The third major section of the book, consisting of Parts 7, 8, and 9, raises the issue of the origin of the ideas used by scientists. Psychology has long tried to formulate a meaningful account of creativity. Many arrangements of the chosen selections appeared possible. We decided to first emphasize two broad classes of activities of the scientist: observation and classification in part 7, experiment and theory construction in part 8. As part 7 demonstrates, consideration of the psychological "theory-ladenness" of observation is not new. Holton's concept of themata (selection 37) represents a powerful summary of this point. In part 8, the specific mechanisms by which new ideas arise in science are directly considered. No final answer to this question is yet possible, but the selections suggest a number of directions for psychological research.

Part 9, the final part of the book, addresses the role of imagery in science. Most striking to us is Einstein's own brief account of the role of imagery in his own thought (selection 48). Brief as this part is, it is a measure of how much is yet to be learned concerning a fundamental issue in the cognitive psychology of science.

The book closes with an epilogue in which we assess the current status of the cognitive psychology of science, and discuss some of its prospects. Questions of methodology are addressed, and some speculation is included on future theories of scientific activity. Finally, we raise the possibility of the development of an "empirical epistemology," of the development of norms for scientific conduct based upon the findings of the psychology of science. The possibility suggests the ultimate fruitfulness of the new research domain that is the topic of this book.

## Note

1. Scholarship generally, and Merton's in particular, is exemplified and parodied with brilliant wit in Merton's *On the Shoulders of Giants* (1965).

# Psychology

# and the Science

# of Science

# Introduction

FAITH IN THE power of science is not new, nor is faith in the potential applicability of psychology to an understanding of science. The selections in this chapter were written as general conceptions of the nature of science, but all share the belief that an understanding of human mental processes is propaedeutic to an understanding of science.

The first selection, by Francis Bacon (1620), serves as a reminder that the problems we are raising were seen at the very beginning of the Scientific Revolution. Bacon anticipated many of the central concerns of the present book: confirmatory tendencies in inference, the psychological creativity implicated in theory construction, the origin of hypotheses. Believing, as he did, that knowledge would give man new power over nature, he emphasized the importance of an understanding of human nature as essential to the growth of that knowledge.

By the time Ernst Mach wrote *Knowledge and Error*, in 1905, the power of science was no longer doubted, but the role of psychology had diminished. The founder of positivism, Auguste Comte, had rather arbitrarily ruled psychology out of the sciences (Comte 1830–1842). Many shared Kant's view that psychology could not be a true science, but could only be "merely" empirical (Kant 1798; Mischel 1967). Though experimental psychology was quickly gaining ground as an independent discipline, some of its greatest practitioners doubted the applicability of experimental methods to such complex human processes as science itself (e.g., Wundt 1896). In particular, even though Wundt had pioneered the use of experimental techniques in psychology, he felt that the higher mental processes could only be studied using nonexperimental techniques such as comparative linguistic analysis (see Bringmann and Tweney 1980; Danziger 1980 for recent views on this point). Mach's view, which differed from Wundt's, was positivistic in its reliance upon sense data for all knowledge, and in its rejection of unobservable constructs. More important (for our purpose) was Mach's belief that there was no real distinction between physics and psychology, and that the same methods ought to be applicable in both. This is certainly no longer a common view, but, in the selection we have chosen, the reader will note how it led Mach to view scientific thinking as a special case of ordinary thinking. It is the same creature doing the thinking, and whatever the nature of the special cultural adaptations of science, this commonality can not be overlooked.

A radical departure from the philosophical position exemplified by Mach is represented by Jean Piaget (selection 3). Piaget's epistemology derives from Kant and from evolutionary conceptions, rather than from positivism, but he shares with Mach the view that science is an extension of ordinary thinking. The understanding of science has been, in certain respects, an ultimate goal for Piaget (as it was for Kant). The research carried out (and in progress) by Piaget and his followers has not yet been applied extensively to the understanding of scientific thinking in working scientists. Instead, the focus has been upon the development of scientific concepts (mass, velocity, causality, etc.) in children and adolescents (e.g., Inhelder and Piaget 1958; Karmiloff-Smith and Inhelder 1975). At this date, however, only a few attempts have been made to extend the theory toward scientific thought as such, most notably by Gruber (1974).

Thomas Kuhn (selection 4) is a physicist-turned-historian whose views of the nature of scientific revolutions have themselves been revolutionary. In the selection chosen, Kuhn argues against Sir Karl Popper's rejection of the role of "psychology" in the understanding of science, and in doing so, elaborates the part that psychological knowledge must play in that understanding. The new emphasis on the psychology of science derives in part from the reorientation that Kuhn's work has forced.

The last selection, by Herbert Simon, was written from a theoretical position of recent vintage, the information-processing approach. The application of psychology to complex cognitive processes has been a major theme in the work of Simon and his colleagues (cf. Simon 1969; Newell and Simon 1972; Newell 1973; Bhaskar and Simon 1977). Simon's paper shows that testable hypotheses may be constructed concerning thought processes considered to be implicated in scientific thinking.

FRANCIS BACON

# The Idols

# of Human

# Understanding

XXXVIII.

THE IDOLS AND false notions which are now in possession of the human understanding, and have taken deep root therein, not only so beset men's minds that truth can hardly find entrance, but even after entrance obtained, they will again in the very instauration of the sciences meet and trouble us, unless men being forewarned of the danger fortify themselves as far as may be against their assaults.

XXXIX.

There are four classes of Idols which beset men's minds. To these for distinction's sake I have assigned names,—calling the first class *Idols of the Tribe*; the second, *Idols of the Cave*; the third, *Idols of the Market-place*; the fourth, *Idols of the Theatre.* . . .

The Idols of the Tribe have their foundation in human nature itself, and in the tribe or race of men. For it is a false assertion that the sense of man is the measure of things. On the contrary, all perceptions as well of the sense as of the mind are according to the measure of the individual and not according to the

Excerpt from Francis Bacon, *Novum Organum* (originally published 1620). In J. Spedding, R. L. Ellis, and D. D. Heath, eds. *The Works of Francis Bacon*. Popular Edition. New York: Hurd and Houghton, 1878, pp. 76–80. This translation first published 1857–1859.

measure of the universe. And the human understanding is like
a false mirror, which, receiving rays irregularly, distorts and
discolours the nature of things by mingling its own nature
with it.

### XLII.

The Idols of the Cave are the idols of the individual man. For
every one (besides the errors common to human nature in
general) has a cave or den of his own, which refracts and
discolours the light of nature; owing either to his own proper
and peculiar nature; or to his education and conversation with
others; or to the reading of books, and the authority of those
whom he esteems and admires; or to the differences of impressions, accordingly as they take place in a mind preoccupied
and predisposed or in a mind indifferent and settled; or the
like. So that the spirit of man (according as it is meted out to
different individuals) is in fact a thing variable and full of
perturbation, and governed as it were by chance. Whence it
was well observed by Heraclitus that men look for sciences
in their own lesser worlds, and not in the greater or common
world.

### XLIII.

There are also Idols formed by the intercourse and association
of men with each other, which I call Idols of the Market-place,
on account of the commerce and consort of men there. For it
is by discourse that men associate; and words are imposed
according to the apprehension of the vulgar. And therefore the
ill and unfit choice of words wonderfully obstructs the understanding. . . .

### XLIV.

Lastly, there are Idols which have immigrated into men's
minds from the various dogmas of philosophies, and also from
wrong laws of demonstration. These I call Idols of the Theatre;
because in my judgment all the received systems are but so
many stage-plays, representing worlds of their own creation
after an unreal and scenic fashion. . . .

### XLV.

The human understanding is of its own nature prone to suppose the existence of more order and regularity in the world

than it finds. And though there be many things in nature which are singular and unmatched, yet it devises for them parallels and conjugates and relatives which do not exist. Hence the fiction that all celestial bodies move in perfect circles; spirals and dragons being (except in name) utterly rejected. Hence too the element of Fire with its orb is brought in, to make up the square with the other three which the sense perceives. Hence also the ratio of density of the so-called elements is arbitrarily fixed at ten to one. And so on of other dreams. And these fancies affect not dogmas only, but simple notions also.

XLVI.

The human understanding when it has once adopted an opinion (either as being the received opinion or as being agreeable to itself) draws all things else to support and agree with it. And though there be a greater number and weight of instances to be found on the other side, yet these it either neglects and despises, or else by some distinction sets aside and rejects; in order that by this great and pernicious predetermination the authority of its former conclusions may remain inviolate. And therefore it was a good answer that was made by one who when they showed him hanging in a temple a picture of those who had paid their vows as having escaped shipwreck, and would have him say whether he did not now acknowledge the power of the gods,—"Aye," asked he again, "but where are they painted that were drowned after their vows?" And such is the way of all superstition, whether in astrology, dreams, omens, divine judgments, or the like; wherein men, having a delight in such vanities, mark the events where they are fulfilled, but where they fail, though this happen much oftener, neglect and pass them by. But with far more subtlety does this mischief insinuate itself into philosophy and the sciences; in which the first conclusion colours and brings into conformity with itself all that come after, though far sounder and better. Besides, independently of that delight and vanity which I have described, it is the peculiar and perpetual error of the human intellect to be more moved and excited by affirmatives than by negatives; whereas it ought properly to hold itself indifferently disposed towards both alike. Indeed in the establishment of any true axiom, the negative instance is the more forcible of the two.

XLVII.

The human understanding is moved by those things most which strike and enter the mind simultaneously and suddenly, and so fill the imagination; and then it feigns and supposes all other things to be somehow, though it cannot see how, similar to those few things by which it is surrounded. But for that going to and fro to remote and heterogeneous instances, by which axioms are tried as in the fire, the intellect is altogether slow and unfit, unless it be forced thereto by severe laws and overruling authority.

ERNST MACH

# On the Importance

# of Psychology

# in the Understanding

# of Science

WITHOUT IN THE least being a philosopher, or even wanting to be called one, the scientist has a strong need to fathom the processes by means of which he obtains and extends his knowledge. The most obvious way of doing this is to examine carefully the growth of knowledge in one's own and the more easily accessible neighbouring fields, and above all to detect the specific motives that guide the enquirer. For to the scientist, who has been close to these problems, having often experienced the tension that precedes solution and the relief that comes afterwards, these motives ought to be more visible than to others. Since in almost every new major solution of a problem he will continue to see new features, he will find systematizing and schematizing more difficult and always apparently premature: he therefore likes to leave such aspects to philosophers who have more practice in this. The scientist can be satisfied if he recognizes the conscious mental activity of the enquirer as a variant of the instinctive activity of animal

Excerpts from E. Mach, *Knowledge and Error: Sketches on the Psychology of Inquiry* (first published 1905). Translated by T. J. McCormack and P. Foulkes. Boston: D. Reidel, 1975, pp. xxxi; 1–2. Reprinted by permission.

and man in nature and society—a variant methodically clarified, sharpened and refined. . . .

The goal of the ordinary imagination is the conceptual completion and perfection of a partially observed fact. The hunter imagines the way of life of the prey he has just sighted, in order to choose his own behaviour accordingly. The farmer considers the proper soil, sowing and maturing of the fruit of plants that he intends to cultivate. This trait of mental completion of a fact from partial data is common to ordinary and scientific thought. Galileo, too, merely wants to represent to himself the trajectory as a whole, given the inital speed and direction of a projected stone. However, there is another feature that often very significantly distinguishes scientific from ordinary thought: the latter, at least in its beginnings, serves practical ends, and first of all the satisfaction of bodily needs. The more vigorous mental exercise of scientific thought fashions its own ends and seeks to satisfy itself by removing all intellectual uneasiness: having grown in the service of practical ends, it becomes its own master. Ordinary thought does not serve pure knowledge, and therefore suffers from various defects that at first survive in scientific thought, which is derived from it. Science only very gradually shakes itself free from these flaws. . . .

The representation in thought of facts or the adaption of thought to fact, enables the thinker mentally to complete partially observed facts, insofar as completion is determined by the observed part. Their determination consists in the mutual dependence of factual features, so that thought has to aim at these. Since ordinary thinking and even incipient scientific thought must make do with a rather crude adaption of thoughts to facts, the former do not quite agree amongst each other. Mutual adaptation of thoughts is therefore the further task to be solved in order to attain full intellectual satisfaction. This last endeavour, which involves logical clarification of thinking though reaching far beyond this goal, is the outstanding mark that distinguishes scientific from ordinary thought. . . .

By what means has our knowledge of nature actually grown in the past, and what are the prospects for further growth in the future? The enquirer's behaviour has developed instinctively in practical activity and popular thought, and has merely been transferred to the field of science where in the end it has been developed into a conscious method. To meet our requirements we shall not need to go beyond the empirically given. We shall be satisfied if we can reduce the features of the enquirer's behaviour to actually observable ones in our own physical and mental life (features that recur in practical life and in the action and thought of peoples); and if we can show that this behaviour really leads to practical and intellectual advantages. A natural basis for this purpose is a general consideration of our physical and mental life.

JEAN PIAGET

# Genetic

# Epistemology

# and Science

GENETIC EPISTEMOLOGY ATTEMPTS to explain knowledge, and in particular scientific knowledge, on the basis of
its history, its sociogenesis, and especially the psychological
origins of the notions and operations upon which it is based.
These notions and operations are drawn in large part from
common sense, so that their origins can shed light on their
significance as knowledge of a somewhat higher level. But
genetic epistemology also takes into account, wherever possible, formalization—in particular, logical formalizations applied to equilibrated thought structures and in certain cases
to transformations from one level to another in the development of thought. . . .

We cannot say that on the one hand there is the history of
scientific thinking, and on the other the body of scientific
thought as it is today; there is simply a continual transformation, a continual reorganization. And this fact seems to me
to imply that historical and psychological factors in these
changes are of interest in our attempt to understand the nature
of scientific knowledge.[1]

. . . How is it that Einstein was able to give a new operational
definition of simultaneity at a distance? How was he able to

Excerpts from J. Piaget, *Genetic Epistemology*. Translated by E. Duckworth. New York:
Columbia University Press, 1970, pp. 1; 4–13. Reprinted by permission.

criticize the Newtonian notion of universal time without giving rise to a deep crisis within physics? Of course his critique had its roots in experimental findings, such as the Michelson-Morley experiment—that goes without saying. Nonetheless, if this redefinition of the possibility of events to be simultaneous at great distances from each other went against the grain of our logic, there would have been a considerable crisis within physics. We would have had to accept one of two possibilities: either the physical world is not rational, or else human reason is impotent—incapable of grasping external reality. Well, in fact nothing of this sort happened. There was no such upheaval. A few metaphysicians (I apologize to the philosophers present) such as Bergson or Maritain were appalled by this revolution in physics, but for the most part and among scientists themselves it was not a very drastic crisis. Why in fact was it not a crisis? It was not a crisis because simultaneity is not a primitive notion. It is not a primitive concept, and it is not even a primitive perception. I shall go into this subject further later on, but at the moment I should just like to state that our experimental findings have shown that human beings do not perceive simultaneity with any precision. If we look at two objects moving at different speeds, and they stop at the same time, we do not have an adequate perception that they stopped at the same time. Similarly, when children do not have a very exact idea of what simultaneity is, they do not conceive of it independently of the speed at which objects are traveling. Simultaneity, then, is not a primitive intuition; it is an intellectual construction.

Long before Einstein, Henri Poincaré did a great deal of work in analyzing the notion of simultaneity and revealing its complexities. His studies took him, in fact, almost to the threshold of discovering relativity. Now if we read his essays on this subject, which, by the way, are all the more interesting when considered in the light of Einstein's later work, we see that his reflections were based almost entirely on psychological arguments. Later on I shall show that the notion of time and the notion of simultaneity are based on the notion of speed, which is a more primitive intuition. So there are all sorts of reasons, psychological reasons, that can explain why the crisis brought about by relativity theory was not a fatal one for physics. Rather, it was readjusting, and one can find the psychological routes for this readjustment as well as the experimental and logical basis. In point of fact, Einstein himself recognized the relevance of psychological factors, and when I had the good chance to meet him for the first time in 1928, he suggested to me that it would be of interest to study the origins in children of notions of time and in particular of notions of simultaneity.

What I have said so far may suggest that it can be helpful to make use of psychological data when we are considering the nature of knowledge. I should like now to say that it is more than helpful; it is indispensable. In fact, all epistemologists refer to psychological factors in their analyses, but for the most part their references to psychology are speculative and are not based on psychological research. I am convinced that all epistemology brings up factual problems as well as formal ones, and once factual problems are encountered, psychological findings become relevant and should be taken into account. The unfortunate thing for psychology is that everybody thinks of himself as a psychologist. This is not true for the field of physics, or for the field of philosophy, but it is unfortunately true for psychology. Every man considers himself a psychologist. As a result, when an epistemologist needs to call on some psychological aspect, he does not refer to psychological research and he does not consult psychologists; he depends on his own reflections. He puts together certain ideas and relationships within his own thinking, in his personal attempt to resolve the psychological problem that has arisen. I should like to cite some instances in epistemology where psychological findings can be pertinent, even though they may seem at first sight far removed from the problem.

My first example concerns the school of logical positivism. Logical positivists have never taken psychology into account in their epistemology, but they affirm that logical beings and mathematical beings are nothing but linguistic structures. That is, when we are doing logic or mathematics, we are simply using general syntax, general semantics, or general pragmatics in the sense of Morris, being in this case a rule of the uses of language in general. The position in general is that logical and mathematical reality is derived from language. Logic and mathematics are nothing but specialized linguistic structures. Now here it becomes pertinent to examine factual findings. We can look to see whether there is any logical behavior in

children before language develops. We can look to see whether
the coordinations of their actions reveal a logic of classes,
reveal an ordered system, reveal correspondence structures.
If indeed we find logical structures in the coordinations of
actions in small children even before the development of language, we are not in a position to say that these logical structures are derived from language. This is a question of fact and
should be approached not by speculation but by an experimental methodology with its objective findings.

The first principle of genetic epistemology, then, is this—to
take psychology seriously. Taking psychology seriously means
that, when a question of psychological fact arises, psychological research should be consulted instead of trying to invent
a solution through private speculation.

It is worthwhile pointing out, by the way, that in the field
of linguistics itself, since the golden days of logical positivism,
the theoretical position has been reversed. Bloomfield in his
time adhered completely to the view of the logical positivists,
to this linguistic view of logic. But currently, as you know,
Chomsky maintains the opposite position. Chomsky asserts,
not that logic is based on and derived from language, but, on
the contrary, that language is based on logic, on reason, and
he even considers this reason to be innate. He is perhaps going
too far in maintaining that it is innate; this is once again a
question to be decided by referring to facts, to research. It is
another problem for the field of psychology to determine. Between the rationalism that Chomsky is defending nowadays
(according to which language is based on reason, which is
thought to be innate in man) and the linguistic view of the
positivists (according to which logic is simply a linguistic
convention), there is a whole selection of possible solutions,
and the choice among these solutions must be made on the
basis of fact, that is, on the basis of psychological research.
The problems cannot be resolved by speculation.

I do not want to give the impression that genetic epistemology is based exclusively on psychology. On the contrary,
logical formalization is absolutely essential every time that we
can carry out some formalization; every time that we come
upon some completed structure in the course of the development of thought, we make an effort, with the collaboration
of logicians or of specialists within the field that we are con-

sidering, to formalize this structure. Our hypothesis is that there will be a correspondence between the psychological formation on the one hand, and the formalization on the other hand. But although we recognize the importance of formalization in epistemology, we also realize that formalization cannot be sufficient by itself. We have been attempting to point out areas in which psychological experimentation is indispensable to shed light on certain epistemological problems, but even on its own grounds there are a number of reasons why formalization can never be sufficient by itself. I should like to discuss three of these reasons.

The first reason is that there are many different logics, and not just a single logic. This means that no single logic is strong enough to support the total construction of human knowledge. But it also means that, when all the different logics are taken together, they are not sufficiently coherent with one another to serve as the foundation for human knowledge. Any one logic, then, is too weak, but all the logics taken together are too rich to enable logic to form a single value basis for knowledge. That is the first reason why formalization alone is not sufficient.

The second reason is found in Gödel's theorem. It is the fact that there are limits to formalization. Any consistent system sufficiently rich to contain elementary arithmetic cannot prove its own consistency. So the following questions arise: logic is a formalization, an axiomatization of something, but of what exactly? What does logic formalize? This is a considerable problem. There are even two problems here. Any axiomatic system contains the undemonstrable propositions or the axioms, at the outset, from which the other propositions can be demonstrated, and also the undefinable, fundamental notions on the basis of which the other notions can be defined. Now in the case of logic what lies underneath the undemonstrable axioms and the undefinable notions? This is the problem of structuralism in logic, and it is a problem that shows the inadequacy of formalization as the fundamental basis. It shows the necessity for considering thought itself as well as considering axiomatized logical systems, since it is from human thought that the logical systems develop and remain still intuitive.

The third reason why formalization is not enough is that

epistemology sets out to explain knowledge as it actually is within the areas of science, and this knowledge is, in fact not purely formal: there are other aspects to it. In this context I should like to quote a logician friend of mine, the late Evert W. Beth. For a very long time he was a strong adversary of psychology in general and the introduction of psychological observations into the field of epistemology, and by that token an adversary of my own work, since my work was based on psychology. Nonetheless, in the interests of an intellectual confrontation, Beth did us the honor of coming to one of our symposia on genetic epistemology and looking more closely at the questions that were concerning us. At the end of the symposium he agreed to co-author with me, in spite of his fear of psychologists, a work that we called *Mathematical and Psychological Epistemology*. This has appeared in French and is being translated into English [Beth & Piaget 1966]. In his conclusion to this volume, Beth wrote as follows: "The problem of epistemology is to explain how real human thought is capable of producing scientific knowledge. In order to do that we must establish a certain coordination between logic and psychology." This declaration does not suggest that psychology ought to interfere directly in logic—that is of course not true—but it does maintain that in epistemology both logic and psychology should be taken into account, since it is important to deal with both the formal aspects and the empirical aspects of human knowledge.

So, in sum, genetic epistemology deals with both the formation and the meaning of knowledge. We can formulate our problem in the following terms: by what means does the human mind go from a state of less sufficient knowledge to a state of higher knowledge? The decision of what is lower or less adequate knowledge, and what is higher knowledge, has of course formal and normative aspects. It is not up to psychologists to determine whether or not a certain state of knowledge is superior to another state. That decision is one for logicians or for specialists within a given realm of science. For instance, in the area of physics, it is up to physicists to decide whether or not a given theory shows some progress over another theory. Our problem, from the point of view of psychology and from the point of view of genetic epistemology, is to explain how the transition is made from a lower level of knowledge to a level that is judged to be higher. The nature of these transitions is a factual question.

## Note

1. Another opinion, often quoted in philosophical circles, is that the theory of knowledge studies essentially the question of the validity of science, the criteria of this validity and its justification. If we accept this viewpoint, it is then argued that the study of science *as it is*, as a fact, is fundamentally irrelevant. Genetic epistemology, as we see it, reflects most decidedly this separation of norm and fact, of valuation and description. We believe that, to the contrary, only in the real development of the sciences can we discover the implicit values and norms that guide, inspire, and regulate them. Any other attitude, it seems to us, reduces to the rather arbitrary imposition on knowledge of the personal views of an isolated observer. This we want to avoid.

# Selection 4

THOMAS KUHN

# Unanswered Questions about Science

I MUST FIRST ask what it is that still requires explanation. Not that scientists discover the truth about nature, nor that they approach ever closer to the truth. Unless, as one of my critics suggests [Hawkins 1963], we simply define the approach to truth as the result of what scientists do, we cannot recognize progress towards that goal. Rather we must explain why science—our surest example of sound knowledge—progresses as it does, and we must first find out how, in fact, it does progress.

Surprisingly little is yet known about the answer to that descriptive question. A vast amount of thoughtful empirical investigation is still required. With the passage of time, scientific theories taken as a group are obviously more and more articulated. In the process, they are matched to nature at an increasing number of points and with increasing precision. Or again, the number of subject matters to which the puzzle-solving approach can be applied clearly grows with time. There is a continuing proliferation of scientific specialties, partly by an extension of the boundaries of science and partly by the subdivision of existing fields.

Those generalizations are, however, only a beginning. We

Excerpt from I. Lakatos and A. Musgrave, eds., *Criticism and the Growth of Knowledge.* Cambridge: Cambridge University Press, 1970, pp. 18–21. Reprinted by permission.

know, for example, almost nothing about what a group of scientists will sacrifice in order to achieve the gains that a new theory invariably offers. My own impression, though it is no more than that, is that a scientific community will seldom or never embrace a new theory unless it solves all or almost all the quantitative, numerical puzzles that have been treated by its predecessor [see Kuhn 1958]. They will, on the other hand, occasionally sacrifice explanatory power, however reluctantly, sometimes leaving previously resolved questions open and sometimes declaring them altogether unscientific [see Kuhn 1962: 102–8]. Turning to another area, we know little about historical changes in the unity of the sciences. Despite occasional spectacular successes, communication across the boundaries between scientific specialties becomes worse and worse. Does the number of incompatible viewpoints employed by the increasing number of communities of specialists grow with time? Unity of the sciences is clearly a value for scientists, but for what will they give it up? Or again, though the bulk of scientific knowledge clearly increases with time, what are we to say about ignorance? The problems solved during the last thirty years did not exist as open questions a century ago. In any age, the scientific knowledge already at hand virtually exhausts what there is to know, leaving visible puzzles only at the horizon of existing knowledge. Is it not possible, or perhaps even likely, that contemporary scientists know less of what there is to know about their world than the scientists of the eighteenth century knew of theirs? Scientific theories, it must be remembered, attach to nature only here and there. Are the interstices between those points of attachment perhaps now larger and more numerous than ever before?

Until we can answer more questions like these, we shall not know quite what scientific progress is and cannot therefore quite hope to explain it. On the other hand, answers to those questions will very nearly provide the explanation sought. The two come almost together. Already it should be clear that the explanation must, in the final analysis, be psychological or sociological. It must, that is, be a description of a value system, an ideology, together with an analysis of the institutions through which that system is transmitted and enforced. Knowing what scientists value, we may hope to understand what problems they will undertake and what choices they will make in particular circumstances of conflict. I doubt that there is another sort of answer to be found.

What form that answer will take is, of course, another matter. At this point, too, my sense that I control my subject matter ends. But again, some sample generalizations will illustrate the sorts of answers which must be sought. For a scientist, the solution of a difficult conceptual or instrumental puzzle is a principal goal. His success in that endeavour is rewarded through recognition by other members of his professional group and by them alone. The practical merit of his solution is at best a secondary value, and the approval of men outside the specialist group is a negative value or none at all. These values, which do much to dictate the form of normal science, are also significant at times when a choice must be made between theories. A man trained as a puzzle-solver will wish to preserve as many as possible of the prior puzzle-solutions obtained by his group, and he will also wish to maximize the number of puzzles that can be solved. But even these values frequently conflict, and there are others which make the problem of choice still more difficult. It is just in this connection that a study of what scientists will give up would be most significant. Simplicity, precision, and congruence with the theories used in other specialties are all significant values for the scientists, but they do not all dictate the same choice nor will they all be applied in the same way. That being the case, it is also important that group unanimity be a paramount value, causing the group to minimize the occasions for conflict and to reunite quickly about a single set of rules for puzzle solving even at the price of subdividing the specialty or excluding a formerly productive member [see Kuhn 1962: 161–69].

HERBERT S. SIMON

# The Psychology

# of Scientific

# Problem Solving

THE THESIS OF the present chapter is that scientific discovery is a form of problem solving, and that the processes whereby science is carried on can be explained in the terms that have been used to explain the processes of problem solving. . . .

To explain scientific discovery is to describe a set of processes that is sufficient—and *just* sufficient—to account for the amounts and directions of scientific progress that have actually occurred. For a variety of reasons, perhaps best understood·by psychoanalysis, when we talk or write about scientific discovery, we tend to dwell lovingly on the great names and the great events—Galileo and uniform acceleration, Newton and universal gravitation, Einstein and relativity, and so on. We insist that a theory of discovery postulate processes sufficiently powerful to produce these events. It is right to so insist, but we must not forget how rare such events are, and we must not postulate processes so powerful that they predict a discovery of first magnitude as a daily matter. . . .

The theory of scientific discovery I propose to set forth rests

Excerpt from H. A. Simon, Scientific discovery and the psychology of problem solving. In R. G. Colodny, ed., *Mind and Cosmos: Essays in Contemporary Science and Philosophy*, pp. 22–25, 28–34, 38–39 (footnotes deleted). Pittsburgh: University of Pittsburgh Press, 1966. Copyright 1966, reprinted by permission University of Pittsburgh Press. The complete article is reprinted as chapter 5.2 in H. A. Simon, ed., *Models of Discovery*. Boston: D. Reidel, 1977.

on the hypothesis that there are no qualitative differences between the *processes* of revolutionary science and of normal
science, between work of high creativity and journeyman
work. I shall not claim that the case can be proven conclusively. My main evidence will be data indicating that the processes that show up in relatively simple and humdrum forms
of human problem solving are also the ones that show up
when great scientists try to describe how they do their work.
How convincing the evidence is can better be judged at the
end of the chapter.

Let us return, then, to the problem-solving theory proposed
in the last chapter and confront that theory with the recorded
phenomena of scientific discovery.

The problem-solving theory asserted that thinking is an organization of elementary information processes, organized
hierarchically and executed serially. In overall organization,
the processes exhibit large amounts of highly selective trial-
and-error search using rules of thumb, or heuristics, as bases
for their selectivity. Among the prominent heuristic schemes
are means-end analysis, planning and abstraction, factoriza-
tion, and satisficing. Our task is to show how a system with
these characteristics can behave like a scientist. . . .

## Incubation

## and Unconscious Processes

## in Discovery

The phenomena of incubation and sudden illumination have
held immense fascination for those who have written on scientific discovery. Poincaré's experience on boarding the bus
at Coutances takes its place in the annals of illumination along
with Proust's madeleine dipped in tea:

> Just at this time I left Caen, where I was then living, to go on
> a geological excursion under the auspices of the school of
> mines. The changes of travel made me forget my mathematical
> work. Having reached Coutances, we entered an omnibus to go
> some place or other. At the moment when I put my foot on the
> step the idea came to me, without anything in my former

thoughts seeming to have paved the way for it, that the transformations I had used to define the Fuchsian functions were
identical with those of non-Euclidean geometry.

Hadamard places particular emphasis on the role of the unconscious in mathematical invention. While he proposes no
specific theory of the processes that go on during incubation,
he argues strongly that these are active processes and not
merely a forgetting of material generated during conscious
work that is inhibiting the problem solution.

The theory of problem solving proposed in the last chapter
does not assign any special role to the unconscious—or, for that
matter, to the conscious. It assumes, implicitly, that the information processes that occur without consciousness of them
are of the same kinds as the processes of which the thinker
is aware. It assumes, further, that the organization of the totality of processes, conscious and unconscious, is fundamentally serial rather than parallel in time. . . .

I should like to describe two mechanisms currently employed in the information-processing theories that appear to
go a long way toward accounting for these phenomena. The
first of these mechanisms is called *familiarization*, the second
is called *selective forgetting*. The familiarization mechanism
emerged in the course of constructing a theory of human rote
memory, the forgetting mechanism in the course of trying to
discover why the organization of the first theorem-proving
program, the Logic Theorist, was more effective in solving
problems than the organization of early versions of the General
Problem Solver. Neither mechanism was devised, then, with
incubation and illumination in mind; they were introduced
into the theory to meet other requirements imposed by the
data on problem solving.

## 1. *FAMILIARIZATION*

Thinking processes make use of certain means in the central
nervous system for holding symbols in short-term or "immediate" memory. Little is known of the neurophysiological
substrate of immediate memory, but a good deal is known
about its phenomenal characteristics. Most important, the
number of symbols that can be stored in immediate memory

is severely limited—in George Miller's words, "seven, plus or minus two." But a "symbol" can serve as the name for anything that can be recognized as familiar and that has information associated with it in permanent memory. Thus "*a*" is a symbol; so is "Lincoln's Gettysburg Address." For most native speakers of English "criminal lawyer" is a symbol, but for a person just learning the language, the phrase may constitute a pair of symbols denoting a lawyer with certain antisocial tendencies.

The important facts are (1) that only about seven symbols can be held and manipulated in immediate memory at one time and (2) that anything can become a symbol through repeated exposure to it, or familiarization. Familiarization involves storing in *permanent* memory information that allows the symbol to be recognized and a single symbol or "name" to be substituted for it. . . .

## 2. SELECTIVE FORGETTING

A second mechanism to be found in information-processing theories of problem solving that is essential to our proposed explanation of incubation and illumination involves more rapid forgetting of some memory contents than of others. The selective forgetting rests, in turn, on the distinction between forms of short-term and long-term memory.

In the typical organization of a problem-solving program, the solution efforts are guided and controlled by a hierarchy or "tree" of goals and subgoals. Thus, the subject starts out with the goal of solving the original problem. In trying to reach this goal, he generates a subgoal that will take him part of the way (if it is achieved) and addresses himself to that subgoal. If the subgoal is achieved, he may then return to the now-modified original goal. If difficulties arise in achieving the subgoal, sub-subgoals may be erected to deal with them.

The operation of such a process requires the goal hierarchy to be held in memory. If a subgoal is achieved, it can be forgotten, but the tree of unattained goals must be retained. In human problem solvers this retention is not always perfect, of course. When part of the structure is lost, the subject says, "Where am I?" or "Now why was I trying to get that result?" and may have to go over some of the same ground to get back into context—i.e., to locate himself in that part of the tree that has been retained in memory. If we were designing such a system, instead of probing the one that human beings possess, we would specify that the goal tree be held in some kind of temporary memory, since it is a dynamic structure, whose function is to guide search, and it is not needed (or certainly not all of it) when the problem solution has been found. Our hypothesis is that human beings are also constructed in this way—that the goal tree is held in a relatively short-term memory.

During the course of problem solving, a second memory structure is being built up. First of all, new complexes are being familiarized, so that they can be handled by the processing system as units. In addition, the problem solver is noticing various features of the problem environment and is storing some of these in memory. If he is studying a chess position, for example, in the course of his explorations he may notice that a particular piece is undefended or that another piece is pinned against the queen.

This kind of information is perceived while the problem solver is addressing himself to particular subgoals. What use is made of it at the time it is noted depends on what subgoal is directing attention at that moment. But some of this information is also transferred to more permanent forms of memory and is associated with the problem environment—in this example, with the chess position. This information about the environment is used, in turn, in the processes that erect new subgoals and that work toward subgoal achievement. Hence, over the longer run, this information influences the growth of the subgoal tree. To have a short name for it (since it is now a familiar unit for us!), I will call the information about the task environment that is noticed in the course of problem solution and fixated in permanent (or relatively long-term) memory the "blackboard."

The course of problem solving, then, involves continuous interaction between goal tree and blackboard. In the course of pursuing goals, information is added to the blackboard. This information, in turn, helps to determine what new goals and subgoals will be set up. During periods of persistent activity, the problem solver will always be working in local goal contexts, and information added to the blackboard will be used, in the short run, only if it is relevant in those contexts.

What happens, now, if the problem solver removes himself from the task for a time? Information he has been holding in relatively short-term memory will begin to disappear, and to disappear more rapidly than information in long-term memory. But we have hypothesized that the goal tree is held in short-term memory, the blackboard in long-term memory. Hence, when the problem solver next takes up the task, many or most of the finer twigs and branches of the goal tree will have disappeared. He will begin again, with one of the higher level goals, to reconstruct that tree—but now with the help of a very different set of information, on the blackboard, than he had the first time he went down the tree.

In general, we would expect the problem solver, in his renewed examination of the problem, to follow a quite different path than he did originally. Since his blackboard now has better information about the problem environment than it did the first time, he has better cues to find the correct path. Under these circumstances (and remembering the tremendous differences a few hints can produce in problem solution), solutions may appear quickly that had previously eluded him in protracted search.

There is almost no direct evidence at the present time for the validity of this explanation of incubation and illumination. (I have been able, introspectively, to account for my most recent illumination experience quite simply in these terms, but perhaps my introspections are compromised as witnesses.) It invokes, however, only mechanisms that have already been incorporated in problem-solving theories. It does leave one aspect of the phenomena unaccounted for—it does not explain how the problem that the problem solver has temporarily (consciously) abandoned is put back on the agenda by unconscious processes. It does, however, account for the suddenness of solution without calling on the subconscious to perform elaborate processes, or processes different from those it and the conscious perform in the normal course of problem-solving activity. Nor does it postulate that the unconscious is capable of random searches through immense problem spaces for the solution.

## Conclusion

Theories are now available that incorporate mechanisms sufficient to account for some of the principal phenomena of problem solving in at least certain relatively well-structured situations. The aim of this chapter has been to ask how much these theories need to be modified or extended in order to account for problem solving in science. The general tenor of the argument has been that problem solving in science, like problem solving in the psychological laboratory, is a tedious, painstaking process of selective trial and error. Our knowledge of it does not suggest the presence of completely unknown processes far more powerful than those that have been observed in the laboratory.

Several kinds of objections can be raised, and have been, against this "minimalist" theory. One objection is that it does not account for striking phenomena like incubation and illumination. To meet this objection, a mechanism has been proposed that is believed sufficient to produce exactly these kinds of phenomena.

Another objection is that the theory only explains how problems are solved that have already been stated and for which there exist well-defined representations. This objection has not been answered in detail, but an answer has been sketched in terms of the broader social environment within which scientific work takes place. Most scientific activity goes on within the framework of established paradigms. Even in revolutionary science, which creates those paradigms, the problems and representations are rooted in the past; they are not created out of whole cloth.

We are still very far from a complete understanding of the whole structure of the psychological processes involved in making scientific discoveries. But perhaps our analysis makes somewhat more plausible the hypothesis that at the core of this structure is the same kind of selective trial-and-error search that has already been shown to constitute the basis for human problem-solving activity in the psychological laboratory.

**Part II** _____

# Hypotheses

# in Science

# Introduction

FEW ACTIVITIES OF the modern scientist are more highly esteemed than the generation of fruitful hypotheses. It is the great theoreticians who achieve lasting recognition. Discoverers of "mere facts," experimenters, and those who apply the results of science may have their moment of glory, but they are more likely to be relegated to footnotes and obscure passages in outdated textbooks. "Baconian science" has become a pejorative term. Consider Albert Einstein, who conducted no experiments (Frank 1947). More striking, however, is the fact that most of his important early papers were based on little or no new empirical evidence. Yet Einstein is the quintessential scientist, widely admired by scientists and nonscientists alike.

These values have not always been in vogue; fashion changes in science as elsewhere. Three hundred years ago speculation and bold hypotheses were seen in less favorable light than today. A prominent illustration can be found in Isaac Newton's assertion "I frame no hypotheses" (see selection 7).[1] What Newton apparently meant was that he did not employ speculative conjectures, unsupported by evidence. In his view, the proper approach was to derive general principles from experiments and phenomena, leaving ultimate explanations of phenomena (e.g., gravitational attraction) to later inquiry.

Newton proposed a data-oriented science, one that used hypotheses in a careful, controlled fashion. His method of analysis and composition (or synthesis) was used to great effect, particularly in the experiments in optics. In his famous prism experiments, Newton first showed that rays of sunlight produced a spectrum of colors when passed through a prism, each color being refracted at a particular angle. He concluded (by "induction") that sunlight is composed of rays of different colors. From this general principle he then deduced new predicted phenomena—for example, that monochrome rays of light should refract at particular angles and not produce spectra. The first part of the procedure, the induction of the general rule from specific facts, is the method of analysis; the second part, the deduction of new effects, is the method of synthesis. Newton deliberately eschewed reference to unseen or "occult" causes, or to unobserved entities of a Cartesian sort.

In all cases where it is necessary or desirable to formulate causal hypotheses, such hypotheses, according to Newton, should be closely scrutinized and regulated. Above all, hypotheses should be subordinated to evidence: "This rule we must follow, that the argument of induction may not be evaded by hypotheses. Similar positions were held by other seventeenth- and eighteenth-century writers, among them John Locke (1632–1704), who was even more skeptical than Newton about "general maxims, precarious principles, and hypotheses laid down at pleasure" (selection 8).

The twentieth-century view of the value of hypotheses is epitomized in the writing of Karl Popper (selection 9). In Popper's opinion, science cannot be based upon inductive procedures such as the method of analysis. There are of course philosophers and scientists who still stress the importance of inductive procedures. Carnap (1962), for example, attempted to explicate the role of inductive evidence in justifying scientific propositions. Under the influence of logical positivism, many psychologists (e.g., Stevens 1939; Skinner 1953; Sidman 1960) have argued that inductively derived empirical generalizations are necessary prerequisites for the development of more abstract theories and hypotheses. However, in Popper's view "pure," theory-free, observations do not exist; true induction is a myth. Thus, belief in scientific propositions cannot be justified by inductive evidence.

Beyond the fact that different people at different times have had radically opposed notions about the value of hypotheses vis-à-vis data, of what relevance is this to the psychology of science? First of all, it is important to note that Newton, Locke, and Popper put forth prescriptive maxims, making assertions about how science ought to be done, specifying procedures,

rules, and general guidelines for the conduct of scientific inquiry. For Newton and Locke, hypotheses are to be employed
(if at all) only when well supported by evidence. While Newton speculated freely, sometimes wildly, in his notes and letters, he carefully circumscribed the use of hypotheses in his
writings. For Popper, any hypothesis, no matter how speculative, may be legitimately entertained, as long as it is potentially testable. Some recent philosophers (e.g., Feyerabend
1975) have dropped even the requirement of testability.

Can empirical evidence be adduced to support one or another of these approaches? Locke's readers are invited to consider how little speculative hypotheses had "helped to satisfy
the inquiries of rational men." Newton pointed to his own
successful use of the methods of analysis and synthesis.
Popper, citing the example of the man who recorded all his
observations and left them to the Royal Society, concluded
that "though beetles may profitably be collected, observations
may not" (selection 9).

What is missing in all these arguments is any systematic,
empirical evidence concerning their correctness. To support
such claims rigorously one would need to show that the prescribed approaches in fact produce more progress than would
an alternative approach, with variables such as the difficulty
of the problem, type of hypothesis employed, and competence
of the scientist held more or less constant. Neither Popper nor
Locke nor Newton (nor so far as we know anyone else) has
ever tried to provide such evidence. Yet the issues they raise
cannot be resolved solely on the basis of historical, anecdotal,
or self-report data; they are simply too complex. It is possible
that the use of speculative hypotheses may be desirable under
some circumstances but not others, that only some types of
speculative hypotheses may be useful, or that the type of hypothesis and type of situation interact to determine what is
optimal. Such possibilities are not easily dealt with by fiat.
On the other hand, it is conceivable that much could be learned
by careful and thoughtful experimentation. The method of laboratory science can here be applied to science itself.

Questions of evidential support for their utility aside, to
what extent do scientists actually follow maxims about the use
of hypotheses? The answer, to the extent that one can be given,
seems to be "very little." A particularly instructive example

can be drawn from Newton. Consider his laws of motion:

1. Every body continues in its state of rest, or of uniform motion in a right line, unless it is compelled to change that state by forces impressed upon it.

2. The change of motion is proportional to the motive force impressed; and is made in the direction of the right line in which that force is impressed.

3. To every action there is always opposed an equal reaction: or, the natural actions of two bodies upon each other are always equal, and directed to contrary parts. (Newton 1687/ 1947:13)

Newton claimed that these principles had been derived from phenomena via the method of analysis. However, the inference process used to arrive at the laws of motion was very different from that used for the laws of optics. In the latter case, Newton generalized from a few instances (*this* prism and *this* light ray) to a universal proposition (*all* prisms and *all* light rays). The phenomena covered by the general law are identical to those observed in the specific instances. The laws of motion, on the other hand, are not just assertions that some observed phenomena will occur generally. Neither Newton nor anyone else had ever observed a body continuing in "uniform motion in a right line." The laws of motion are not literal extrapolations from observed events. A much more "hypothetical" sort of inference was involved. The observed behavior of actual objects—cannon balls, falling apples, and planets—can be successfully predicted and explained if the laws of motion are assumed to be true, but the laws themselves refer to events which have not and cannot be observed. Thus, Newton actually made considerable use of hypotheses, but in a complex fashion (Butts and Davis 1970). Newton clearly employed what today might be called hypothetical constructs.[2]

These examples make it clear that two quite different approaches to research can be found in Newton's publications, one corresponding to experimental inquiry, as in the 1704 *Opticks*, and one corresponding to theoretical inquiry, as in the *Principia Mathematica* of 1687 (Cohen 1956). In the *Principia*, the three laws of motion are taken as an axiomatic starting point, from which consequences are deduced. In this respect, even the *Principia* can be seen as a test of a single grand hypothesis. This is not to take Newton to task for failing to follow a particular prescription about hypotheses (in this instance his own). Rather it is to suggest that actual scientific practice may show little relation to "rules" of scientific practices or to autobiographical accounts. Good scientists, and certainly great scientists like Newton, often show remarkable intuition in ignoring such rules.

All this supports the previous point that whether or not to employ speculative hypotheses is a question too complex to be adequately dealt with by simple prescriptive rules. Newton and many other scientists somehow seem to sense when such hypotheses are necessary and when they are impediments. How such judgments are made is unknown, and virtually unexplored. Experimental research might go a long way toward providing an answer, especially if a typology of problems could be formulated. Scientists' "intuitions" about when and how to employ speculative hypotheses might then turn out to be much less mysterious than they now appear.

Whether such problems can be addressed experimentally is very much an open question, and at present we can point to little relevant research. There is, however, a fairly large research literature on hypothesis testing, a not totally dissimilar problem, and some reason for believing that the results of this research generalize to actual science. This material is discussed extensively in part 4, and in our view it suggests strongly that questions about how hypotheses are used are in fact researchable questions.

## Notes

1. This statement appeared first in the 1712 edition of the *Principia*. Hanson (1970) argued that it was written in response to Cartesian and Aristotelian criticism of the 1687 edition.

2. Newton even postulated some very "occult" hypothetical notions, especially in connection with his theory of the ether. But he carefully cricumscribed his use of such speculations. See the collection of his papers edited by I. B. Cohen (Newton 1958).

# Selection 6

ISAAC NEWTON

# On the Use

# of Hypotheses

ALL THESE THINGS being consider'd, it seems probable to me, that God in the Beginning form'd Matter in solid, massy, hard, impenetrable, moveable Particles, of such Sizes and Figures, and with such other Properties, and in such Proportion to Space, as most conduced to the End for which he form'd them; and that these primitive Particles being Solids, are incomparably harder than any porous Bodies compounded of them; even so very hard, as never to wear or break in pieces; no ordinary Power being able to divide what God himself made one in the first Creation. While the Particles continue entire, they may compose Bodies of one and the same Nature and Texture in all Ages: But should they wear away, or break in pieces, the Nature of Things depending on them, would be changed. Water and Earth, composed of old worn Particles and Fragments of Particles, would not be of the same Nature and Texture now, with Water and Earth composed of entire Particles in the Beginning. And therefore, that Nature may be lasting, the Changes of corporeal Things are to be placed only in the various Separations and new Associations and Motions of these permanent Particles; compound Bodies being apt to break, not in the midst of solid Particles, but where those Particles are laid together, and only touch in a few Points.

It seems to me farther, that these Particles have not only a Vis *inertiae*, accompanied with such passive Laws of Motion

Excerpt from I. Newton, *Opticks* (originally published 1704). London: Bell & Hyman, 1931, pp. 400–6.

as naturally result from that Force, but also that they are moved by certain active Principles, such as is that of Gravity, and that which causes Fermentation, and the Cohesion of Bodies. These Principles I consider, not as occult Qualities, supposed to result from the specifick Forms of Things, but as general Laws of Nature, by which the Things themselves are form'd; their Truth appearing to us by Phaenomena, though their Causes be not yet discover'd. For these are manifest Qualities, and their Causes only are occult. And the *Aristotelians* gave the Name of occult Qualities, not to manifest Qualities, but to such Qualities only as they supposed to lie hid in Bodies, and to be the unknown Causes of manifest Effects: Such as would be the Causes of Gravity, and of magnetick and electrick Attractions, and of Fermentations, if we should suppose that these Forces or Actions arose from Qualities unknown to us, and uncapable of being discovered and made manifest. Such occult Qualities put a stop to the Improvement of natural Philosophy, and therefore of late Years have been rejected. To tell us that every Species of Things is endow'd with an occult specifick Quality by which it acts and produces manifest Effects, is to tell us nothing: But to derive two or three general Principles of Motion from Phaenomena, and afterwards to tell us how the Properties and Actions of all corporeal Things follow from those manifest Principles, would be a very great step in Philosophy, though the Causes of those Principles were not yet discover'd: And therefore I scruple not to propose the Principles of Motion above-mention'd, they being of very general Extent, and leave their Causes to be found out. . . .

As in Mathematicks, so in Natural Philosophy, the Investigation of difficult Things by the Method of Analysis, ought ever to precede the Method of Composition. This Analysis consists in making Experiments and Observations, and in drawing general Conclusions from them by Induction, and admitting of no Objections against the Conclusions, but such as are taken from Experiments, or other certain Truths. For Hypotheses are not to be regarded in experimental Philosophy. And although the arguing from Experiments and Observations by Induction be no Demonstration of general Conclusions; yet it is the best way of arguing which the Nature of Things admits of, and may be looked upon as so much the stronger, by how much the Induction is more general. And if no Exception occur from Phaenomena, the Conclusion may be pronounced generally. But if at any time afterwards any Exception shall occur from Experiments, it may then begin to be pronounced with such Exceptions as occur. By this way of Analysis we may proceed from Compounds to Ingredients, and from Motions to the Forces producing them; and in general, from Effects to their Causes, and from particular Causes to more general ones, till the Argument end in the most general. This is the Method of Analysis: And the Synthesis consists in assuming the Causes discover'd, and establish'd as Principles, and by them explaining the Phaenomena proceeding from them, and proving the Explanations.

In the two first Books of these Opticks, I proceeded by this Analysis to discover and prove the original Differences of the Rays of Light in respect of Refrangibility, Reflexibility, and Colour, and their alternate Fits of easy Reflexion and easy Transmission, and the Properties of Bodies, both opake and pellucid, on which their Reflexions and Colours depend. And these Discoveries being proved, may be assumed in the Method of Composition for explaining the Phaenomena arising from them: An Instance of which Method I gave in the End of the first Book. In this third Book I have only begun the Analysis of what remains to be discover'd about Light and its Effects upon the Frame of Nature, hinting several things about it, and leaving the Hints to be examin'd and improv'd by the farther Experiments and Observations of such as are inquisitive. And if natural Philosophy in all its Parts, by pursuing this Method, shall at length be perfected, the Bounds of Moral Philosophy will be also enlarged. For so far as we can know by natural Philosophy what is the first Cause, what Power he has over us, and what Benefits we receive from him, so far our Duty towards him, as well as that towards one another, will appear to us by the Light of Nature.

# Selection 7

## ISAAC NEWTON

# The Rules

# of Hypothesizing

RULE I

*We are to admit no more causes of natural things than such as are both true and sufficient to explain their appearances.*

To this purpose the philosophers say that Nature does nothing in vain, and more is in vain when less will serve; for Nature is pleased with simplicity, and affects not the pomp of superfluous causes.

RULE II

*Therefore to the same natural effects we must, as far as possible, assign the same causes.*

As to respiration in a man and in a beast; the descent of stones in *Europe* and in *America*; the light of our culinary fire and of the sun; the reflection of light in the earth, and in the planets.

RULE III

*The qualities of bodies, which admit neither intensification nor remission of degrees, and which are found to belong to all bodies within the reach of our experiments, are to be esteemed the universal qualities of all bodies whatsoever.*

For since the qualities of bodies are only known to us by experiments, we are to hold for universal all such as univer-

Excerpts from I. Newton, *Mathematical Principles of Natural Philosophy* (originally published 1687, translated by A. Motte in 1729, revised by F. Cajori). Berkeley: University of California Press, 1934, pp. 398–400; 546–47 (footnotes deleted).

sally agree with experiments; and such as are not liable to diminution can never be quite taken away. We are certainly not to relinquish the evidence of experiments for the sake of dreams and vain fictions of our own devising; nor are we to recede from the analogy of Nature, which is wont to be simple, and always consonant to itself. We no other way know the extension of bodies than by our senses, nor do these reach it in all bodies; but because we perceive extension in all that are sensible, therefore we ascribe it universally to all others also. That abundance of bodies are hard, we learn by experience; and because the hardness of the whole arises from the hardness of the parts, we therefore justly infer the hardness of the undivided particles not only of the bodies we feel but of all others. That all bodies are impenetrable, we gather not from reason, but from sensation. The bodies which we handle we find impenetrable, and thence conclude impenetrability to be an universal property of all bodies whatsoever. That all bodies are movable, and endowed with certain powers (which we call the inertia) of persevering in their motion, or in their rest, we only infer from the like properties observed in the bodies which we have seen. The extension, hardness, impenetrability, mobility, and inertia of the whole, result from the extension, hardness, impenetrability, mobility, and inertia of the parts; and hence we conclude the least particles of all bodies to be also extended, and hard and impenetrable, and movable, and endowed with their proper inertia. And this is the foundation of all philosophy. Moreover, that the divided but contiguous particles of bodies may be separated from one another, is matter of observation; and, in the particles that remain undivided, our minds are able to distinguish yet lesser parts, as is mathematically demonstrated. But whether the parts so distinguished, and not yet divided, may, by the powers of Nature, be actually divided and separated from one another, we cannot certainly determine. Yet, had we the proof of but one experiment that any undivided particle, in breaking a hard and solid body, suffered a division, we might by virtue of this rule conclude that the undivided as well as the divided particles may be divided and actually separated to infinity.

Lastly, if it universally appears, by experiments and astronomical observations, that all bodies about the earth gravitate towards the earth, and that in proportion to the quantity of matter which they severally contain; that the moon likewise, according to the quantity of its matter, gravitates towards the earth; that, on the other hand, our sea gravitates towards the moon; and all the planets one towards another; and the comets in like manner towards the sun; we must, in consequence of this rule, universally allow that all bodies whatsoever are endowed with a principle of mutual gravitation. For the argument from the appearances concludes with more force for the universal gravitation of all bodies than for their impenetrability; of which, among those in the celestial regions, we have no experiments, nor any manner of observation. Not that I affirm gravity to be essential to bodies: by their *vis insita* I mean nothing but their inertia. This is immutable. Their gravity is diminished as they recede from the earth.

RULE IV

*In experimental philosophy we are to look upon propositions inferred by general induction from phenomena as accurately or very nearly true, notwithstanding any contrary hypotheses that may be imagined, till such time as other phenomena occur, by which they may either be made more accurate, or liable to exceptions.*

This rule we must follow, that the argument of induction may not be evaded by hypotheses. . . .

Hitherto we have explained the phenomena of the heavens and of our sea by the power of gravity, but have not yet assigned the cause of this power. This is certain, that it must proceed from a cause that penetrates to the very centres of the sun and planets, without suffering the least diminution of its force; that operates not according to the quantity of the surfaces of the particles upon which it acts (as mechanical causes used to do), but according to the quantity of the solid matter which they contain, and propagates its virtue on all sides to immense distances, decreasing always as the inverse square of the distances. Gravitation towards the sun is made up out of the gravitations towards the several particles of which the body of the sun is composed; and in receding from the sun decreases accurately as the inverse square of the distances as far as the orbit of Saturn, as evidently appears from the quiescence of the aphelion of the planets; nay, and even to the remotest aphelion of the comets, if those aphelions are also quiescent. But hitherto I have not been able to discover the cause of those properties of gravity from phenomena, and I frame no hypotheses, for whatever is not deduced from the phenomena is to be called an hypothesis; and hypotheses, whether metaphysical or physical, whether of occult qualities or mechanical, have no place in experimental philosophy. In this philosophy particular propositions are inferred from the phenomena, and afterwards rendered general by induction. Thus it was that the impenetrability, the mobility, and the impulsive force of bodies, and the laws of motion and of gravitation, were discovered. And to us it is enough that gravity does really exist, and act according to the laws which we have explained, and abundantly serves to account for all the motions of the celestial bodies, and of our sea.

JOHN LOCKE

# On the Use

# and Misuse

# of Hypotheses

I WOULD NOT therefore be thought to disesteem or dissuade the study of nature. I readily agree the contemplation of his works gives us occasion to admire, revere, and glorify their Author and, if rightly directed, may be of greater benefit to mankind than the monuments of examplary [sic] charity that have at so great charge been raised by the founders of hospitals and almshouses. He that first invented printing, discovered the use of the compass, or made public the virtue and right use of kin kina[1], did more for the propagation of knowledge, for the supplying and increase of useful commodities, and saved more from the grave than those who built colleges, work-houses, and hospitals. All that I would say is that we should not be too forwardly possessed with the opinion or expectation of knowledge where it is not to be had, or by ways that will not attain it; that we should not take doubtful systems for complete sciences, nor unintelligible notions for scientifical demonstrations. In the knowledge of bodies, we must be content to glean what we can from particular experiments, since we cannot from a discovery of their real essences grasp at a time whole sheaves, and in bundles comprehend the nature and properties of whole species together. Where our inquiry

Excerpt from J. Locke, *An Essay concerning Human Understanding*, book 4, chapter 12 (originally published 1690). London: John Bumpus, 1824, pp. 589–90.

is concerning co-existence of repugnancy to co-exist, which by contemplation of our ideas we cannot discover, there experience, observation, and natural history must give us by our senses and by retail an insight into corporeal substances. The knowledge of bodies we must get by our senses warily employed in taking notice of their qualities and operations on one another; and what we hope to know of separate spirits in this world we must, I think, expect only from revelation. He that shall consider how little general maxims, precarious principles, and hypotheses laid down at pleasure have promoted true knowledge or helped to satisfy the inquiries of rational men after real improvements, how little, I say, the setting out at that end has for many ages together advanced men's progress towards the knowledge of natural philosophy, will think we have reason to thank those who in this latter age have taken another course and have trod out to us, though not an easier way to learned ignorance, yet a surer way to profitable knowledge.

Not that we may not, to explain any phenomena of nature, make use of any probable hypothesis whatsoever: hypotheses, if they are well made, are at least great helps to the memory and often direct us to new discoveries. But my meaning is that we should not take up any one too hastily (which the mind, that would always penetrate into the causes of things and have principles to rest on, is very apt to do) till we have very well examined particulars and made several experiments in that thing which we would explain by our hypothesis and see whether it will agree to them all, whether our principles will carry us quite through and not be as inconsistent with one phenomenon of nature, as they seem to accommodate and explain another. And at least that we take care that the name of principles deceive us not, nor impose on us, by making us receive that for an unquestionable truth which is really at best but a very doubtful conjecture, such as are most (I had almost said all) of the hypotheses in natural philosophy.

# Notes

1. *Kin kina* is a Peruvian bark, the source of quinine. [Eds.]

KARL POPPER

# The Myth

# of Inductive

# Hypothesis

# Generation

THE BELIEF THAT science proceeds from observation to theory is still so widely and so firmly held that my denial of it is often met with incredulity. I have even been suspected of being insincere—of denying what nobody in his senses can doubt.

But in fact the belief that we can start with pure observations alone, without anything in the nature of a theory, is absurd; as may be illustrated by the story of the man who dedicated his life to natural science, wrote down everything he could observe, and bequeathed his priceless collection of observations to the Royal Society to be used as inductive evidence. This story should show us that though beetles may profitably be collected, observations may not.

Twenty-five years ago I tried to bring home the same point to a group of physics students in Vienna by beginning a lecture with the following instructions: "Take pencil and paper; carefully observe, and write down what you have observed!" They

Excerpts from K. Popper, *Conjectures and Refutations* (first published in slightly different form in 1962). London: Routledge & Kegan Paul, 1978, pp. 46–48, 52–53 (footnotes deleted). Reprinted by permission of Sir Karl Popper.

asked, of course, *what* I wanted them to observe. Clearly the instruction, "Observe!" is absurd. (It is not even idiomatic, unless the object of the transitive verb can be taken as understood.) Observation is always selective. It needs a chosen object, a definite task, an interest, a point of view, a problem. And its description presupposes a descriptive language, with property words; it presupposes similarity and classification, which in their turn presuppose interests, points of view, and problems. "A hungry animal," writes Katz (1937) "divides the environment into edible and inedible things. An animal in flight sees roads to escape and hiding places. . . . Quite generally objects change . . . according to the needs of the animal." We may add that objects can be classified, and can become similar or dissimilar, *only* in this way—by being related to needs and interests. This rule applies not only to animals but also to scientists. For the animal a point of view is provided by its needs, the task of the moment, and its expectations; for the scientist by his theoretical interests, the special problem under investigation, his conjectures and anticipations, and the theories which he accepts as a kind of background: his frame of reference, his "horizon of expectations."

The problem "Which comes first, the hypothesis (H) or the observation (O)?" is soluble; as is the problem, "Which comes first, the hen (H) or the egg (O)?". The reply to the latter is, "An earlier kind of egg"; to the former, "An earlier kind of hypothesis." Is is quite true that any particular hypothesis we choose will have been preceded by observations—the observations, for example, which it is designed to explain. But these observations, in their turn, presupposed the adoption of a frame of reference: a frame of expectations: a frame of theories. If they were significant, if they created a need for explanation and thus gave rise to the invention of a hypothesis, it was because they could not be explained within the old theoretical framework, the old horizon of expectations. There is no danger here of an infinite regress. Going back to more and more primitive theories and myths we shall in the end find unconscious, *inborn* expectations.

The theory of inborn *ideas* is absurd, I think; but every organism has inborn *reactions* or *responses*; and among them, responses adapted to impending events. These responses we may describe as "expectations" without implying that these "expectations" are conscious. The newborn baby "expects," in this sense, to be fed (and, one could even argue, to be protected and loved). In view of the close relation between expectation and knowledge we may even speak in quite a reasonable sense of "inborn knowledge." This "knowledge," however, is not *valid a priori*; an inborn expectation, no matter how strong and specific, may be mistaken. (The newborn child may be abandoned, and starve.)

Thus we are born with expectations; with "knowledge" which, although not *valid a priori*, is *psychologically* or *genetically a priori*, i.e. prior to all observational experience. One of the most important of these expectations is the expectation of finding a regularity. It is connected with an inborn propensity to look out for regularities, or with a *need* to *find* regularities, as we may see from the pleasure of the child who satisfies this need.

This "instinctive" expectation of finding regularities, which is psychologically *a priori*, corresponds very closely to the "law of causality" which Kant believed to be part of our mental outfit and to be *a priori* valid. One might thus be inclined to say that Kant failed to distinguish between psychologically *a priori* ways of thinking or responding and *a priori* valid beliefs. But I do not think that his mistake was quite as crude as that. For the expectation of finding regularities is not only psychologically *a priori*, but also logically *a priori*: it is logically prior to all observational experience, for it is prior to any recognition of similarities, as we have seen; and all observation involves the recognition of similarities (or dissimilarities). But in spite of being logically *a priori* in this sense the expectation is not valid *a priori*. For it may fail: we can easily construct an environment (it would be a lethal one) which, compared with our ordinary environment, is so chaotic that we completely fail to find regularities. (All natural laws could remain valid: environments of this kind have been used in the animal experiments mentioned in the next section.)

Thus Kant's reply to Hume came near to being right; for the distinction between an *a priori* valid expectation and one which is both genetically *and* logically prior to observation, but not *a priori* valid, is really somewhat subtle. But Kant proved too much. In trying to show how knowledge is possible, he proposed a theory which had the unavoidable consequence that our quest for knowledge must necessarily succeed, which is clearly mistaken. When Kant said, "Our intellect does not draw its laws from nature but imposes its laws upon nature," he was right. But in thinking that these laws are necessarily true, or that we necessarily succeed in imposing them upon nature, he was wrong. Nature very often resists quite successfully, forcing us to discard our laws as refuted; but if we live we may try again. . . .

From what I have said it is obvious that there was a close link between the two problems which interested me at that time: demarcation, and induction or scientific method. It was easy to see that the method of science is criticism, i.e. attempted falsifications. Yet it took me a few years to notice that the two problems—of demarcation and of induction—were in a sense one.

Why, I asked, do so many scientists believe in induction? I found they did so because they believed natural science to be characterized by the inductive method—by a method starting from, and relying upon, long sequences of observations and experiments. They believed that the difference between genuine science and metaphysical or pseudo-scientific speculation depended solely upon whether or not the inductive method was employed. They believed (to put it in my own terminology) that only the inductive method could provide a satisfactory *criterion of demarcation.*

I recently came across an interesting formulation of this belief in a remarkable philosophical book by a great physicist—Max Born's *Natural Philosophy of Cause and Chance* (1949:7). He writes: "Induction allows us to generalize a number of observations into a general rule: that night follows day and day follows night. . . . But while everyday life has no definite criterion for the validity of an induction, . . . science has worked out a code, or rule of craft, for its application." Born nowhere reveals the contents of this inductive code (which, as his wording shows, contains a "definite criterion for the validity of an induction"); but he stresses that "there is no logical argument" for its acceptance: "it is a question of faith"; and he is therefore "willing to call induction a metaphysical principle." But why does he believe that such a code of valid inductive rules must exist? This become clear when he speaks of the "vast communities of people ignorant of, or rejecting, the rule of science, among them the members of anti-vaccination societies and believers in astrology. It is useless to argue with them; I cannot compel them to accept the same criteria of valid induction in which I believe: the code of scientific rules." This makes it quite clear that *"valid induction" was here meant to serve as a criterion of demarcation between science and pseudo-science.*

But it is obvious that this rule or craft of "valid induction" is not even metaphysical: it simply does not exist. No rule can ever guarantee that a generalization inferred from true observations, however often repeated, is true. (Born himself does not believe in the truth of Newtonian physics, in spite of its success, although he believes that it is based on induction.) And the success of science is not based upon rules of induction, but depends upon luck, ingenuity, and the purely deductive rules of critical argument.

I may summarize some of my conclusions as follows:

(1) Induction, i.e. inference based on many observations, is a myth. It is neither a psychological fact, nor a fact of ordinary life, nor one of scientific procedure.

(2) The actual procedure of science is to operate with conjectures: to jump to conclusions—often after one single observation (as noticed for example by Hume and Born).

(3) Repeated observations and experiments function in science as *tests* of our conjectures or hypotheses, i.e. as attempted refutations.

(4) The mistaken belief in induction is fortified by the need for a criterion of demarcation which, it is traditionally but wrongly believed, only the inductive method can provide.

(5) The conception of such an inductive method, like the criterion of verifiability, implies a faulty demarcation.

(6) None of this is altered in the least if we say that induction makes theories only probable rather than certain.

# Part III

# Hypothesis
# Testing:
# The Role of
# Disconfirmation

# Introduction

IT HAS LONG been recognized that even the most carefully formulated hypotheses may turn out to be wrong. Indeed, it can be argued that eventually all scientific propositions will be found to be in error to a greater or lesser degree. It may take decades or centuries for this to occur, but even the most powerful and widely accepted scientific theories have failed or will fail. Even so, many scientists have held tenaciously to the most fanciful ideas, often in the face of large amounts of anomalous evidence. They have seemed to act as if absolute truth exists and as if they could possess it. This has suggested to many observers that healthy skepticism is an important ingredient in successful science; that, paradoxically, doubt is an invaluable aid in seeking knowledge.

Skepticism can be manifested in various ways. Selection 10, a brief passage by Descartes (1637) illustrates a rationalist approach. Descartes's search for indisputable first principles upon which to build both philosophy and science led him to apply systematic doubt to all of his ideas, to see which, if any, would survive. He concluded that there were indeed some things beyond doubt: that he, who was doing the doubting, must exist, and that a Perfect Being must exist. From these premises Descartes attempted to deduce a number of consequences, for example, that one could in general trust one's senses since a Perfect Being would not deceive. He further concluded that other ideas equally "clear and distinct" could be held with the same certainty. Such ideas could then serve as a priori axioms for the physical sciences. It was not Descartes's view that detailed physical theories could be derived from such axioms alone, since empirical testing was necessary in order to choose between those hypotheses that were compatible with the axioms. The axioms themselves, however, were not seen as open to test. Instead, they were justified on nonempirical grounds, on their "clarity" and "distinctiveness." Descartes' attempt to base physical science on such axioms ultimately failed when pitted against the more empirical Newtonian systems. His ideas in other areas were more fruitful—his mechanistic physiology for example, was a precursor of reflex theory (Fearing 1930)—but given his essentially rationalist epistomology, it is not surprising that his lasting contributions were to philosophy and mathematics, not to science.

Descartes's procedure failed because his "clear and distinct" ideas were not, by themselves, immutable. Why not doubt such ideas as well, a procedure which results in complete skepticism? Thus, when David Hume (1748) used the Lockean theory of ideas as the starting point for a critique of causal reasoning, he created a dilemma for philosophy which has, in certain respects, never been resolved. In the eyes of scientists, the effect was to render irrelevant the reasonings of philosophy about the ultimate sources of knowledge (cf. Reichenbach 1951). While eighteenth- and nineteenth-century philosophers debated the origin of ideas, most scientists simply shrugged off such questions and assumed the existence of a knowable reality. Doubt then shifted from the ontological level to the operational—it became a working tool. In particular, the widespread use of controlled experimentation made it possible to rule out, i.e., to "rigorously doubt," large numbers of hypotheses. When Anton Mesmer created a public stir in Paris in the 1780s with his "animal magnetism," a simple experiment (devised by Benjamin Franklin, Lavoisier, and others) disproved the claim that "magnetic fluids" were responsible for Mesmer's bizarre effects[1] (Darnton 1968; the relevant documents are reprinted in Hunter and Macalpine 1963).

To illustrate how dramatically the role of doubt in science changed, we have chosen an unusual treatment of that role in selection 11, written in 1749 by David Hartley. He was concerned with the empirical rather than the rational consequences of doubt, and proposed that using the "rule of false," the conscious attempt at disproof, would "much abate that unreasonable fondness, which those who make few or no distinct hypotheses, have for such confused ones as occur accidently to their imaginations" Clear as this sounds, it should be pointed out that Hartley was usually incapable of applying such doubt to his own ideas (Tweney in press). Thus, though sometimes regarded as the first physiological psychologist (Boring 1950), he is more often remembered for his fanciful invention of unseen "nervous vibrations" (Schofield 1969). Nevertheless, Hartley's prescription has a contemporary ring. He recognized the impossibility of conducting science without hypotheses—"It is vain to bid an inquirer form no hypotheses." The proper approach is not to disdain hypotheses, but rather to subject them to rigorous testing, weeding out those which fail. In spite of his own inability to apply such methods, his claims were in advance of most scientists of his era, who followed Newton, and widely disdained hypotheses (Laudan 1969).

Similar views on the usefulness of disconfirmation had been put forth more than a century earlier by Francis Bacon, who placed great emphasis on what he called the "Instance of the Fingerpost; borrowing the term from the fingerposts which are set up where roads part, to indicate the several directions" (*Novum Organum*, book 2, aphorism 36, in Bacon 1620/1878:253); in other words evidence that allowed a choice to be made between competing hypotheses. Bacon actually sketched out several "crucial experiments" on such phenomena as gravitational effects, magnetism, and tidal motion.

Disconfirmation is at the heart of Popper's analysis of scientific inquiry (selection 12). His analysis is based upon the logical asymmetry between confirmatory and disconfirmatory evidence, which is apparent when scientific propositions are represented as conditional relationships of the form "If $p$ then $q$," where $p$ is some theory and $q$ a logically derived prediction. If $p$ is a universal generalization, no finite amount of confirmatory evidence is sufficient to establish its truth. Asserting that $p$ is true on the basis of the occurrence of $q$ is a logical fallacy. On the other hand, if the predicted event $q$ does not occur, then $p$ is false. Instances of confirmation, no matter how numerous, can never verify a scientific proposition, but even one disconfirmation is sufficient, in principle, to falsify it.

In actual practice, matters are not so simple. In particular, when falsification occurs, one does not know which aspect of $p$ is incorrect. Most theories consist of complex networks of

assumptions, hypothetical constructs, interpretation statements, operational definitions, and so forth, some of which
are explicit, others implicit. Some or all of the network may
be in error: a single disconfirmed prediction does not provide
the information needed to determine what exactly is wrong.
Furthermore, the falsification itself may be erroneous. Since
all measurement has an error component, the falsifying observation, "not $q$," may actually be "$q$ with a large error component." Finally, disconfirmatory evidence can almost always
be incorporated into a theory by means of ex post facto modifications of the theory. It would obviously be unproductive
to abandon powerful, well-confirmed theories every time an
anomaly crops up. But when should one stop "patching up"
a disconfirmed theory and reject it instead? The question does
not admit of an obvious answer.

In spite of these and other problems, Popper has vigorously
applied his ideas about falsification to a number of issues concerning the nature of science, most notably the demarcation
between "science" and "nonscience." In Popper's view, what
demarcates science from other inquiry is that scientific propositions are falsifiable, while "pseudoscientific" propositions
are not. Relativity theory, Popper's paradigmatic example of
good science, made a number of explicit and "risky" predictions, one of which concerned the existence of a gravitational
effect on light (a prediction that was confirmed by Eddington
in 1919—see Dyson, Eddington and Davidson 1919). By contrast,
Marxist economics and Freudian psychoanalytic theory, which
can "explain" virtually any event after it happens, have little
or no predictive power of this "risky" sort. The more explicit
the predictions a theory or hypothesis makes, the more potentially falsifiable it is, and hence the more scientific it is.

One consequence is that it makes no sense to speak of "degrees of confirmation": to say that one theory is "more confirmed" than another. What counts in evaluating, say, relativity theory is not the amount of confirmation it has received,
but rather the lack of disconfirmation of highly precise predictions. Psychoanalytic theory has had more confirmation
than relativity theory, but is nevertheless pseudoscience because it does not make falsifiable predictions (Popper 1978).
When a theory makes falsifiable predictions but is not falsified,
Popper speaks of "corroboration," the extent to which a theory

had performed successfully in the past. Popper stresses that no amount of corroboration can serve as a guide to how a theory will perform in the future (Popper 1959).

That confirmation counts, logically, for nothing, that induction has no role in science, and that scientists ought to try actively to disprove their theories are surprising and counterintuitive conclusions. Some scientists consider them irrelevant to the practice of science (see Mitroff 1974a, and the introduction to part 4). Nevertheless, Popper's ideas have been influential, not only among philosophers of science, but among certain scientists as well, who have taken them as norms for the conduct of science. One clear instance of this is presented in selection 14, in which Sir John Eccles describes how his beliefs about scientific investigation were changed by Popper. Eccles "experienced a great liberation" in escaping the view that falsification of one's hypotheses was a sign of failure, and even came "to rejoice in the refutation of a cherished hypothesis." While such expressions are rare, it is perhaps true that scientists will seek deliberate falsification in the early stages of thinking about a problem. The scientist may conjure up a hypothesis to explain some data set, but realize that the hypothesis is falsified by some other set of known facts. Such falsification does not ordinarily find its way into print.[2]

The clear falsification of a well-formulated hypothesis is, however, a somewhat rare event in science. Very little research is conducted in such a way that a single prediction makes or breaks a theory. Entire networks of interconnected experiments represent the more common pattern. Can the logic of falsification be extended to incorporate such interconnected hypotheses, predictions, and experimental outcomes? One extension of that logic has been proposed by the biophysicist J. R. Platt (1964), who argued that the spectacular successes of molecular biology and high energy physics during the 1950s and 1960s were due in part to a research strategy that he called "strong inference." It consists of the following procedure:

1. Devising alternative hypotheses;
2. Devising a crucial experiment (or several of them), with alternative possible outcomes, each of which, as nearly as possible, excludes one or more of the hypotheses;
3. Carrying out the experiment to get a clear result;

4. Recycling the procedure, making subhypotheses or sequential hypotheses to refine the possibilities that remain. (Platt 1964:347)

Platt contended that the problem with Popper's version of falsificationism is that it is ". . . a hard doctrine. If you have a hypothesis and I have another hypothesis, evidently one of them must be eliminated. The scientist seems to have no choice but to be either soft-headed or disputatious" (Platt 1964:350). Strong inference removes the difficulty, since multiple hypotheses are no longer "personal property." Conflict occurs between hypotheses, not between scientists. Platt is placing the locus of multiple hypotheses in the individual scientist, rather than in the social institution of science.

In addition to decreasing personal allegiance to pet theories, strong inference can result in a clearer focus on central issues. The scientist employing it is more likely to concentrate on a relatively small number of "crucial experiments." The problem with studies designed to be one more brick in the temple of science is that "Most such bricks just lie around in the brick yard." (Platt 1964:391).

As Platt acknowledged, the essential ideas of strong inference were put forth by the American geologist T. C. Chamberlin,[3] who advocated the method of "multiple working hypotheses." We have used one of Chamberlin's papers, which appeared in 1904 in *Popular Science Monthly* (selection 13), as an example of the advocacy of falsification via the use of multiple hypotheses. Chamberlin's paper eloquently attacks some of the more common distortions of scientific procedure. Thus, what he called "The Method of Colorless Observation," is likely to result in many "unobtrusive but yet vital elements" being overlooked. On the other hand, "The Method of Ruling Theory," while providing a framework for observation and experimentation is also seriously deficient. The problem is that a hypothesis often becomes accepted more quickly than is warranted by the data, and once accepted, comes to have many pernicious effects. The proper approach is neither "colorless observation" nor a "ruling theory," but rather "The Method of Multiple Working Hypotheses," which is virtually identical to Platt's "strong Inference." Chamberlin argued that this approach ensures the scientist a "full and rich" analysis, while at the same time neutralizing "as far as may be, his emotional nature."

Chamberlin noted, as did Platt, that multiple hypotheses are especially suited for investigating complicated, multiply determined, phenomena. It is not surprising, then, that this approach has received attention among psychologists who have been receptive to methodologies promising to make psychological problems more tractable. As an example of this, we have included a comment by Robert Sekuler —selection 15), taken from an account of a series of perceptual experiments. Sekuler and his students serendipitously discovered that when temporally and spatially ordered stimuli are rapidly presented, there is a strong tendency to perceive those in the left part of the visual field as occurring prior to those in the right part of the visual field, even when the actual temporal order is right first, left second. Sekuler's paper describes the self-conscious use of Platt's strong inference strategy to determine what processes account for the "temporal order effect." He was able to rule out a number of alternative methodological and substantive explanations of the effect, until only one viable hypothesis remained.

## Notes

1. The incident is also instructive as an example of the premature use of falsification. In rejecting the physical fluids, the eminent committee also rejected the psychological phenomenon later to be known as hypnotism.

2. We are indebted to Michael Bradie for suggesting this point.

3. In fact, another prominent American geologist, G. K. Gilbert, had advocated a similar method even earlier than Chamberlin. He used it to "disprove" the meteoric orgin of the geological feature in Arizona now known as "Meteor Crater"! (Gilbert 1896; Pyne 1978).

RENÉ DESCARTES

# Doubt as the

# Starting Point

# of Knowledge

MORE ESPECIALLY DID I reflect in each matter that came before me as to anything which could make it subject to suspicion or doubt, and give occasion for mistake, and I rooted out of my mind all the errors which might have formerly crept in. Not that indeed I imitated the sceptics, who only doubt for the sake of doubting, and pretend to be always uncertain; for, on the contrary, my design was only to provide myself with good ground for assurance, and to reject the quicksand and mud in order to find the rock or clay. In this task it seems to me, I succeeded pretty well, since in trying to discover the error or uncertainty of the propositions which I examined, not by feeble conjectures, but by clear and assured reasonings, I encountered nothing so dubious that I could not draw from it some conclusion that was tolerably secure, if this were no more than the inference that it contained in it nothing that was certain. And just as in pulling down an old house we usually preserve the debris to serve in building up another, so in destroying all those opinions which I considered to be ill-founded, I made various observations and acquired many experiences, which have since been of use to me in establishing those which are more certain. . . .

Excerpt from R. Descartes. *The Philosophical Works of Descartes Rendered into English by E. S. Haldane and G. R. T. Ross.* Vol. 1. Cambridge: Cambridge University Press, 1969 (1911), pp. 99–101. (First published 1637).

I do not know that I ought to tell you of the first meditations there made by me, for they are so metaphysical and so unusual that they may perhaps not be acceptable to everyone. And yet at the same time, in order that one may judge whether the foundations which I have laid are sufficiently secure, I find myself constrained in some measure to refer to them. For a long time I had remarked that it is sometimes requisite in common life to follow opinions which one knows to be most uncertain, exactly as though they were indisputable, as has been said above. But because in this case I wished to give myself entirely to the search after Truth, I thought that it was necessary for me to take an apparently opposite course, and to reject as absolutely false everything as to which I could imagine the least ground of doubt, in order to see if afterwards there remained anything in my belief that was entirely certain. Thus, because our senses sometimes deceive us, I wish to suppose that nothing is just as they cause us to imagine it to be; and because there are men who deceive themselves in their reasoning and fall into paralogisms, even concerning the simplest matters of geometry, and judging that I was as subject to error as was any other, I rejected as false all the reasons formerly accepted by me as demonstrations. And since all the same thoughts and conceptions which we have while awake may also come to us in sleep, without any of them being at that time true, I resolved to assume that everything that ever entered into my mind was no more true than the illusions of my dreams. But immediately afterwards I noticed that whilst I thus wished to think all things false, it was absolutely essential that the 'I' who thought this should be somewhat, and remarking that this truth '*I think, therefore I am*' was so certain and so assured that all the most extravagant suppositions brought forward by the sceptics were incapable of shaking it, I came to the conclusion that I could receive it without scruple as the first principle of the Philosophy for which I was seeking.

DAVID HARTLEY

# Hypotheses and the "Rule of the False"

. . . AS ACCORDING TO the rule of false, the arithmetician supposes a certain number to be that which is sought for; treats it as if it was that; and finding the deficiency or overplus in the conclusion, rectifies the error of his first position by a proportional addition or subtraction, and thus solves the problem; so it is useful in inquiries of all kinds to try all such suppositions as occur with any appearance of probability, to endeavour to deduce the real phaenomena from them; and if they do not answer in some tolerable measure, to reject them at once; or if they do, to add, expunge, correct, and improve, till we have brought the hypothesis as near as we can to an agreement with nature. After this it must be left to be farther corrected and improved, or entirely disproved, by the light and evidence reflected upon it from the contiguous, and even, in some measure, from the remote branches of other sciences.

Were this method commonly used, we might soon expect a great advancement in the sciences. It would much abate that

Excerpt from D. Hartley, *Observations on Man: His Frame, His Duty, and His Expectations*. Sixth edition. (first published 1748). London: Thomas Tegg, 1834, pp. 217–19.

unreasonable fondness, which those who make few or no distinct hypotheses, have for such confused ones as occur accidentally to their imaginations, and recur afterwards by association. For the ideas, words, and reasonings, belonging to the
favourite hypothesis, by recurring, and being much agitated
in the brain, heat it, unite with each other, and so coalesce in
the same manner, as genuine truths do from induction and
analogy. Verbal and grammatical analogies and coincidences
are advanced into real ones; and the words which pass often
over the ear, in the form of subject and predicate, are from the
influence of other associations made to adhere together insensibly, like subjects and predicates, that have a natural connexion. It is in vain to bid an inquirer form no hypothesis.
Every phaenomenon will suggest something of this kind: and,
if he do not take care to state such as occur fully and fairly,
and adjust them one to another, he may entertain a confused
inconsistent mixture of all, of fictitious and real, possible and
impossible: and become so persuaded of it, as that counter-
associations shall not be able to break the unnatural bond. But
he that forms hypotheses from the first, and tries them by the
facts, soon rejects the most unlikely ones; and, being freed
from these, is better qualified for the examination of those that
are probable. He will also confute his own positions so often,
as to fluctuate in equilibrio, in respect of prejudices, and so
be at perfect liberty to follow the strongest evidences.

In like manner, the frequent attempts to make an hypothesis
that shall suit the phaenomena, must improve a man in the
method of doing this; and beget in him by degrees an imperfect
practical art, just as algebraists and decypherers, that are much
versed in practice, are possessed of innumerable subordinate
artifices, besides the principal general ones, that are taught by
the established rules of their arts; and these, though of the
greatest use to themselves, can scarce be explained or communicated to others. These artifices may properly be referred
to the head of factitious sagacity, being the result of experience,
and of impressions often repeated, with small variations from
the general resemblance.

Lastly, the frequent making of hypotheses, and arguing from
them synthetically, according to the several variations and
combinations of which they are capable, would suggest numerous phaenomena, that otherwise escape notice, and lead

to *experimenta crucis*, not only in respect of the hypothesis under consideration, but of many others. The variations and combinations just mentioned suggest things to the invention, which the imagination unassisted is far unequal to; just as it would be impossible for a man to write down all the changes upon eight bells, unless he had some method to direct him.

But this method of making definite hypotheses, and trying them, is far too laborious and mortifying for us to hope that inquirers will in general pursue it. It would be of great use to such as intend to pursue it, to make hypotheses for the phaenomena, whose theories are well ascertained; such as those of circulation of the blood, of the pressure of the air, of the different refrangibility of the rays of light, &c. and see how they are gradually compelled into the right road, even from wrong suppositions fairly compared with the phaenomena. This would habituate the mind to a right method, and beget the factitious sagacity above-mentioned.

KARL POPPER

# Science,

# Pseudo-Science,

# and Falsifiability

THE PROBLEM WHICH troubled me at the time was neither, "When is a theory true?" nor, "When is a theory acceptable?" My problem was different. I *wished to distinguish between science and pseudo-science;* knowing very well that science often errs, and that pseudo-science may happen to stumble on the truth.

I knew, of course, the most widely accepted answer to my problem: that science is distinguished from pseudo-science—or from "metaphysics"—by its *empirical method*, which is essentially *inductive*, proceeding from observation or experiment. But this did not satisfy me. On the contrary, I often formulated my problem as one of distinguishing between a genuinely empirical method and a non-empirical or even a pseudo-empirical method—that is to say, a method which, although it appeals to observation and experiment, nevertheless does not come up to scientific standards. The latter method may be exemplified by astrology, with its stupendous mass of empirical evidence based on observation—on horoscopes and on biographies.

But as it was not the example of astrology which led me to

Excerpt from K. Popper, *Conjectures and Refutations* (first published in slightly different form in 1962). London: Routledge & Kegan Paul, 1978, pp. 33–39 (footnotes deleted). Reprinted by permission of Sir Karl Popper.

my problem I should perhaps briefly describe the atmosphere in which my problem arose and the examples by which it was stimulated. After the collapse of the Austrian Empire there had been a revolution in Austria: the air was full of revolutionary slogans and ideas, and new and often wild theories. Among the theories which interested me Einstein's theory of relativity was no doubt by far the most important. Three others were Marx's theory of history, Freud's psycho-analysis, and Alfred Adler's so-called "individual psychology."

There was a lot of popular nonsense talked about these theories, and especially about relativity (as still happens even today), but I was fortunate in those who introduced me to the study of this theory. We all—the small circle of students to which I belonged—were thrilled with the result of Eddington's eclipse observations which in 1919 brought the first important confirmation of Einstein's theory of gravitation. It was a great experience for us, and one which had a lasting influence on my intellectual development.

The three other theories I have mentioned were also widely discussed among students at that time. I myself happened to come into personal contact with Alfred Adler, and even to co-operate with him in his social work among the children and young people in the working-class districts of Vienna where he had established social guidance clinics.

It was during the summer of 1919 that I began to feel more and more dissatisfied with these three theories—the Marxist theory of history, psycho-analysis, and individual psychology; and I began to feel dubious about their claims to scientific status. My problem perhaps first took the simple form, "What is wrong with Marxism, psycho-analysis, and individual psychology? Why are they so different from physical theories, from Newton's theory, and especially from the theory of relativity?"

To make this contrast clear I should explain that few of us at the time would have said that we believed in the *truth* of Einstein's theory of gravitation. This shows that it was not my doubting the *truth* of those other three theories which bothered me, but something else. Yet neither was it that I merely felt mathematical physics to be more *exact* than the sociological or psychological type of theory. Thus what worried me was neither the problem of truth, at that stage at least, nor the problem of exactness or measurability. It was rather that I felt that these other three theories, though posing as sciences, had in fact more in common with primitive myths than with science; that they resembled astrology rather than astronomy.

I found that those of my friends who were admirers of Marx, Freud, and Adler, were impressed by a number of points common to these theories, and especially by their apparent *explanatory power*. These theories appeared to be able to explain practically everything that happened within the fields to which they referred. The study of any of them seemed to have the effect of an intellectual conversion or revelation, opening your eyes to a new truth hidden from those not yet initiated. Once your eyes were thus opened you saw confirming instances everywhere: the world was full of *verifications* of the theory. Whatever happened always confirmed it. Thus its truth appeared manifest; and unbelievers were clearly people who did not want to see the manifest truth; who refused to see it, either because it was against their class interest, or because of their repressions which were still "un-analysed" and crying aloud for treatment.

The most characteristic element in this situation seemed to me the incessant stream of confirmations, of observations which "verified" the theories in question; and this point was constantly emphasized by their adherents. A Marxist could not open a newspaper without finding on every page confirming evidence for his interpretation of history; not only in the news, but also in its presentation—which revealed the class bias of the paper—and especially of course in what the paper did *not* say. The Freudian analysts emphasized that their theories were constantly verified by their "clinical observations." As for Adler, I was much impressed by a personal experience. Once, in 1919, I reported to him a case which to me did not seem particularly Adlerian, but which he found no difficulty in analysing in terms of his theory of inferiority feelings, although he had not even seen the child. Slightly shocked, I asked him how he could be so sure. "Because of my thousandfold experience," he replied; whereupon I could not help saying: "And with this new case, I suppose, your experience has become thousand-and-one-fold."

What I had in mind was that his previous observations may

not have been much sounder than this new one; that each in its turn had been interpreted in the light of "previous experience," and at the same time counted as additional confirmation. What, I asked myself, did it confirm? No more than that a case could be interpreted in the light of the theory. But this meant very little, I reflected, since every conceivable case could be interpreted in the light of Adler's theory, or equally of Freud's. I may illustrate this by two very different examples of human behaviour: that of a man who pushes a child into the water with the intention of drowning him; and that of a man who sacrifices his life in an attempt to save the child. Each of these two cases can be explained with equal ease in Freudian and in Adlerian terms. According to Freud the first man suffered from repression (say, of some component of his Oedipus complex), while the second man had achieved sublimation. According to Adler the first man suffered from feelings of inferiority (producing perhaps the need to prove to himself that he dared to commit some crime), and so did the second man (whose need was to prove to himself that he dared to rescue the child). I could not think of any human behaviour which could not be interpreted in terms of either theory. It was precisely this fact —that they always fitted, that they were always confirmed—which in the eyes of their admirers constituted the strongest argument in favour of these theories. It began to dawn on me that this apparent strength was in fact their weakness.

With Einstein's theory the situation was strikingly different. Take one typical instance—Einstein's prediction, just then confirmed by the findings of Eddington's expedition. Einstein's gravitational theory had led to the result that light must be attracted by heavy bodies (such as the sun), precisely as material bodies were attracted. As a consequence it could be calculated that light from a distant fixed star whose apparent position was close to the sun would reach the earth from such a direction that the star would seem to be slightly shifted away from the sun; or, in other words, that stars close to the sun would look as if they had moved a little away from the sun, and from one another. This is a thing which cannot normally be observed since such stars are rendered invisible in daytime by the sun's overwhelming brightness; but during an eclipse

it is possible to take photographs of them. If the same con-
stellation is photographed at night one can measure the distances on the two photographs, and check the predicted effect.

Now the impressive thing about this case is the *risk* involved
in a prediction of this kind. If observation shows that the
predicted effect is definitely absent, then the theory is simply
refuted. The theory is *incompatible with certain possible results of observation*—in fact with results which everybody before Einstein would have expected. This is quite different from
the situation I have previously described, when it turned out
that the theories in question were compatible with the most
divergent human behaviour, so that it was practically impossible to describe any human behaviour that might not be
claimed to be a verification of these theories.

These considerations led me in the winter of 1919–20 to
conclusions which I may now reformulate as follows.

(1) It is easy to obtain confirmations, or verifications, for
nearly every theory—if we look for confirmations.

(2) Confirmations should count only if they are the result
of *risky predictions*; that is to say, if, unenlightened by the
theory in question, we should have expected an event which
was incompatible with the theory—an event which would have
refuted the theory.

(3) Every "good" scientific theory is a prohibition: it forbids
certain things to happen. The more a theory forbids, the better
it is.

(4) A theory which is not refutable by any conceivable event
is non-scientific. Irrefutability is not a virtue of a theory (as
people often think) but a vice.

(5) Every genuine *test* of a theory is an attempt to falsify it,
or to refute it. Testability is falsifiability; but there are degrees
of testability: some theories are more testable, more exposed
to refutation, than others; they take, as it were, greater risks.

(6) Confirming evidence should not count *except when it
is the result of a genuine test of the theory*; and this means
that it can be presented as a serious but unsuccessful attempt
to falsify the theory. (I now speak in such cases of "corroborating evidence".)

(7) Some genuinely testable theories, when found to be
false, are still upheld by their admirers—for example by intro-

ducing *ad hoc* some auxiliary assumption, or by re-interpreting the theory *ad hoc* in such a way that it escapes refutation. Such a procedure is always possible, but it rescues the theory from refutation only at the price of destroying, or at least lowering, its scientific status. (I later described such a rescuing operation as a *"conventionalist twist"* or a *"conventionalist stratagem."*)

One can sum up all this by saying that *the criterion of the scientific status of a theory is its falsifiability, or refutability, or testability.*

I may perhaps exemplify this with the help of the various theories so far mentioned. Einstein's theory of gravitation clearly satisfied the criterion of falsifiability. Even if our measuring instruments at the time did not allow us to pronounce on the results of the tests with complete assurance, there was clearly a possibility of refuting the theory.

Astrology did not pass the test. Astrologers were greatly impressed, and misled, by what they believed to be confirming evidence—so much so that they were quite unimpressed by any unfavourable evidence. Moreover, by making their interpretations and prophecies sufficiently vague they were able to explain away anything that might have been a refutation of the theory had the theory and the prophecies been more precise. In order to escape falsification they destroyed the testability of their theory. It is a typical soothsayer's trick to predict things so vaguely that the predictions can hardly fail: that they become irrefutable.

The Marxist theory of history, in spite of the serious efforts of some of its founders and followers, ultimately adopted this soothsaying practice. In some of its earlier formulations (for example in Marx's analysis of the character of the "coming social revolution") their predictions were testable, and in fact falsified. Yet instead of accepting the refutations the followers of Marx re-interpreted both the theory and the evidence in order to make them agree. In this way they rescued the theory from refutation; but they did so at the price of adopting a device which made it irrefutable. They thus gave a "conventionalist twist" to the theory; and by this stratagem they destroyed its much advertised claim to scientific status.

The two psycho-analytic theories were in a different class. They were simply non-testable, irrefutable. There was no conceivable human behaviour which could contradict them. This does not mean that Freud and Adler were not seeing certain things correctly: I personally do not doubt that much of what they say is of considerable importance, and may well play its part one day in a psychological science which is testable. But it does mean that those "clinical observations" which analysts naïvely believe confirm their theory cannot do this any more than the daily confirmations which astrologers find in their practice. And as for Freud's epic of the Ego, the Super-ego, and the Id, no substantially stronger claim to scientific status can be made for it than for Homer's collected stories from Olympus. These theories describe some facts, but in the manner of myths. They contain most interesting psychological suggestions, but not in a testable form.

At the same time I realized that such myths may be developed, and become testable; that historically speaking all—or very nearly all—scientific theories originate from myths, and that a myth may contain important anticipations of scientific theories. Examples are Empedocles' theory of evolution by trial and error, or Parmenides' myth of the unchanging block universe in which nothing ever happens and which, if we add another dimension, becomes Einstein's block universe (in which, too, nothing ever happens, since everything is, four-dimensionally speaking, determined and laid down from the beginning). I thus felt that if a theory is found to be non-scientific, or "metaphysical" (as we might say), it is not thereby found to be unimportant, or insignificant, or "meaningless," or "nonsensical." But it cannot claim to be backed by empirical evidence in the scientific sense—although it may easily be, in some genetic sense, the "result of observation."

(There were a great many other theories of this pre-scientific or pseudo-scientific character, some of them, unfortunately, as influential as the Marxist interpretation of history; for example, the racialist interpretation of history—another of those impressive and all-explanatory theories which act upon weak minds like revelations.)

Thus the problem which I tried to solve by proposing the criterion of falsifiability was neither a problem of meaningfulness or significance, nor a problem of truth or acceptability.

It was the problem of drawing a line (as well as this can be done) between the statements, or systems of statements, of the empirical sciences, and all other statements—whether they are of a religious or of a metaphysical character, or simply pseudo-scientific. Years later—it must have been in 1928 or 1929—I called this first problem of mine the *"problem of demarcation."* The criterion of falsifiability is a solution to this problem of demarcation, for it says that statements or systems of statements, in order to be ranked as scientific, must be capable of conflicting with possible, or conceivable, observations.

**Selection 13** ⎯⎯⎯⎯⎯⎯⎯⎯⎯⎯⎯⎯⎯⎯⎯⎯⎯⎯⎯⎯⎯⎯⎯⎯⎯

# THOMAS CHROWDER CHAMBERLIN

# On

# Multiple

# Hypotheses

THAT WHICH PASSES under the name earth-science is not all *science* in the strict sense of the term. Not a little consists of generalizations from incomplete data, of inferences hung on chains of uncertain logic, of interpretations not beyond question, of hypotheses not fully verified and of speculation none too substantial. A part of the mass is true science, a part is philosophy, as I would use the term, a part is speculation, and a part is yet unorganized material. . . . I believe an appropriate atmosphere of philosophy is as necessary to the wholesome intellectual life of our sciences as is the earth's physical atmosphere to the life of the planet. None the less, it must ever be our endeavor to reduce speculation to philosophy, and philosophy to science. For the perpetuation of the necessary philosophic atmosphere, we may safely trust to the evolution of new problems concurrently with the solution of the old.

But granting the importance of the philosophic element, we doubtless agree without hesitation that the solid products of accurate and complete observation, natural or experimental, are the bed-rock of our group of sciences. The first great object

Excerpts from T. C. Chamberlin, The methods of the earth-sciences. *Popular Science Monthly* (1904) 66:66–75 (footnote deleted).

sought by laudable methods is, therefore, the promotion of the most accurate, searching, exhaustive and unbiased observation that is possible. One of the primary efforts in behalf of our sciences, therefore, was naturally directed to the task of promoting the best observational work. It was soon discovered that two chief dangers threatened the worker—bias and incompleteness. To guard against the first there was evolved

## The Method

## of Colorless

## Observation

Under its guidance, the observer endeavors to keep his mind scrupulously free from prepossessions and favored views. However tensely he may strain his observing powers to see what is to be seen, he seeks solely a record of facts uncolored by preferences or prejudices. To this end, he restrains himself from theoretical indulgence, and modestly contents himself with being a recorder of nature. He does not presume to be its interpreter and prophet. At length, in the office, he gathers his observations into an assemblage, with such inferences and interpretations as flow from them spontaneously, but even then he guards himself against the prejudices of theoretical indulgence.

Laudable as this method is in its avoidance of partiality, it is none the less seriously defective. No one who goes into the field with a mind merely receptive, or merely alert to see what presents itself, however nerved to a high effort, will return laden with all that might be seen. Only a part of the elements and aspects of complex phenomena present themselves at once to even the best observational minds. Some parts of the complex are necessarily obscure. Some of the most significant elements are liable to be unimpressive. These unobtrusive but yet vital elements will certainly escape observation unless it is forced to seek them out, and to seek them out diligently, acutely and intensely. To make a reasonably complete set of observations, the mind must not only see what spontaneously arrests its attention, but it must immediately draw out from what it observes inferences, interpretations, and hypotheses to promote further observations. It must at once be seen that if a given inference is correct, certain collateral phenomena must accompany it. If another inference be correct, certain other phenomena must accompany it. If still a third interpretation be the true one, yet other phenomena must be present to give proof of it. Once these suggestions have arisen, the observer seeks out the phenomena that discriminate between them, and, under such stimulus, phenomena that would otherwise have wholly escaped attention at once come into view because the eye has now been focused for them. It may be affirmed with great confidence that without the active and instantaneous use of these concurrent processes the observer will rarely, if ever, record the whole of any one set of significant elements, much less the whole of all sets. His record will contain incomplete parts of different sets of significant elements, *but no complete set of any one.* The obscure factors of each set are quite sure to be overlooked and the obtrusive factors of several sets indiscriminately commingled. The method of colorless observation is thus seriously defective in the completeness of its products, while it successfully guards them from bias.

Standing over against it, in strong contrast, is the method which at once endeavors to seek out and put together the phenomena that are thought to be significant. This leads promptly to the construction of a theory or an explanation which soon comes to guide the work and gives rise to

## The Method
## of the Ruling
## Theory

The chief effort here centers on an elucidation of phenomena, not on an exhaustive determination of the facts. Properly enough the crown of the work is the end, explanation is brought to the forefront and eagerly made the immediate end of endeavor. As soon as a phenomenon is presented, a theory of elucidation is framed. Laudable enough in itself, the theory is liable to be framed before the phenomena are fully and accurately observed. The elucidation is likely to embrace only the more obtrusive phenomena, not the full complement of the obtrusive and the unimpressive. The field is quite likely to present many repetitions of the leading phenomena and a theory framed to fit those that first arrest attention naturally fits the oft-recurrent phenomena of the same class. While there may be really no new evidence, nor any real test, nor any further inquiry into the grounds of the theory, its repeated application with seeming success leads insidiously to the delusion that is has been strengthened by additional investigation. Unconsciously then it begins to direct observation to the facts it so happily elucidates. Unconsciously the facts to which it gives no meaning become less impressive and fall into neglect. Selective observation creeps insidiously in and becomes a persistent habit. Soon also affection is awakened with its blinding influence. The authorship of an original explanation that seems successful easily begets fondness for one's intellectual child. This affection adds its alluring influence to the previous tendency toward an unconscious selection. The mind lingers with pleasure upon the facts that fall happily into the embrace of the theory, and feels a natural indifference toward those that assume a refractory or meaningless attitude. Instinctively, there is a special searching-out of phenomena that support the theory; unwittingly also there is a pressing of the theory to make it fit the facts and a pressing of the facts to make them fit the theory. When these biasing tendencies set in, the mind soon glides into the partiality of paternalism, and the theory rapidly rises to a position of control. Unless it happens to be the true one, all hope of the best results is gone. The defects of this method are obvious and grave.

It is safe to say, however, that under this method, with all its defects, many facts will be gathered that an observer of colorless attitude would have quite overlooked. The reverse may doubtless also be said. An effort to avoid the dangers at once of the colorless Scylla and the biasing Charybdis gave rise to

# The Method

# of the Working

# Hypothesis

This may be regarded as the distinctive feature of the methodology of the last century. This differs from the method of the ruling theory in that the working hypothesis is made a means of determining facts, not primarily a thesis to be established. Its chief function is the suggestion and guidance of lines of inquiry; inquiry not for the sake of the hypothesis, but for the sake of the facts and their final elucidation. The hypothesis is a mode rather than an end. Under the ruling theory, the stimulus is directed to the finding of facts for the support of the theory. Under the working hypothesis, the facts are sought for the purpose of ultimate induction and demonstration, the hypothesis being but a means for the more ready development of facts and their relations, particularly *their relations*.

It will be seen that the distinction is somewhat subtle. It is rarely if ever perfectly sustained. A working hypothesis may glide with the utmost ease into a ruling theory. Affection may as easily cling about a beloved intellectual child under the name of a working hypothesis as under any other, and may become a ruling passion. The moral atmosphere associated with the working hypothesis, however, lends some good influence toward the preservation of its integrity. The author of a working hypothesis is not presumed to father or defend it, but merely to use it for what it is worth.

Conscientiously followed, the method of the working hypothesis is an incalculable advance upon the method of the ruling theory, as it is also upon the method of colorless observation; but it also has serious defects. As already implied, it is not an adequate protection against a biased attitude. Even if it avoids this, it tends to narrow the scope of inquiry and direct it solely along the lines of the hypothesis. It undoubtedly gives acuteness, incisiveness and thoroughness in its own lines, but it inevitably turns inquiry away from other lines. It has dangers therefore akin to its predecessor, the ruling theory.

A remedy for these dangers and defects has been sought in

### The Method

### of Multiple Working

### Hypotheses

This differs from the method of the simple working hypothesis in that it distributes the effort and divides the affections. It is thus in some measure protected against the radical defects of the two previous methods. The effort is to bring up into distinct view every rational explanation of the phenomenon in hand and to develop into working form every tenable hypothesis of its nature, cause or origin, and to give to each of these a due place in the inquiry. The investigator thus becomes the parent of a family of hypotheses; and by his paternal relations to all is morally forbidden to fasten his affections unduly upon any one. In the very nature of the case, the chief danger that springs from affection is counteracted. Where some of the hypotheses have been already proposed and used, while others are the investigator's own creation, a natural tendency to bias arises, but the right use of the method requires the impartial adoption of all into the working family. The investigator thus at the outset puts himself in cordial sympathy and in the parental relations of adoption, if not of authorship, with every hypothesis that is at all applicable to the case under investigation. Having thus neutralized, so far as may be, the partialities of his emotional nature, he proceeds with a certain natural and enforced erectness of mental attitude to the inquiry, knowing well that some of the family of hypotheses must needs perish in the ordeal of crucial research, but with a reasonable expectation that more than one of them may survive, since it often proves in the end that several agencies were conjoined in the production of the phenomenon. Honors must often be divided between hypotheses. In following a single hypothesis, the mind is biased by the presumptions of the method toward a single explanatory conception. But an adequate explanation often involves the coordination of several causes. This is especially true when the research deals with complicated phenomena such as prevail in the field of the earth-sciences. Not only do several agencies often participate, but their proportions and relative importance vary from instance to instance in the same class of phenomena. The true explanation is therefore necessarily multiple, and often involves an estimate of the measure of participation of each factor. For this the simultaneous use of a full staff of working hypotheses is demanded. The method of the single working hypothesis is here incompetent.

The reaction of one hypothesis upon another leads to a fuller and sharper recognition of the scope of each. Every added hypothesis is quite sure to call forth into clear recognition neglected aspects of the phenomena. The mutual conflicts of hypotheses whet the discriminative edge of each. The sharp competition of hypotheses provokes keenness in the analytic processes and acuteness in differentiating criteria. Fertility in investigative devices is a natural sequence. If therefore an ample group of hypotheses encompass the subject on all sides, the total outcome of observation, of discrimination and of recognition of significance and relationship is full and rich. . . .

## The Method

## of Regenerative

## Hypotheses

In the method of multiple hypotheses, the members of the group are used simultaneously and are more or less mutually exclusive, or even antagonistic. Supplementary to this method is the use of a *succession* of hypotheses related genetically to one another. In this the results of an inquiry under the first hypothesis give rise to the assumptions of the succeeding hypothesis. The precise conclusions of the first inquiry are not made the assumptions of the second, for the process would then be little more than repetitive, but the revelations and intimations, perhaps the incongruities and incompatibilities of the first results beget, by their suggestiveness, the basis of the second. The latter is the offspring of the former, but between parent and offspring there is mutation with an evolutionary purpose. A cruder first attempt generates a more highly organized and specialized working scheme fitted to the new state of knowledge developed. . . .

As our working basis, we assume that our perceptions represent reality, when duly directed and corrected, but that error and illusion lurk on all sides and must be scrupulously avoided. We assume that we are capable of detecting error and of demonstrating truth; and that, as requisite means, we have choice, and some measure of volitional command over ourselves and over nature. . . .

The most serious source of error in the development of the earth-sciences, in my judgment, is our relative neglect to probe fundamental conceptions and to recognize the extent to which they influence the most common observations and interpretations. We need a method of thought that shall keep us alive to these basal considerations. To this end I believe it to be conducive to soundness of intellectual procedure to regard our whole system of interpretation as but an effort to develop a consistent system of workable hypotheses. I think we should do well to abandon all claims that we are reaching absolute truth, in the severest sense of that phrase, and content ourselves with the more modest effort to work out a system of interpretation which shall approve itself in practise under such tests as human powers can devise. Wherein lie

### The Basal

### Criteria of

### Our Sciences?

I believe they lie essentially in *the working quality*. Whatever conforms thoroughly to the working requirements of nature probably corresponds essentially to the absolute truth, though it may be much short of the full truth. That may be accepted, for the time being, as true which duly approves itself under all tests, *as though it were true*. Whenever it seems to fail under test in any degree, confidence is to be withdrawn in equal degree, and a rectification of conceptions sought. This may well hold for all conceptions, however fundamental, whether they relate to the physical, the vital, or the mental phenomena which the earth presents. Let us entirely abandon the historic effort of the metaphysicians to build an inverted pyramid on an apex of axioms assumed to be incontestable truth, and let us rear our super-structure on the results of working trials applied as widely and as severely as possible. Let us seek our foundation in the broadest possible contact with phenomena. I hold that *the working test* when brought to bear in its fullest, most intimate and severest forms is *the supreme criterion* of that which should stand to us for truth. Our interpretative effort should, therefore, be to organize a complete set of working hypotheses for all phenomena, physical, vital and mental, so far as appropriate to our sphere of research. These should be at once the basis of our philosophy and of our science. These hypotheses should be constantly revised, extended and elaborated by all available means, and should be tested continually by every new relation which comes into view, until the crucial trials shall become as the sands of the sea for multitude and their severity shall have no bounds but the limits of human capacity. That which under this prolonged ordeal shall give the highest grounds of assurance may stand to us for science, that which shall rest more upon inference than upon the firmer modes of determination may stand to us for our philosophy, while that which lies beyond these, as something doubtless always will, may stand to us for the working material of the future.

JOHN C. ECCLES

# In Praise

# of Falsification

DURING THE FIRST two of the eight years (1944–1951) I spent in Dunedin, New Zealand, I had the good fortune to be associated with the eminent philosopher of science Karl Popper. I learned from him what for me is the essence of scientific investigation—how to be speculative and imaginative in the creation of hypotheses, and then to challenge them with the utmost rigor, both by utilizing all existing knowledge and by mounting the most searching experimental attacks. In fact I learned from him even to rejoice in the refutation of a cherished hypothesis, because that, too, is a scientific achievement and because much has been learned by the refutation.

Through my association with Popper I experienced a great liberation in escaping from the rigid conventions that are generally held with respect to scientific research. Until 1944 I held the following conventional ideas about the nature of research: First, that hypotheses grow out of the careful and methodical collection of experimental data. (This is the inductive idea of science that we attribute to Bacon and Mill.) Second, that the excellence of a scientist can be judged by the reliability of his developed hypotheses, which, no doubt, need elaboration as more data accumulate, but which, it is hoped, stand as a firm and secure foundation for further conceptual development. Finally, and this is the important point, that it is in

Excerpt from J. C. Eccles, Under the spell of the synapse. In F. G. Worden, J. P. Swazey, and G. Adelman, eds., The Neurosciences: Paths of Discovery. Cambridge: MIT Press, 1975, pp. 162–63. Copyright © 1975, The MIT Press, Cambridge, Massachusetts. Reprinted by permission.

the highest degree regrettable and a sign of failure if a scientist espouses an hypothesis that is falsified by new data so that it has to be scrapped altogether. When one is liberated from these restrictive dogmas, scientific investigation becomes an exciting adventure opening up new visions; and this attitude has, I think, been reflected in my own scientific life since that time.

My prolonged isolation in the Antipodes was relieved for three weeks early in 1946 by my first visit to the United States. In those immediate postwar weeks the journey from New Zealand to New York was certainly an unpredictable and exciting experience, but that is another story! The occasion was a meeting of the New York Academy of Sciences organized by David Nachmansohn, and I was most grateful to him for the opportunity of meeting so many of the leading neurophysiologists of America and also the international visitors from Europe. On that occasion I developed still further my story of electrical synaptic transmission. . . . I had been encouraged by Karl Popper to make my hypothesis as precise as possible, so that it would call for experimental attack and falsification. It turned out that it was I who was to succeed in this falsification by the discoveries that came towards the end of my stay in New Zealand.

ROBERT SEKULER

# In Praise

# of Strong

# Inference

AS YOU PROBABLY know already, it is the scientist's lot that one experiment begets another. Since we enjoy doing research, this is pleasurable rather than upsetting. I believe, however, that the genealogy and pedigree of experiments is an important business. There should be vigilant population control. We do not want random and unplanned begetting of experiments. There are optimal ways to organize experiments so that you get the most information in the shortest possible time. One of the guiding principles which scientists ought to use is that of "strong inference" (Platt 1964). The series of experiments I'm about to describe makes pretty fair use of strong inference—for a psychologist.

Look at the problem at hand. We came upon a new and peculiar phenomenon, the left-right TOE,[1] quite by accident. We decided to try and explain it, since that's what science is all about. Now what exactly is involved in *explaining* a perceptual phenomenon? First, we must acknowledge that there are many competing, alternative hypotheses (or "stories," as I like to call them at this immature stage of their development) which we could cook up to explain the TOE. Each experiment

Excerpt from R. Sekuler, Seeing and the nick in time. In M. H. Siegel and H. P. Zeigler, eds., *Psychological Research: The Inside Story*. New York: Harper & Row, 1976, pp. 182–83. Reprinted by permission of Harper & Row.

we do ought to eliminate one possible explanation. If we're really clever, an experiment will eliminate more than just one alternative.

In this view, science is a treelike structure—a decision tree. Each experiment provides a choice point. The results of each experiment direct us out along the limb in the appropriate direction. However, if you subscribe to this view you can't sit down at the beginning of the study and say with certainty exactly what experiments you'll do. The sequence of experiments is determined as we go along, mandated by the pattern of the outcomes we have. Hopefully, the results of well-designed and executed experiments will give us better successive approximations to truth.

## Note

1. The TOE (temporal order effect) is a phenomenon which occurs when two stimuli are presented simultaneously (or nearly simultaneously) in a left-to-right spatial order. The left stimulus appears to precede the right, even if the right stimulus is actually presented temporally first. [Eds.]

# Part IV

# Hypothesis Testing: The Role of Confirmation

# Introduction

THE DOCTRINE OF falsification has a compelling logical power which has attracted many adherents. Nevertheless, it is not hard to find historical examples which are at variance with this model. Kelvin's unwillingness to accept Roentgen's discovery of X-rays, Mesmer's reluctance to test alternatives to his belief in animal magnetism, the resistance to Harvey's discovery of the circulation; in each case, we can find practicing scientists—sometimes great scientists—turning away from empirical data just as irrationally (and perhaps as confidently!) as the bishops turned away from Galileo's telescope (Santillana 1955).

A recent body of psychological research similarly suggests that prescriptive falsification rules of the sort advocated by Popper, Chamberlin, and Platt are frequently violated. Most of this research has not employed scientists as subjects; but the pervasive nature of the effects obtained, together with the results of studies of scientists, indicates that many people, including scientists, manifest a bias to confirm. They do so by their failure to do one or more of the following:

1. Seek disconfirmatory evidence.
2. Utilize disconfirmatory evidence when it is available.
3. Test alternative hypotheses.
4. Consider whether evidence supporting a favored hypothesis supports alternative hypotheses as well.

Each is a slightly different manifestation of an excessive focus on a favored hypothesis—a confirmation bias. The term "bias" is not used here in a pejorative sense. It is a descriptive term which refers to a general tendency not to give up a favored hypothesis.

Each of the four types of confirmation bias can be demonstrated by pointing to a number of studies. We shall review these, as well as other, related, studies, and discuss some of the implications of confirmation bias.

**Failure**

**to Seek**

**Disconfirmatory**

**Evidence**

This is probably the most pervasive of the four manifestations of confirmation bias. A good deal of relevant experimental work has been carried out on this problem, much of it using a cognitive task developed by Peter Wason.

Wason (1968b) required college student subjects to select evidence to test the truth or falsity of a conditional rule. Specifically, subjects were shown four cards, on each of which a single letter or number was printed. For example:

A   B   3   4

After being told that the cards with numbers on one side had letters on the other side, and vice versa, they were instructed to select those cards and only those cards which needed to be turned over to test a conditional rule of the form, "If a card has an A on one side, then it has a 3 on the other side." ("If $p$, then $q$.") Logically, the combination of $p$ and $not$-$q$ is of critical importance, since it alone can show the rule to be false (via the *modus tollens* syllogism). In Wason's experiment, subjects overwhelmingly chose cards which corresponded to $p$ and $q$ (A and 3 in the example above). The $p$ card can either confirm or disconfirm the rule, depending on whether 3 or 4 is on the reverse side, but the $q$ card cannot disconfirm the rule, no matter what letter is on the other side. The $not$-$q$ card (4), which could disconfirm the rule if A were on the other side, was rarely chosen. Wason and his colleagues employed several different instructional and training procedures in an attempt to induce subjects to choose the $not$-$q$ card but had only limited success (Wason & Johnson-Laird 1972). Even subjects who could verbalize the correct strategy experienced great difficulty in actually using it. Wason and Shapiro (1971) and Johnson-Laird, Legrenzi, and Legrenzi (1972) were able to induce most subjects to use the correct strategy by presenting the task in a realistic context. Unfortunately, the general principles that govern whether or not subjects will respond correctly have not been determined and the "realism effect" has not proved easy to replicate (Manktelow & Evans, 1979; Yachanin, 1980). Further, Evans and Lynch (1973) have shown that the pattern of cards chosen may result either from logical reflection or (more often) from nonlogical attempts to match the chosen cards with the cards named in the rule. Complete understanding of subjects' performance in the four-card task is, thus, likely to require fairly complex models. Nonetheless, for whatever reason, there is clearly a failure to search for disconfirmatory evidence.

Bean (1979) presented psychologists, biologists, and physicists, as well as humanities faculty and undergraduates, with a series of logical reasoning problems. Although nearly all the scientists correctly recognized the validity or invalidity of some types of abstract logical constructions (e.g., modus ponens and denial of the antecedent), fewer than half recognized the validity of modus tollens, a performance about equal to that of humanities faculty and undergraduates. Einhorn and Hogarth (1978) gave a group of statisticians a concrete version of Wason's four-card task. The major finding was that ". . . a majority of analytically sophisticated subjects failed to make the appropriate response. In particular, half the subjects chose to examine the same piece of confirmatory information" (p. 400). Bean (1979) obtained similar results in her study.

Wason (1960, 1968a) developed another task, described in selections 16 and 17, which requires that subjects play a more active role. Here also confirmation bias is frequently manifested. In this task, subjects try to discover an experimenter-defined numerical rule by seeking evidence to evaluate their hypotheses. Wason presented subjects with an initial three-digit sequence, 2 4 6, which was compatible with the rule, and instructed the subjects to generate their own test sequences. Following each sequence, the subjects were told whether or not their example conformed to the rule. After a series of such tasks, subjects could announce what they thought the rule was, and would be told if they were correct. In all cases, the experimenter's rule was deceptively simple: "any three ascending numbers." Mahoney and DeMonbreun (selection 17) administered the same task to psychologists, physical scientists, and Protestant ministers. They argued that the scientists, as a group, should use disconfirmation more than ministers, since scientists have been trained in the use of "critical thinking skills," presumably including the virtues of falsification, while ministers have not. The results of the study indicated, however, that the majority of subjects, including the scientists, did not seek disconfirmatory evidence and placed little value on it when it occurred.

## Failure

## to Utilize

## Disconfirmatory

## Evidence

Even when disconfirmatory evidence is present, it is not always used. For example, Pitz and his colleagues (Pitz, Downing, & Reinhold, 1967; Geller & Pitz, 1968; Pitz 1969) have repeatedly found evidence of an "inertia effect" in studies of sequential decision making. In these experiments, undergraduate subjects were presented with several pieces of evidence confirming and disconfirming two hypotheses. After each item in the sequence, the subjects made likelihood judgments about one, or both, of the hypotheses. Geller and Pitz (1968) summarize this research by stating ". . . a resistance to revising opinion following disconfirming information appears to be a general phenomenon in studies of confidence revision, probability estimation, decision making, and information buying" (p. 199). The same general conclusion was reached by Einhorn and Hogarth (1978). In developing a model of judgmental confidence, they suggested that confidence generally increases more following positive feedback than it decreases following negative feedback (p. 402). Ross (1977) and Snyder and Swann (1978) found the same effect in person-perception studies.

Even when an initial hypothesis about a stranger was completely discredited, subjects persevered in believing the hypothesis.

In order to study such effects under conditions more closely approximating those of actual scientific inquiry, we constructed an artificial research environment using a computer-controlled plasma display screen. Subjects were presented with various "universes" consisting of fixed shapes, such as disks and triangles, and small moving dots or "particles" (Mynatt, Doherty, & Tweney 1978, selection 18). Advanced science majors were instructed to attempt to discover the "laws" of motion in the universe by conducting particle firing "experiments."

All of the subjects initially generated many incorrect hypotheses, tests of which rapidly led to disconfirmatory evidence. Following disconfirmation, most subjects either abandoned an hypothesis temporarily but returned to it later, revised an hypothesis *ex post facto* to accommodate the anomalous evidence, or simply ignored the disconfirmation and went on testing the same hypothesis. The results were consistent with the data from Wason's and Mahoney and De-Monbreun's 2 4 6 studies. In both investigations, more than half of the subjects who obtained evidence inconsistent with their hypothesized rule continued to test the same rule.

It is, of course, not the case that subjects always ignore disconfirmatory evidence. In our first study (1977), we found that subjects were often able to use disconfirmatory evidence when it was fortuitously available. The exact conditions which determine whether or not such data will be used are not yet known (see also Tweney et al. 1980).

**Failure**

**to Test**

**Alternative**

**Hypotheses**

In the first study using a computer simulated "universe" (Mynatt, Doherty, & Tweney 1977) we gave undergraduate subjects two different displays designed to induce a particular incorrect hypothesis. Subjects were then given the opportunity to test their hypotheses on additional screens which enabled them to fire more particles at either (a) confirmatory targets, or (b) at other targets which could disconfirm the induced hypotheses. Only screens of the second type allowed alternatives to the induced hypothesis to be tested. Subjects were asked to make forced choices between these two types of test screens. Of the chosen screens, 71 percent were of the first type. This occurred even though some subjects had been explicitly instructed to attempt to disconfirm their hypotheses.

Some of the subjects were given instructions on multiple hypothesis testing, but there was no effect due to instructions. Further, there was almost no evidence of consideration of alternative hypotheses in the subject protocols. The study clearly implies that people do not test alternative hypotheses, even after being explicitly told to do so.

## Failure to

## Evaluate Evidence

## in the Light of

## Alternative Hypotheses

Even when clear alternative hypotheses are available, subjects often fail to evaluate evidence against all possibilities. Normative behavior in certain instances is simple to describe in terms of Bayes' theorem, which specifies that the import of data that are compatible with one hypothesis cannot be determined independently of the extent to which the same data are compatible with *other* hypotheses as well. For example, an archeologist who accepts a theory which states that a certain group of extinct primates were carnivores might count as confirmatory evidence a finding that many fossilized antelope bones had been found in caves known to have been inhabited by these primates. However, the conclusion is unwarranted unless the archeologist considers the likelihood of finding such evidence, even if the theory in question were incorrect—

if, for example, antelope were brought into the cave by other predators during periods when the primates were absent.

Doherty, Mynatt, Tweney, and Schiavo (1979; selection 19) found that many people do not seek data relevant to an alternative hypothesis, even when such data are easily available.[1] When subjects' responses were constrained in a subsequent study so that they had to choose some data relevant to the alternative, later unconstrained responses showed a substantial increase in such choices, but only if their original choices had resulted in incorrect conclusions. (M. Doherty, personal communication). Without such experience, there seems to be a very strong tendency not to consider information relevant to alternative hypothesis.

## Confirmation Bias
## Among
## Scientists

All of the above research suggests the existence of a general and pervasive tendency to avoid giving up a favored hypothesis. Most subjects neither sought nor utilized information so as to make disconfirmation likely. Nor did they seek or utilize information in such a way that would make acceptance of an alternative hypothesis likely. People to not behave in accordance with prescriptive falsification rules.

It could be argued that this says nothing about what Popper, Chamberlin, and Platt were concerned with; namely, the conduct of *scientific* inquiry. The subjects in many of the reviewed experiments were college undergraduates. Equally important, all of the studies used tasks that were substantially different from "real science" in many important ways. Surely one must be cautious, to say the least, in extrapolating results from one population and set of tasks. In the present case, this is an especially serious concern, since much scientific training could be seen as directed toward the elimination of confirmation bias.

However, the results of an interview study of NASA geophysicists conducted by Mitroff (1974b; selection 20) suggest that confirmation bias is indeed prevalant among scientists. Mitroff began interviewing high-level NASA research scientists three months before the first Apollo moon landing. Many of the scientists regarded themselves and their colleagues as strongly committed advocates of particular theories, rather than as impartial seekers after truth. For at least some of the scientists, this commitment persisted even in the face of contradictory evidence obtained during the 1969 moon landing. Mitroff selected out of his original sample several scientists who were acknowledged by their peers to be especially articulate and prestigious spokesmen for various theories. The other scientists in the sample were asked to rate the degree of commitment of these theorists to their theories both before the Apollo flights and after, when relevant, frequently disconfirmatory, data had become available for the first time. Little or no change was perceived.

Gruber (1974; selection 21), in his examination of the development of Darwin's thinking on evolution, found little evidence for use of anything like Platt's strong inference strategy. "The picture of scientific ·thought is often painted as being carried forward by the construction of alternative hypotheses followed by the rational choice between them. Darwin's notebooks do not support this rationalist myth" (p. 146).

An extraordinary example of confirmatory thinking was cited by Holton (1972). It occurred during a conversation between Einstein and one of his students, Ilse Rosenthal-Schneider. She recalls:

> Once when I was with Einstein in order to read with him a work that contained many objections against his theory . . . he suddenly interrupted the discussion of the book, reached for a telegram that was lying on the windowsill, and handed it to me with the words, "Here, this will perhaps interest you." It was Eddington's cable with the results of measurement of the eclipse expedition (1919). When I was giving expression to my joy that the results coincided with his calculations, he said quite unmoved, "But I knew that the theory is correct"; and when I asked, what if there had been no confirmation of his prediction, he countered: "Then I would have been sorry for the dear Lord (Eddington)—the theory is correct." (cited in Holton 1972:361)

The most striking feature of this incident is the difference between Einstein's and Popper's perception of the importance of the Eddington observations for relativity theory. For Popper, relativity theory is the example, par excellence, of a falsifiable theory, and this is nowhere more evident than in the precise predictions of gravitational effects on light, which were confirmed by Eddington. Popper contended this was the crucial difference between pseudoscience, such as Marxist economics or Freudian psychodynamics, and true science. Had Eddington's data been different, relativity theory would have been disconfirmed. No such disconfirmatory experiment would be acknowledged by the Marxist or the Freudian. Quite clearly, in this instance Einstein was also unwilling to acknowledge any such crucial experiment.

## The Utility

## of Confirmation

## and Disconfirmation

Thus, there seems to be a good deal of evidence that falsification strategies are *not* employed in a wide variety of inference tasks, including scientific inquiry. But is this necessarily a bad thing? Are falsification strategies, in fact, optimal inference techniques?

There are two distinct ways in which the term "optimal" can be used. First, it can be argued that falsification strategies ought to be employed in order to act *rationally*; to fail to employ them is to violate accepted canons of reason. Second, it can be argued that falsification strategies ought to be employed in order to act *efficiently*; to fail to employ them is to decrease the likelihood of obtaining true theories and laws. In one sense the two kinds of optimality are independent of each other. One can envision, for example, highly efficient but nonrational procedures, such as consulting an infallible oracle. In another sense, however, they are not independent. If a significant portion of successful scientific activity proceeds in violation of philosophical models of rationality, this could make the philosophical canons irrelevant, regardless of their formal validity, or their power to attract assent.

Our major concern is optimality of the second, empirical type. Several types of evidence suggest that falsification strategies are not necessarily efficient.

Many of Mitroff's (1974a) scientists saw strong, even extreme, commitment as not only present, but as desirable in scientists. In our laboratory even the occasional subjects who did employ falsification strategies were not conspicuously successful (Mynatt, Doherty, & Tweney 1978; selection 18). In fact, subjects who quickly and permanently abandoned disconfirmed hypotheses were further away from the correct solution at the end of the experiment than they had been during the earlier stages.

In an effort to control the use of falsification strategies, Tweney and colleagues (1980) taught one group of college students to seek disconfirmation and another group to test multiple hypotheses when attempting to solve Wason's 2 4 6 task. No improvement in any of several measures of solution efficiency resulted. In fact, excessive use of disconfirmation and requiring subjects to use multiple hypotheses, actually worsened performance.[2]

Several philosophers of science, for example Lakatos (1970, 1978), have argued that Popperian falsification simply does not occur:

> Popper's criterion ignores the remarkable tenacity of scientific theories. Scientists have thick skins. They do not abandon a theory merely because facts contradict it. They normally either invent some rescue hypothesis to explain what they then call a mere anomaly, or, if they cannot explain the anomaly, they ignore it, and direct their attention to other problems. Note that scientists talk about anomalies, recalcitrant instances, not refutations. History of science, of course, is full of accounts of how critical experiments allegedly killed theories. But such accounts are fabricated long after the theory had been abandoned. Had Popper ever asked a Newtonian scientist under what experimental conditions he would abandon Newtonian theory, some Newtonian scientists would have been exactly as nonplussed as are some Marxists. (1978:3–4)

It is confirmation, rather than disconfirmation, which is crucial for Lakatos, but only confirmation of a special type.

> . . . let us take Einstein's programme. This programme made the stunning prediction that if one measures the distance be-

tween two stars in the night and if one measures the distance between them during the day (when they are visible during an eclipse of the sun), the two measurements will be different. Nobody had thought to make such an observation before Einstein's programme. Thus, in a progressive research programme, theory leads to the discovery of hitherto unknown novel facts. In degenerating programmes, however, theories are fabricated only in order to accommodate known facts. (Lakatos 1978:5–6)

Finally, one must be especially careful in judging new theories according to the harsh doctrine of falsification (a point made by Feyerabend, 1975, as well as Lakatos).

One must treat budding programmes leniently: programmes may take decades before they get off the ground and become empirically progressive. Criticism is not a Popperian quick kill, by refutation. Important criticism is always constructive: there is no refutation without a better theory. (Lakatos 1978:6)

These arguments, together with the data from Mynatt, Doherty, and Tweney (1978), Tweney et al. (1980), and Mitroff (1974b) imply that falsification strategies are not always effective. Further, they suggest that a good deal of careful inquiry is needed into whether, or under what conditions, falsification might be an optimal inference technique. Specifically, it would appear reasonable to try to delimit more precisely the "boundary conditions" of falsification; that is, to specify those circumstances under which it is likely to prove successful and those circumstances under which it is not.

We believe that it is primarily the quality of the relationship between data and hypotheses that determines whether use of falsification will be productive or not. Unless there is a clear deductive link between a set of data and the hypothesis under test, the import of the data for the hypothesis will be ambiguous. Disconfirmation of a hypothesis can arise from any of several conditions:

1. The hypothesis under test is fundamentally wrong.
2. The hypothesis under test is fundamentally right, but is incomplete or incorrect in one or more specific details.
3. The hypothesis under test is fundamentally right, but data quality is low (e.g., there is a high degree of measurement error).
4. The relation between the hypothesis and data is not rigorously elaborated.

Only under condition 1 would it be appropriate to abandon the hypothesis completely and permanently. Only when the quality of the hypothesis is high (i.e., it is relatively complete and specific) and the quality of the data is high (i.e., there is no ambiguity about the data) can it be concluded that condition 1 is, indeed, the case. Otherwise, disconfirmation might be due to lack of precision in either the hypothesis or the data, or both, and not to the fundamental falsity of the hypothesis.

The scientist working in a relatively underdeveloped field might therefore be well advised to pay little heed to occasional disconfirmation, and not to waste time seeking disconfirmation. It might be more productive to concentrate on development of the hypothesis in hand and on refinement of data collection techniques. The process of developing a good hypothesis may require systematic use of **confirmatory** procedures and may be hindered by use of disconfirmatory procedures; on the other hand, disconfirmation might be used very productively when both hypothesis and data quality are high enough to allow unambiguous inferences to be made.

This analysis can be schematized as shown in the accompanying diagram.

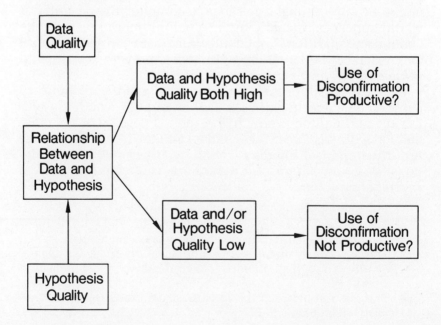

There is some evidence supporting the validity of this analysis. In the Mynatt, Doherty, and Tweney (1978) study, subjects were put into a completely novel situation and forced to develop hypotheses "from scratch" about the underlying laws of motion in the artificial universe. Not surprisingly, nearly all the initial hypotheses were in one respect or another incorrect or incomplete; that is, hypothesis quality was low. Significantly, the most successful subject developed a relatively cogent but not completely correct hypothesis which received a good deal of initial confirmation. When disconfirmation later occurred, he did not discard his hypothesis entirely. Rather, he used the disconfirmatory evidence productively by modifying those parts of his hypothesis which were, in fact, incorrect. This finding is consistent with the results of a recent study on problem solving in children by Karmiloff-Smith and Inhelder (1975). They concluded that only when a hypothesis is sufficiently "consolidated and generalized is (a child) ready to recognize some form of unifying principle for the counter-examples which were earlier rejected as mere exceptions" (p. 204).[3]

Of course falsification may also be counterproductive at later stages of inquiry as well, particularly if there is no reasonable alternative to the falsified hypothesis. It seems absurd to give up a theory which accounts for, say, 99 percent of the relevant data and replace it with nothing.

Many of these ideas about the appropriate conditions for falsification can be illustrated and summarized by an examination of Watson and Crick's (1953) research on DNA, which is cited by Platt (1964) as an example of the successful application of multiple-hypothesis testing. In this instance both hypothesis and data quality were high. There was a fairly well developed inductive data base. It was known that DNA consisted of only certain nucleic acids, and that these could be linked to form the DNA molecule only in certain ways. Furthermore, reliable techniques for collecting relevant data were available, and the import of these data with respect to the various alternative structural models was reasonably clear. There was not, however, an overwhelmingly dominant theory which admitted of no disconfirmation at all. Thus, it is not surprising that Watson and Crick found falsification and multiple-hypothesis testing to be highly effective procedures. Contrast this to the situation in lunar geology in the late 1960s and early 1970s as described by Mitroff. The field was a new one, so that hypotheses were relatively undeveloped and nonspecific. Data quality was also relatively low due to the extreme constraints on data collecting. Not surprisingly, and perhaps not ineffectively, confirmation bias flourished.

The selections in part 3 showed a profound understanding of the human tendency toward confirmation. The words of Newton, Locke, Hartley, and the rest, evidenced good psychology in this regard. Popper showed the same sensitivity to this issue and juxtaposed to it a logical analysis. The result was a demarcation criterion which has been taken by many (perhaps erroneously) as a prescription for how to do science. But it is manifest that the issue of how to do good science is a tangled one, with deep psychological implications. Certainly at least two broad psychological questions are involved. One is "Can people falsify?" The second is "Should they?"

# NOTES

1. In a recent study, D. Adsit and R. Tweney found that a majority of research mathematicians and statisticians, many of whom were thoroughly familiar with Bayes' theorem, also failed to seek data relevant to alternative hypotheses when given the Doherty et al. task (R. Tweney, personal communication).

2. Tweney et al. (1980) found one manipulation that appeared to enhance use of evidence relevant to alternative hypotheses. When subjects were required to determine what rules governed construction of DAX triples and MED triples, they performed far more efficiently than subjects who sought rules that governed DAX triples and non-DAX triples, the usual form of the 246 task, even when the MED and non-MED triples were identical. The conditions under which such purely structural changes in problems can lead to better performance are not well understood.

3. Rose and Tweney (1980) manipulated the kind of evidence given children in a balance-beam task, and found that confirmatory evidence was necessary before advancement to higher-order rules was possible.

PETER WASON
and P. N. JOHNSON-LAIRD

# The Discovery

# of a General

# Rule

IN THE LAST three chapters the deductive component of scientific inference was investigated in a miniature task. The
subjects were presented with a hypothesis, and had to decide
what items of evidence were relevant for testing its truth. In
spite of the difficulties of the task, it could be argued that, as
an analogy to scientific research, half the work had already
been done for them. In doing research one has to postulate a
hypothesis, collect the evidence which would be relevant to
it, and then evaluate this evidence in order to modify the
initial hypothesis. The experiments reported in this chapter
simulate this activity in an abstract task. More specifically,
they seek to examine the adequacy of the tests to which hypotheses about an unknown rule are subjected. At least two
factors may be important: (a) the ability to think up hypotheses, and (b) the ability to relinquish hypotheses. These
factors are likely to interact. The ability to conceive a hypothesis which seems to "work," in the sense that it fits the data,
may make the individual reluctant to abandon it. On the contrary, he may simply try to strengthen it by adducing more

Excerpts from P. Wason and P. N. Johnson-Laird, *Psychology of Reasoning: Structure
and Content*. Cambridge: Harvard University Press, 1972, pp. 202–14. Reprinted by
permission of B. T. Batsford Ltd.

confirming evidence for it. The experiments were designed to investigate the propensity of individuals to offer premature solutions based on such evidence. In this way, they are related to tolerance for uncertainty. . . .

In our first experiment (Wason 1960) the subjects were told that the three numbers, 2 4 6, conformed to a simple relational rule which the experimenter had in mind, and that their task was to try to discover it by generating successive triads of numbers. After each triad they were told whether, or not, the numbers conformed to the rule. They were allowed to keep a written record of their numbers, and their hypotheses, but were strictly instructed to show the rule to the experimenter ("announce the rule") only when they were highly confident that they had discovered it. If they announced a hypothesis other than the rule, they were told they were wrong and instructed to proceed with the task. The rule was: *numbers increasing in order of magnitude.* But the real point of the experiment was not to see whether the subjects discovered it, but to see how they set about trying to discover it. A very general rule was deliberately chosen so that positive instances of it would also tend to be positive instances of the more restricted hypotheses which would be likely to occur first of all to the subjects, e.g. "intervals of two between increasing numbers." In a strict sense, of course, the task is impossible if "high confidence" were to be equated with proof. The rule cannot be proved, but it will be demonstrated subsequently how any more specific (sufficient) hypotheses can be disproved. This task, which relaxes the constraints in the conventional concept attainment task, has three distinctive features.

First, the subject is not presented with all the available evidence at the start. He has to generate both his own instances and his own hypotheses.

Second, the universe of instances is potentially infinite, and hence the number of instances which exemplify any hypothesis can never be exhausted. For example, an endless number of instances, exemplifying a sufficient hypothesis such as "intervals of two between increasing numbers," can be generated without forcing the subject to generate an instance which does not exemplify it. It follows that if a subject only verifies, or confirms, his hypothesis, he will be forced to announce it to the experimenter as the only way of finding out whether it is the rule. As we have seen, this does not necessarily occur in Bruner's task because the universe of instances is finite, and the degree of generality of the possible concepts is correlated with the number of instances which satisfy them.

Third, the subjects do not have to remember either their previous instances, or their previous hypotheses, but can refer to their record sheet. In nearly all traditional concept attainment tasks memory has been a factor. It may, of course, be of considerable interest to investigate the interaction between reasoning and memory (e.g., Whitfield 1951), but often memory has been a gratuitous variable without specific predictions being made about it. The use of paper and pencil is not denied in real life as an aid to thinking and planning.

It is evident that there are three strategies which a subject could adopt in the present task. The first one is to try to *verify* a hypothesis, and, if it is confirmed, to announce it as the rule. As we have seen, the task is biased so that this strategy will almost certainly lead to plausible but wrong conclusions.

The second strategy is to try to *falsify* a hypothesis. The strictest criterion for this strategy is when a subject generates an instance which is *inconsistent* with the hypothesis he is entertaining. For example, he might generate the instance, 3 6 10, and write down on the record sheet "to see if the rule is successive multiples of the first number."

The third strategy is to *vary* his current hypothesis instead of trying to confirm it, or deliberately falsify it. Unlike the second strategy, this one is particularly interesting because it is consistent with Kuhn's (1962) thesis that scientists only relinquish a theory when an alternative theory, or hypothesis, is available. In the present context it entails that the subject will only abandon a hypothesis when another one is conceived.

Both the second and third strategies share the common factor that the rule will probably be discovered without announcing any incorrect hypotheses, and the aim of the experiment was primarily to investigate the extent to which procedures of this kind were used. It is simplest to show how the third strategy might achieve its end.

Suppose a subject has entertained the hypothesis, "intervals of two between ascending numbers," on the strength of the positive instances, 8 10 12, 14 16 18, he might then consider the more general hypothesis," an equal interval between ascending numbers" (arithmetic progression). And in doing this, he might generate 7 11 15. Since that instance turns out to be a positive instance of the rule, he has disproved at a stroke his former hypothesis, "intervals of two between ascending numbers." Similarly, his new hypothesis of arithmetic progression could be disproved by finding out that an instance such as 1 6 7 is also positive. Thus the strategy of varying a hypothesis, which has been confirmed, will gradually purge the initial instance, 2 4 6, of all the surplus meaning inherent in it, and inexorably tend to lead the subject towards a consideration of the rule. It should be noted that there is a pertinent psychological distinction between the strategy of varying (falsifying) hypotheses on the one hand, and the strategy of verifying hypotheses on the other hand. When hypotheses are varied the work involved in discovery is internalized; but when hypotheses are only verified on the basis of confirming evidence the subject is forced to appeal to an external authority (the experimenter) to find out if his conclusions are correct.

In the exploratory experiment (Wason 1960) only 21 per cent of the subjects discovered the rule without making any announcements of incorrect hypotheses, and the results showed that these subjects varied their hypotheses significantly more often than the remainder. They did not hit on the rule by chance. There was, however, very little evidence for the use of the falsification strategy, and hence the main interest lies between the variation of hypotheses (which of course, frequently leads to implicit falsification of former hypotheses), and the verification of hypotheses by piling up confirming evidence for them. This provides some corroboration, in an artificial task, for Kuhn's (1962) view that beliefs, or hypotheses, are only abandoned (if at all) when more adequate alternatives become available. But even the spontaneous variation of hypotheses is rare in the present experiment. The majority of subjects announced at least one incorrect hypothesis, and one subject only discovered the rule after making four incorrect announcements. However, only 28 per cent failed to discover the rule by whatever strategy. The two protocols which follow illustrate the difference between trying out alternative hypotheses and the failure to do so. The first is derived from a subject who made no incorrect announcement ("immediate correct announcement"), and the second from a subject who made four incorrect announcements.

No. 1. Immediate correct announcement. Female. Psychology undergraduate.

| Instances | Hypotheses |
|-----------|-----------|
| 2 4 6 ( + ) (Given) | |
| 3 6 9 ( + ) | Three goes into the second figure twice and into the third figure three times. |
| 2 4 8 ( + ) | Perhaps the figures have to have an L.C.D. |
| 2 4 10 ( + ) | Same reason. |
| 2 5 10 ( + ) | The second number does not have to be divided by the first one. |
| 10 6 4 ( − ) | The highest number must go last. |
| 4 6 10 ( + ) | The first number must be the lowest. |
| 2 3 5 ( + ) | It is only the order that counts. |
| 4 5 6 ( + ) | Same reason. |
| 1 7 13 ( + ) | Same reason. |

Announcement: *The rule is that the figures must be in numerical order.* (Correct: 16 min.)

It is apparent that this subject does spontaneously vary her hypotheses, and hence avoids getting entangled in her own thought. It is also clear that with one instance she does attempt the falsification strategy: she generates the instance 10 6 4, and cites a hypothesis inconsistent with it, "the highest number must go last."

No. 2. Four incorrect announcements. Female. Psychology undergraduate.

| Instances | Hypotheses |
|-----------|-----------|
| 2 4 6 ( + ) (Given) | |
| 8 10 12 ( + ) | Two added each time. |
| 14 16 18 ( + ) | Even numbers in order of magnitude. |
| 20 22 24 ( + ) | Same reason. |
| 1 3 5 ( + ) | Two added to preceding number. |

Announcement: *The rule is that by starting with any number two is added each time to form the next number.* (Incorrect)

| | |
|-----------|-----------|
| 2 6 10 ( + ) | The middle number is the arithmetic mean of the other two. |
| 1 50 99 ( + ) | Same reason. |

Announcement: *The rule is that the middle number is the arithmetic mean of the other two.* (Incorrect)

3 10 17 (+)                Same number, seven, added each time.
0   3   6 (+)                Three added each time.

Announcement: *The rule is that the difference between two numbers next to each other is the same.* (Incorrect)

12   8   4 (−)                The same number is subtracted each time to form the next number.

Announcement: *The rule is adding a number, always the same one, to form the next number.* (Incorrect)

1   4   9 (+)                Any three numbers in order of magnitude.

Announcement: *The rule is any three numbers in order of magnitude.* (Correct: 17 min.) . . .

This protocol is an extreme example of a trend which was apparent throughout the experiment. In fact, over all the subjects, as many as 51.6 percent of the instances, which were generated immediately after an incorrect announcement, remained consistent, rather than inconsistent, with the hypothesis just announced. On more than half the possible occasions the hypothesis is not relinquished, even when it is known to be wrong. Time is needed to find a new idea in a large number of cases—a point which is again in conformity with Kuhn's (1962) views. Typical incorrect announcements were fairly stereotyped, e.g. "arithmetic progression" (constant interval between ascending numbers), "increasing intervals of two," "successive multiples of the first number," "consecutive even numbers," etc. A few were more idiosyncratic, e.g. "arithmetic or geometric progression." . . .

It seems a reasonable conclusion from this exploratory experiment that most highly intelligent adults, in an abstract task, tend to use only a verification strategy in attempting to discover an unknown rule. It is supported by a similar study carried out by G. A. Miller (1967) in which the subjects had to discover the rules governing artificial grammars. According to Miller (personal communication): "Once a subject finds a rule that seems to work, he is unlikely to suspect the existence of other positive instances that lie beyond the scope of his particular rule."

The general conclusion from our experiment, however,

should be viewed with some caution because it is open to a considerable number of objections, and until these have been met not much progress has really been made. First, it could be claimed that a positive instance of the rule would reinforce the hypothesis which is also exemplified by it. The conflation between hypothesis and rule, wedded by the same positive instance, would tend to reward the generation of incorrect hypotheses. Miller (1967) has argued cogently, and at length, against this criticism, and we were able to adduce some empirical evidence to support his arguments. In a subsequent experiment, Angela Fine (reported in Wason 1971) delayed telling her subjects whether their instances were positive or negative until they had been generated in blocks of varying sizes. If a reinforcement principle had been responsible for the failure to vary hypotheses, then it would be predicted that generating (say) eight instances in a block, before receiving feedback about them, would lead to more variation of hypotheses than would immediate feedback, given after each instance. In fact, no increase in variation of hypotheses was observed. Feedback does not function simply as reward; if subjects are going to stick to their hypotheses, they will do so in any case. Fine's study, incidentally, also provided the comforting evidence that 75 percent of the subjects (27 out of 36) rated the task subsequently as either "enjoyable," or "very enjoyable." Furthermore, all the subjects said afterwards that they considered the rule was perfectly fair: they did not feel cheated.

Second, it could be claimed that the results were a function of the material used in the task. More specifically, the charge might be that students would be familiar with the "number series" type of intelligence test item, in which a unique continuation is assumed correct, and hence extrapolate this knowledge to the supposition that only the most "fitting" rule was correct. However, J. Penrose (1962) substituted verbal material for numbers, and devised a sort of inverted game of 20 questions. The subjects were given an instance, e.g. "Siamese cat," and had to discover a class under which it fell ("living things"), and which the experimenter had in mind. They did this by generating instances, and were told each time whether they were included, or not included, in this class. Very similar effects were obtained. For example, one subject only changed his hypothesis from "domestic pets" to "animals" after generating 12 instances. Tirril Gatty (reported in Wason 1968a) adopted a different approach. She retained numerical material, but tried to alert the subject to a variety of different possible rules. The main instruction was to discover "rules which *could be* the correct one, but were not the one the experimenter had in mind," and a subsidiary instruction was to announce the correct rule if this was discovered, as a byproduct of the main task. The results showed no difference in the number of incorrect hypotheses, announced as the rule, compared with a control group who were instructed to discover the rule in the ordinary way. In the experimental group, six out of the 11 subjects announced at least one incorrect hypothesis as the rule, and only two apparently appreciated that a single instance was sufficient to prove conclusively that a hypothesis was incorrect. A particularly interesting finding was that five out of the 11 subjects first of all generated instances to confirm a hypothesis (in contradiction to their instructions), and only then attempted to eliminate them. This confirms, yet again, Kuhn's (1962) argument that the scientist carries out research with reference to a pre-existing "paradigm." Even in an artificial task, in which no particular values are invested, the technique of deliberately trying to disconfirm a hypothesis, in accordance with explicit instructions, would seem to be a logically possible, but deviant procedure. Penrose's and Gatty's studies suggest respectively that neither the material used in our original experiment, nor the possible belief that only one rule is appropriate, could explain our results.

Third, it could be claimed that the subjects in the original experiment were merely announcing incorrect hypotheses in order to remove themselves from the experimental situation as quickly as possible. This seems unlikely because they were highly motivated, and were very surprised when told that their first incorrect announcement was not the rule. However, a modest study (reported in Wason 1968a) was carried out to try to determine whether subjects were unwilling, or unable, to adopt the appropriate strategy. One group was given ten shillings (50p) initially, and told that they would lose half a crown (12.5p) for every incorrect announcement. A control group was not given this incentive. There were no differences between the groups in the number of incorrect announcements made; the incentive only had the effect of increasing the number of confirming instances generated before making an announcement. The study suggested that the strategy of varying hypotheses does not exist in the repertoire of most subjects, rather than that the effort of using this strategy is too onerous.

Fourth, it was pointed out that if a subject can only use the verifying strategy, then he must announce that hypothesis, which he verifies, in order to find out whether it is the rule. The present task provides no objective information about whether a subject "really" believes his hypothesis is the rule. Strategy and belief are confounded, and subjective confidence ratings could be merely artifacts. They would be expected to increase as a function of the need to find out about a hypothesis rather than provide an independent criterion of the strength of belief. An experiment carried out in collaboration with Martin Katzman (reported in Wason 1968a) modified the task in order to try to clarify this issue. The subjects were told that they would be given only one opportunity to announce the rule. If an incorrect hypothesis was announced, they were not told it was wrong but were asked: "If you were wrong, how could you find out?" In response to this question, nine out of the 16 subjects, to whom it was asked, replied that they would continue to generate instances consistent with their hypothesis, and wait for one to be a negative instance of the rule. Four replied that they would either vary their hypotheses, or generate instances inconsistent with them. Three gave the revealing answer that no other rules were possible, e.g. "I can't be wrong since my rule is correct for those numbers"; "Rules are relative—if you were the subject and I were the experimenter, then I should be right." It is evident that some subjects at any rate, not only believe that their hypotheses are the rule, but also believe that they have logical reasons for supposing that they cannot be mistaken. This phenomenon, in which subjects seem to identify with their hypotheses, is presumably the most extreme consequence of the verifying strategy.

MICHAEL J. MAHONEY
and BOBBY G. DeMONBREUN

# Problem-Solving

# Bias

# in Scientists

THE PRESENT STUDY was designed to provide a . . . direct and controlled assessment of scientists' problem-solving skills. Using the same numerical task employed by Wason and his colleagues, 30 academic scientists were tested. Their problem-solving skills and, in particular, their confirmatory bias were compared to those of 15 ministers from conservative Protestant churches. The Wason task was deemed particularly appropriate due to its emphasis on both the collection (or generation) of experimental data and their subsequent interpretation.

## Method

The subjects were 15 Ph.D. psychologists, 15 Ph.D. physical scientists, and 15 ministers representing conservative Protestant religious denominations. All subjects were male. The psychologists and physical scientists (physicists and engineers) were faculty members of a large eastern university. The ministers were drawn from the same geographical area (radius—

Excerpts from M. J. Mahoney and B. G. DeMonbreun, Psychology of the scientist: An analysis of problem-solving bias. Cognitive Therapy and Research (1978), 1(3):229–38. Reprinted by permission of Plenum Publishing Corporation.

20 kilometers) and represented the following denominations: Church of Christ (2), Baptist (2), Independent Christian (1), Methodist (1), Free Methodist (1), Calvary Bible Church (1), Pentecostal (1), Christian Missionary Alliance (2), Nazarene (1), Assembly of God (2), Church of God (1). Three of the ministers had master's degrees, one in education and two in theology. The formal educational level of the remainder of the ministers ranged from high school through the bachelor's of theology degree. Ministers were chosen as the comparison group because of the widespread assumption that they are more dogmatic and less critical in their reasoning skills (Rokeach 1960).

The sample of psychologists were initially 19 but 3 declined participation and 1 more had to be eliminated due to equipment (tape recorder) malfunction. Seventeen physical scientists were originally contacted; 1 declined to participate and 1 was eliminated by equipment malfunction. None of the ministers contacted refused to participate, although 2 of this group were also eliminated by equipment failures.

The experimenter was an advanced female undergraduate (psychology major) who was blind to the experimental hypotheses. Using a standardized interview format, she introduced herself, briefly explained the problem-solving nature of the experiment, and invited each subject's participation. If they agreed, she activated a portable tape recorder and presented standardized instructions regarding the experimental task. Subjects were informed that their task was to discover a simple relational rule of which 2 4 6 was an example. They were told to generate other triads of numbers to test their hypotheses about the rule. After each triad, the experimenter informed them as to whether it conformed to the rule. Participants were instructed to announce their hypothesis only when they felt confident in its accuracy. If the announced hypothesis was wrong, they were told to continue their experimentation. Task performance continued until the subject announced the correct hypothesis, gave up, or completed a standardized 10-minute interval (measured by the experimenter).

During the 10-minute experimental phase, the interviewer followed a structured and rehearsed interaction pattern, which was designed to minimize bias and interview variance. If subjects requested further clarification of the task, the experimenter simply restated the instructions. Any other requests were deferred or declined with a verbal reference to the need for a standardized and objective interview sequence. When the interview was completed, each subject was asked to verbally respond to three questions on the difficulty of the task, its appropriateness as a measure of problem-solving skills, and the subject's preexperimental familiarity with this particular task. Subjects were then thanked for their participation and asked not to discuss its content with any of their colleagues until completion of the study. All subjects were interviewed within a total time interval of 17 days. Interviews took place in either their office or their home.

## Results

Recordings of experimental interviews were scored on 11 dependent variables. These data are summarized in table 17.1. Reliability checks were performed on three of the primary variables—announcement of the correct hypothesis, disconfirmation, and number of hypotheses offered. Inter-observer reliabilities were 100 percent, 100 percent, and 85 percent, respectively.

Replicating the previous findings of Wason and his colleagues, the present results showed that disconfirmation was a less popular problem-solving strategy than confirmation. Only 17 of the 45 subjects ever tested any of their announced hypotheses via a falsifying experiment and the average percentage of trials that were disconfirmatory was 13.3 (binomial $z = 9.76$, $p < .001$). In similar congruity with the prior finding that disconfirmation is an efficient reasoning strategy, those few subjects who did disconfirm were significantly more successful in attaining the correct hypothesis ($\chi^2 = 5.16$, $p < .05$, 2-tailed).

Differences among the three experimental groups were analyzed via nonparametric statistics. For dichotomous variables, chi-square tests were performed. Remaining data were analyzed via Kruskal-Wallis and Mann-Whitney statistics. These were chosen both because of their relative conservatism and due to the questionable appropriateness of parametric statis-

**Table 17.1** The Relative Performance of Psychologists, Physical Scientists, and Ministers on Each of the Dependent Variables

| Variable | Group performance | | |
|---|---|---|---|
| | Psychologists | Physical scientists | Ministers |
| Percent who announce the correct hypothesis | 47 | 27 | 40 |
| Percent who disconfirmed | 40 | 40 | 33 |
| Percentage of trials that were disconfirmatory | 9 | 12 | 19 |
| Average number of hypotheses generated | 3.07 (SD = 1.53) | 3.33 (SD = 1.84) | 2.00 (SD = 1.25) |
| Average number of experiments generated | 7.53 (SD = 2.76) | 7.60 (SD = 5.97) | 8.33 (SD = 7.64) |
| Average ratio of experiments to hypotheses | 3.03 (SD = 1.86) | 2.01 (SD = 1.48) | 6.22 (SD = 7.79) |
| Average number of experiments before first hypothesis | 1.93 (SD = 1.11) | 1.07 (SD = 1.62) | 5.40 (SD = 8.00) |
| Latency in seconds to first hypothesis | 35.4 (SD = 21.76) | 37.7 (SD = 39.62) | 132.9 (SD = 161.50) |
| Percentage who reconfirmed a previously falsified hypothesis | 93 | 93 | 53 |
| Average difficulty rating | 4.04 (SD = 4.39) | 4.33 (SD = 2.26) | 5.87 (SD = 2.16) |
| Average appropriateness rating | 5.50 (SD = 2.50) | 4.67 (SD = 2.63) | 5.46 (SD = 2.10) |

tics for some of the variates in question. Since the null hypothesis in this study predicted a superiority of scientists over ministers, reported levels of significance are one-tailed. As will be described in the Discussion, a second epistemological consideration also favored the use of a directional null hypothesis.

There were no significant differences among the three groups in successful attainment of the correct hypothesis ($\chi^2 = 1.32$). Similarly, looking dichotomously at whether they ever disconfirmed any of their announced hypotheses, the scientists and ministers did not differ ($\chi^2 = .19$). Even using the more general, correct hypothesis as a disconfirmatory reference point (rather than the hypothesis announced by the subject), no group differences emerged ($\chi^2 = 1.27$). Moreover, an analysis of trials found the ministers using disconfirmation with slightly higher frequency than either of the scientist groups, although the difference did not attain statistical significance ($\chi^2$ = 3.96). In terms of efficient use of disconfirmation in hypothesis testing, analyses were performed on the success of falsifying subjects in each group. Again, there were no significant differences ($\chi^2$ = 1.32).

The psychologists, physical scientists, and ministers did not vary in the total number of triads (experiments) they generated ($H$ = .08). However, there were substantial differences in their respective rates of hypothesizing ($H$ = 6.23, $p$ < .025). The ministers generated significantly fewer hypotheses than either the psychologists ($U$ = 60.5, $p$ < .025) or the physical scientists ($U$ = 64.4, $p$ < .03). The average number of triads (experiments) generated per hypothesis also suggested some possible group differences ($H$ = 4.53, $p$ < .06). Ministers had the highest ratio of experiments to hypotheses (table 17.1). This ratio was significantly greater than that of the physical scientists ($U$ = 65, $p$ < .03), but the comparison with psychologists failed to attain statistical significance ($U$ = 91.5).

The intergroup differences in speculative behavior were further illustrated by factors surrounding their first hypothesis. The three groups differed significantly in the number of experiments they generated prior to their first announced hypothesis ($H$ = 9.66, $p$ < .005). This difference was apparently due to the fact that ministers experimented much more than physical scientists at this stage of the task ($U$ = 48, $p$ < .005). The minister-psychologist difference was not significant ($U$ = 89). Assessing latency (in seconds) to the first hypothesis, the overall analysis again revealed substantial group variation ($H$ = 8.84, $p$ < .01). Ministers were much slower to hypothesize than either the physical scientists ($U$ = 52, $p$ < .01) or the psychologists ($U$ = 53.5, $p$ < .01). Two subjects were "errorless" in that their first and only hypothesis was the correct one; both were ministers.

To evaluate subjects' tenacity and cognitive redundancy, analyses were performed on the frequency with which individuals reconfirmed a hypothesis that had already been designated as incorrect by the task administrator. The three groups showed sizable differences ($\chi^2$ = 9.22, $p$ < .005) with the ministers displaying this error less often than either of the scientist groups (both $\chi^2$ = 4.75, $p$ < .02).

Although ministers rated the experimental task as being somewhat more difficult (table 17.1), the overall analysis failed to yield significant intergroup differences ($H = 4.09$). The three groups did not vary in their ratings of task appropriateness ($H = 1.38$). . . .

## Discussion

The present results must, of course, be interpreted with qualifications regarding both the subjects employed and the experimental task. They do not necessarily bear on all scientists or on alternate aspects of scientist behavior. Within the sample and task constraints, however, some tentative generalizations appear warranted.

First, in terms of problem-solving efficiency, the present study found no significant intergroup differences. Physical scientists and psychologists were no more successful than ministers in attaining the correct hypothesis. They also failed to demonstrate any differential reliance on disconfirmatory logic. Both scientists and nonscientists showed marked tendencies to confirm rather than disconfirm their hypotheses and, contrary to the popular image, they did not differ in this respect. The canons of "null hypothesis testing" would halt interpretation here—a nonsignificant result neither rejects nor supports the null hypothesis. However, there are more than a few epistemological and practical problems with null hypothesis testing (Greenwald 1975), some of which are illustrated in the present study. The "nonsignificant" difference between scientists and nonscientists is a conceptually "significant" finding, even when restricted to this task and sample. Unfortunately, conventional models of hypothesis testing are often inadequate for evaluating the importance of a nondifference. The canons of null hypothesis testing also force an investigator to "take sides" when a nondirectional (two-tailed) test yields a statistically significant difference. While more liberal in their probability levels, the one-tailed tests here reported allow a more conservative data interpretation—namely, the additional possibility of equivalence (no difference). Thus, in the present study, significant statistical tests simply recommend the rejection of a "scientist superior" hypothesis—

they do not convey a necessary superiority on the part of nonscientists. In terms of both the external validity and the interpretation of this study, the above qualification is critically important.

With the above constraints in mind, it is noteworthy that the scientists in this study appeared to be strongly inclined toward early speculation with relatively little experimentation. Moreover, within the brief analogue task employed, they appeared to be somewhat more committed to their hypotheses, often reconfirming them even after they had been conclusively rejected by the task administrator. Both of these phenomena— the apparent penchant for quick speculation and tenacious fidelity to a hypothesis—have been observed as relatively common phenomena in the scientific culture (Mitroff 1974a; Mahoney 1976). Their implications for the conduct and refinement of scientific inquiry bear careful consideration.

In summary, the present findings raise serious questions about the presumed superiority of at least some scientists' reasoning and problem-solving skills. In the study reported here, their performance was neither superior nor remarkably efficient. Both scientists and nonscientists were relatively unsuccessful in their logical performance. Moreover, both groups demonstrated considerable bias toward confirming rather than disconfirming their hypotheses. The critical importance of this practice in scientific research warrants continued examination. Additional studies with broader samples and alternate tasks will, it is hoped, shed more light on this area. Until the scientist's behavior is critically scrutinized by the scientists' tools, however, one can hardly expect either clarification or correction.

CLIFFORD R. MYNATT,
MICHAEL E. DOHERTY,
and RYAN D. TWENEY

# A Simulated

# Research

# Environment

RECENT LABORATORY STUDIES on inference indicate that hypothesis testing is subject to a pervasive confirmation bias. Mynatt, Doherty, and Tweney (1977) presented 45 subjects with a set of computer displays consisting of stationary geometric objects and moving particles whose motion was influenced by the objects. Subjects were asked to discover the laws which determined particle motion. After observing events in this "universe" for a period of time, each subject generated an initial hypothesis and then chose among several potential experiments to test that hypothesis. A significant majority of the chosen experiments were ones in which the subjects could only confirm, but not test alternatives to, their original hypothesis. The results were similar to those of Wason (1960, 1968b), who found that subjects did not try to eliminate hypotheses when attempting to discover an experimenter-defined numerical rule and did not seek disconfirmatory evidence when testing a conditional rule of the form "If P, then Q."

Excerpts from C. R. Mynatt, M. E. Doherty, and R. D. Tweney, Consequences of confirmation and disconfirmation in a simulated research environment. *Quarterly Journal of Experimental Psychology* (1978), 30:395–406.

These empirical findings contrast with widely accepted normative views of scientific inference in which disconfirmation and tests of alternative hypotheses play major roles. For example, Popper's (1959, 1962) entire philosophical analysis of science centers around the idea of disconfirmation. Platt (1964) has proposed that scientists should develop and seek to disconfirm successive generations of alternative hypotheses—a strategy he labels "strong inference." The findings of Mynatt, Doherty, and Tweney (1977) and Wason (1960, 1968a), although not based upon data from scientists, do suggest that strategies involving disconfirmation may be difficult for most people to employ.

An even more fundamental problem is raised by Mitroff (1974a). Both Popper and Platt contend that inference techniques employing falsification should be particularly effective in reaching some desired end ("well corroborated theories" in Popper's case and "rapid advances" in Platt's). Mitroff, however, presents evidence that practicing scientists may not agree with this contention. In a series of extensive interviews with 40 NASA scientists, Mitroff found that a surprising number were not only highly committed to confirming their own theoretical positions but saw this commitment as desirable and even necessary. They argued that without it, many good, new, but undeveloped ideas would die as a result of premature falsification. Even more interestingly, the scientists who held such views tended to be those who were rated by their peers as especially prominent and successful.

The question of whether falsification strategies are, in fact, effective inference techniques (irrespective of how difficult it may be to get people to use them) has not received much attention in the psychological research literature. Some of Wason's (1960) subjects did attempt to eliminate alternative hypotheses and were more successful than other subjects, but the inference task employed in the study was relatively simple. Using a more complex task, Mynatt, Doherty, and Tweney (1977) used brief instructions in an attempt to induce a confirmation strategy in some subjects and a falsification strategy in others, but the manipulation produced little effect; subjects neither sought disconfirmation nor tested alternative hypotheses. One purpose then of the present study was to attempt again to induce subjects to use an inference technique based upon falsification—specifically Platt's strong inference strategy—and to assess its effectiveness. . . .

# Method

## SUBJECTS

Sixteen upper division (Junior or Senior) students at Bowling Green State University served as subjects. Eleven were males and five were females. There were two physics, three chemistry, four biology, two mathematics, and three psychology majors. . . .

## APPARATUS

A more complicated version of the simulated research environment described in Mynatt, Doherty, and Tweney (1977) was used. It was controlled a NOVA 1220 computer and visually displayed on a 21.6 cm sq. Owens-Illinois Digivue plasma screen. Objects of three different shapes (triangles, squares, or disks), three different sizes (0.5, 2 or 9 cm$^2$) and three different brightness levels (all points enclosed in a figure lit, one-half of all points lit, or one-fourth of all points lit) could be drawn on the screen. . . .

Subjects could place a "firing grid" on a screen at any point they chose. From this grid, they could "shoot" a small lighted dot or "particle" in any direction at a constant speed of approximately 0.3 cm/s. Circular, non-visible boundaries with a radius of 8.4 cm were centered around each object. When a particle encountered one of these boundaries it was deflected away from the object. Deflection occurred only once as a particle approached the object. No deflection occurred if a particle encountered the boundary while traveling away from the object or if it was fired toward the object from within the boundary. The angle of deflection, $\theta$, was determined by three parameters, $\alpha$, $\beta$ and $\chi$. The value of $\alpha$ was the angle of incidence between the initial particle path and a radius drawn from the center of the object to the point at which the particle encountered the boundary. Objects with brightness levels 1.0, 0.5 and 0.25 were assigned $\beta$ values of 0, 0.5 and 1.0, respectively. Large, Medium and Small objects were assigned $\chi$ values of

0.5, 1.0 and 0.5, respectively. The angle θ was given by θ = (90° − α)βχ. Note that θ is a curvilinear function of size and an inverse linear function of both brightness and angle of incidence. Note also that since the brightness level 1.0 is assigned the β value of 0, any object of this brightness will show no deflection.

Eight different "universes" obeying these "laws" were programmed, each consisting of nine screens in a 3 × 3 array. Each universe contained one object of each possible combination of shape, size, and brightness for a total of 27 objects. The objects were randomly distributed in each universe with the constraint that each screen have three objects. An example of one such universe together with the boundaries (which were not visible to subjects) is shown in Fig. 18.1. Subjects could look at any of the nine screens but only one at a time. The boundaries around the objects could overlap onto adjacent screens so that particles could be deflected by objects not on the screen a subject happened to be observing. A "tracker" was programmed into the system which allowed subjects to determine (though laboriously) the numerical coordinates of any point on a screen. This allowed, potentially, the precise determination of deflection points, angles and locations of objects.

Each universe, while based upon a few very simple rules, was from a subject's point of view quite complex. Several aspects of the system contributed to this complexity. First of all, the invisible boundaries were very large. Thus their effects were not easily seen to be localized around a particular object. Often two or more boundaries overlapped making this problem even more difficult. Double or even triple deflections could occur on a single screen under such conditions. Secondly, since the boundaries overlapped from one screen to another, particles frequently deflected as a result of hitting boundaries of objects that were not on the particular screen a subject was observing. Finally, very bright objects caused no deflection. Even if a subject arrived at the more or less correct conclusion that objects had circular deflective boundaries around them of a certain size which overlapped from screen to screen, an important regular exception, capable of producing striking disconfirmation, still existed. All of these problems are illus-

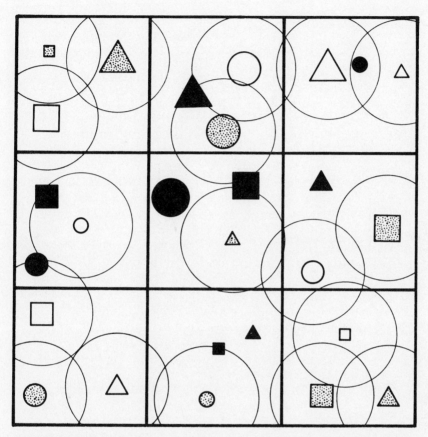

FIGURE 18.1. An example of a nine-screen universe. Black figures represent
1.0 brightness objects, stippled figures represent 0.5 brightness objects, and
outlined figures represent 0.25 brightness objects. Boundaries, which were
not visible to the subjects, have been drawn in around those figures which
caused deflections (i.e. 0.5 and 0.25 objects).

trated in the subject protocols which are presented later in the
paper.

PROCEDURE
The subjects were individually instructed in how to fire particles, change screens and use the tracker and were told that

their task was to discover the laws of particle motion in the artificial universe. It was emphasized that the laws were deterministic and would not change during the course of the experiment. Subjects were paid $2.50 per hour for up to 10 hours of participation but told that if they discovered the laws sooner, they would be paid the full $25. Subjects were randomly assigned to one of two groups: a Strong Inference Instructions group or a No Instructions Control group. Subjects in the Strong Inference group received extensive written instructions stressing the value of falsification and multiple hypothesis testing, were given a copy of Platt's (1964) article to read, and were required at periodic intervals during the experiment to write down alternative hypotheses. Control group subjects were given instructions only in the use of the system. It was assumed that they would probably employ some sort of confirmation strategy, but no instructions encouraging this were given. Each of the eight subjects in both groups was assigned to a different universe in order to minimize effects due to particular combinations of objects.

Subjects were run at their convenience in 1- or 2-hour sessions, usually spaced a day or two apart. They were encouraged to continue working on the problem at home but instructed not to discuss their work with others. During each session, they were asked to "think aloud" frequently into a tape recorder about what they were doing and why, especially when testing hypotheses. At the end of each session, they were asked to write down what they had done during that session, again with particular emphasis on hypothesis testing. All subject input to the system (screen changes, particle firing, and tracker usage) was automatically stored by the computer.

## Results

### QUANTITATIVE ANALYSES

*Solutions.* No subject solved the system. In fact, no subject came even close to recovering its quantitative relationships (i.e., the equation for particle deflection). A few, however, did achieve a fairly good qualitative understanding.

*Hypothesis testing.* Each transcript was matched with the computer record of that subject's input. Two judges then independently used these records to identify hypotheses which the subjects were testing, whether outcomes of the tests were confirmatory or disconfirmatory, and what happened to the hypotheses as a result of the tests. If a hypothesis was disconfirmed, the result of the disconfirmation was categorized in one of four ways: (1) Hypothesis permanently abandoned; (2) Hypothesis temporarily abandoned; (3) Hypothesis revised and retested; and (4) Hypothesis retested without revision. If a hypothesis was confirmed, the result of the confirmation was categorized in one of two ways: (1) Hypotheses revised and retested; and (2) Hypothesis retested without revision. The results of this analysis are shown in table 18.1.

There are several interesting features of these data. First many more disconfirmations occurred than confirmations. This was almost inevitable given the complexity of the system. However, there was no clear pattern of differences between the two groups. Subjects who had been carefully instructed in strong inference were no more likely to get disconfirmation than were subjects who were not so instructed. Furthermore, when disconfirmation occurred, subjects in the two conditions did not react differently to it. Subjects in the Strong Inference

**Table 18.1** Results of hypothesis tests

|  |  | Condition | | |
|---|---|---|---|---|
|  |  | Strong Inference | Control | Total |
| Responses following disconfirmation | Permanently abandoned hypothesis | 15 | 11 | 26 |
|  | Temporarily abandoned hypothesis | 8 | 3 | 11 |
|  | Revised hypothesis and retested | 14 | 19 | 33 |
|  | Retested hypothesis without revision | 8 | 10 | 18 |
|  | Total | 45 | 43 | 88 |
| Responses following confirmation | Revised hypothesis and retested | 4 | 1 | 5 |
|  | Retested hypothesis without revision | 13 | 11 | 24 |
|  | Total | 17 | 12 | 29 |

group permanently or temporarily abandoned their hypothesis following disconfirmation 51 percent of the time; Control condition subjects permanently or temporarily abandoned their hypotheses only 33 percent of the time following disconfirmation. This difference was not, however, significant ($t = 1.46$, $df = 13$, n.s.). This failure of the Strong Inference Instructions to produce effects was further evidenced by the fact that there were only three clear instances of multiple hypothesis testing among the Strong Inference subjects (none occurred among Control subjects).

Thus, most subjects showed a strong tendency to discount or disregard falsification. Of the 88 hypothesis tests in both groups which resulted in disconfirmation, only 26 led to the permanent abandonment of the hypothesis. Following the remaining 62 tests, the subjects either abandoned the hypothesis but returned to it later, revised it to attempt to account for the anomalous data, or simply ignored the disconfirmation and went on testing the same hypothesis.


QUALITATIVE RESULTS
The quantitative analysis presented so far does not fully capture the behavior of many of the subjects. In order to provide a more complete description we selected three subjects whose responses were especially interesting. Portions of these subjects' transcribed protocols appear below. Presented in figure 18.2 are drawings of the screens which the subjects were observing while the protocols were being recorded.

*Subject No. 7: An example of confirmation bias*
Some subjects returned time and again to the same falsified hypothesis, seemingly unable or unwilling to give it up. Subject 7 provided an especially clear, but not atypical example of this response pattern.

After some initial exploratory firings, this subject generated a "line hypothesis":

> My pet hypothesis, I guess you could call it—my first one anyway—is that it looks like it was deflected on a line that would be determined by the triangle in the corner—by extending a line from the one point which is closest to the place where I fired

the particle and another point determined by the opposite side and the sidepoint of the opposite side. So, I'm going to start trying to figure out if that's feasible.

The subject tested this hypothesis and got confirmatory results. (See fig. 18.2, screen 7:1. The particle was actually deflecting off the boundary of the upper square.)

It looks like it's exactly between—that is, on the line that uses the upper right-hand corner of the lower left-hand square and the lower left-hand corner of the upper right-hand triangle.

He then tested the hypothesis again and got disconfirmatory evidence. (See fig. 18.2, screen 7:2. The particle was deflecting off the boundary of the triangle.)

It didn't work out like I thought. I thought it would deflect on the line between the aforementioned points, but it doesn't. I have to allow a little variation though because the program isn't exact and also my data isn't that accurate, but that's too much error.

Later the subject returned to essentially the same hypothesis.

Another hypothesis is that they deflect when they hit the line determined by the two points of the objects closest to them.

He tested this hypothesis and again got disconfirmatory data. (See fig. 18.2, screen 7:3.)

I don't know what to think—guess I'll keep firing until something else happens.

Later he returned again to the same hypothesis.

Well, I'm going back to my "pet hypothesis" and try to figure out if any of these deflection points are on any of the lines determined by any of the points.

Later, after more disconfirmatory evidence he returned again to the same hypothesis.

The old line theory just won't let me alone right now.

Still later, near the end of the experiment, and following even more disconfirmatory evidence he generated a new, much vaguer, and probably unfalsifiable version of this hypothesis.

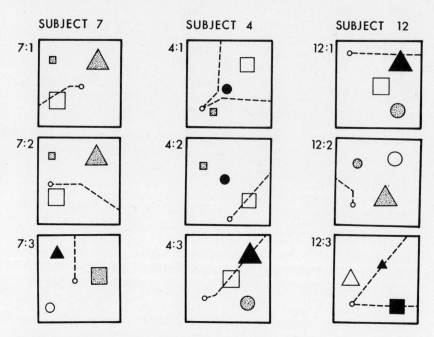

FIGURE 18.2. Examples of hypothesis tests on various screens for three subjects. Dotted lines represent particle paths.

I'm thinking that if you take all the lines from the corners and midpoint of an object, take all the lines you can possibly draw, you'll come up with a bunch of lines that really mess up the diagram; but maybe all of the points of deflection are on those lines.

*Subject No. 4: An example of the use of disconfirmatory evidence leading a subject astray*

Paradoxically, while an overconcern with confirmation caused many subjects to concentrate on incorrect hypotheses, a few subjects were led seriously astray by the way in which they used *disconfirmation*. Occasionally, a subject would hit on a promising, but partially incorrect, hypothesis. Since the hypothesis was only partially correct, it would sooner or later receive clear disconfirmation. Three different subjects quickly abandoned promising hypotheses and began testing others which were, in fact, much less close to the "truth." The responses of Subject 4 provided a clear example of this phenomenon.

This subject first generated and tested a "repulsion" hypothesis. He initially received some confirmatory evidence. (See fig. 18.2, screen 4:1. The particles were actually deflecting off the boundary of the upper square.)

> It seems that the planets are causing repulsion of the particle.
> It's the area or position, or a combination of these.

He then tested the hypothesis again and got disconfirmatory results. (See fig. 18.2, screen 4:2. The firing grid was inside the boundary of the lower square.)

He immediately retested the hypothesis on a new screen and got more disconfirmatory evidence. (See fig. 18.2, screen 4:3. The particle appeared to be attracted toward the square and the triangle. It was actually deflecting off the boundary of the disk.)

> It's obvious that I'm going to have to divorce myself completely from the knowledge that I have, since it's not working like it should.
> I will have to discard the repulsion theory. This is very confusing.

*Subject No. 12: An example of the productive use of both confirmation and disconfirmation*

Subject 12, who was the most successful subject in the experiment, used *both* confirmatory and disconfirmatory evidence. This subject generated a "magnetic field" hypothesis after some initial observations. His early tests were confirmatory, but after switching to a new screen he got a disconfirmatory result. (See fig. 18.2, screen 12:1. The triangle is a 1.0 brightness object which caused no deflection and the particle just missed the boundary of the square.)

At this point the subject went back to an earlier screen, retested his hypothesis and received confirmatory evidence. (See fig. 18.2, screen 12:2.)

> It should be deflected by the field around the small circle.

He then revised his original hypothesis so that it now took the brightness of the objects into account.

Hypothesis: Fully lit objects won't deflect the particle and other ones will.

Following this revision he tested his new hypothesis on another screen, which contained two 1.0 brightness objects, and received clear confirmation. (See fig. 18.2, screen 12:3.) . . .

## Discussion

The robustness of the phenomenon of confirmation bias was again shown by its resistance to instructional manipulation. As in Mynatt, Doherty, and Tweney (1977), instructions to falsify had little or no effect, even though the instructional manipulations were much more extensive than in the earlier study. This result is consistent with the findings of Wason and Johnson-Laird (1972), who report striking inabilities to alter confirmatory strategies in a variety of inference tasks. The repeated failure to produce instructional effects seems to indicate that future efforts to modify confirmation bias should proceed along other lines, such as manipulation of task parameters or training.

The lack of an instructional effect undercut one of the primary purposes of the study—the attempt to assess the relative effectiveness of a strong inference strategy. The Strong Inference group did not differ significantly from the Control group on any of the quantitative indices used. Furthermore, there was very little evidence in the protocols that instructed subjects were using strong inference. Thus, it was not possible to make any overall group comparisons of the effectiveness of the strong inference strategy. . . .

While most subjects did not use disconfirmatory evidence, a few did seem to be strongly influenced by it. These cases make possible some comparison between confirmation and disconfirmation strategies. The three subjects who quickly abandoned disconfirmed hypotheses did not rapidly progress. In fact, they were further from discovering the laws of the system at the conclusion of the study than they had been earlier (e.g., Subject 4). This result, although based on only a few subjects, nevertheless suggests that the use of disconfirmation may *not* be a universally effective inference technique. This may be especially true in the early stages of a complex inference task. When still groping for a means of dealing with open-ended inference problems, a disconfirmation strategy may simply overload the cognitive capacity of most people. It may not be psychologically possible to generate good, alternative hypotheses and to establish low-level, inductive regularities at the same time. In fact, it may not be possible to establish such inductive regularities without some theory, even an incorrect one. The most successful subject in the study, Subject 12, first developed a relatively well-confirmed hypothesis. When disconfirmation later occurred, he was so confident in his hypothesis that he did not discard it entirely. Rather, he used the disconfirmatory evidence productively by modifying those parts which were, in fact, incorrect.

The suggestion that at least some confirmatory evidence is necessary for successful inference is consistent with the reports of Mitroff's (1974a) NASA scientists and with some "unorthodox" philosophical analyses of science (e.g., Feyerabend 1975; Lakatos 1970). Interestingly, it is also consistent with the results of a recent study on problem solving in young children by Karmiloff-Smith and Inhelder (1975). They concluded that only when a hypothesis is sufficiently "consolidated and generalized is [a child] ready to recognize some form of unifying principle for the counterexamples which [were] earlier rejected as mere exceptions" (p. 204).

Thus, confirmation bias may not be completely counterproductive. This conclusion suggests that traditional philosophical analysis of falsification may not have recognized important psychological realities. It further suggests that empirical investigation into the roles of confirmation and disconfirmation at different stages during inference processes may be richly rewarding.

MICHAEL E. DOHERTY,
CLIFFORD R. MYNATT,
RYAN D. TWENEY,
and MICHAEL D. SCHIAVO

# Pseudodiagnosticity

COGNITION INVOLVES A complex interplay between mechanisms of information input, information processing and information search. While this has long been clear, too little is known about the interaction of these mechanisms. In particular, behavioral scientists have discovered much about the limits on cognitive processes imposed by, for example, the limited capacity to store information in short-term memory. But little is known about mechanisms that people use to overcome these limitations. The present study was designed to explore limitations of another sort, limitations which operate when subjects seek information needed to make a judgment under uncertainty. The task was such that three basic cognitive limitations were manifested: (1) failure to use statistical base-rate information (Tversky and Kahneman 1974); (2) confirmation bias (Wason 1960; Mynatt, Doherty, and Tweney 1977); and (3) failure to indentify and select diagnostically relevant information. The latter phenomenon, which we have called pseudodiagnosticity, has not previously been described.

While the present paper is not a test of the adequacy of Bayes' theorem as a descriptive model of how subjects revise their opinions, it will be convenient to define the three limitations of interest in terms of Bayes' theorem. Consider the

From M. E. Doherty, C. R. Mynatt, R. D. Tweney, and M. D. Schiavo, Pseudodiagnosticity. *Acta Psychologica* (1979), 43:11–21.

equation

$$P(H_1/D_i) = \frac{P(H_1)\, P(D_i/H_1)}{P(H_1)\, P(D_i/H_1)\, +\, P(H_2)\, P(D_i/H_2)} \qquad (1)$$

in which $H$ and $D$ stand for hypothesis and data, respectively, the subscripts 1 and 2 label two mutually exclusive and exhaustive hypotheses, and the subscript $i$ indexes a set of data. Assume further that the probability ($P$) values are defined as relative frequencies. The base rate for $H_1$ is simply the prior probability that $H_1$ is true, and is denoted by $P(H_1)$. Note that very high or low values of $P(H_1)$ can overwhelm the diagnostic effects of data, even when those data are highly diagnostic. Whether behavior is influenced appropriately (i.e., in accordance with the normative model) by $P(H_1)$ is one way of determining how well people use base-rate data. Confirmation bias, the second limitation, may reflect itself in several ways, one being when a subject favoring $H_1$ does not search for data which would be likely to favor $H_2$, or does not seek any information about $H_2$ at all. Finally, Bayes' theorem points to a strong qualitative test of whether subjects possess a working knowledge of the concept of diagnosticity. Suppose a subject in a two hypothesis task is given the opportunity to choose information revealing $P(D_1/H_1)$ and $P(D_1/H_2)$ or $P(D_2/H_1)$ and $P(D_2/H_2)$. If the subject instead chooses information revealing $P(D_1/H_1)$ and $P(D_2/H_1)$ in order to make an inference, then such a subject cannot be said to have a working understanding of diagnosticity. That is, the posterior probability of either hypothesis, say $P(H_1/D_1)$, can be calculated only if *both* the probability of a piece of data given the hypothesis under test, $P(D_1/H_1)$, *and* the probability of the same piece of data given the alternative hypothesis, $P(D_1/H_2)$, are known.

A compelling demonstration of people's failure to use base rate data appropriately is provided by Tversky and Kahneman (1974). Subjects asked to classify university students ignored radical base-rate differentials as soon as they were given very weak individuating data. That is, though values of $P(D_1/H_1)$ and $P(D_1/H_2)$ were much too close to each other to overturn the base rates, subjects were quick to abandon the hypothesis favored by the base rates. Since failure to consider base-rate information is of both theoretical and practical consequence, the phenomenon merits further attention, including attempts at replication in other situations.

Confirmation bias can be inferred from two different forms of behavior: (a) failure to change one's opinion in the face of nonsupporting or even contradictory evidence and (b) selection of data favoring one's hypothesis while ignoring sources of data which might be likely to contradict one's hypothesis. Confirmation bias seems to be a pervasive phenomenon. Pitz's "inertia effect" (Pitz, Downing, and Reinhold 1967; Geller and Pitz 1968) is an instance in a bookbag and poker chip task, as is "freezing" of hypotheses in the clinical judgment setting (Wiggins 1973; see also Wallsten 1976). A tendency to look only for confirmatory evidence was found in Wason's four-card selection task and 2 4 6 studies (Wason 1960, 1968b). Mynatt, Doherty, and Tweney (1978) observed the effect in simulations of scientific research, and Mitroff (1974a) noted that top-grade scientists not only admit to a confirmation bias but assert that it plays a necessary role in scientific progress. Confirmation bias thus seems to be a phenomenon meriting further exploration, especially given the logical asymmetry (Popper 1959) between confirmation and disconfirmation in inductive inference.

Much research on human inference has compared opinion revision with the degree of such revision prescribed by Bayes' theorem (Pitz 1975). One recent experiment is of special interest. Troutman and Shanteau (1977) explored a phenomenon called the "nondiagnosticity effect." They provided subjects with samples from one of two boxes of beads. One box had 70 red, 30 white, and 50 blue beads. The other box had 30 red, 70 white, and 50 blue beads. Thus a blue bead was noninformative. Yet subjects significantly revised their subjective probabilities when they were given blue beads, in the direction of less extreme judgments. Wallsten's (1976) subjects were also influenced by neutral information. There is other evidence that subjects are not properly sensitive to diagnosticity (e.g., Steinmann and Doherty 1972). The fundamental issue we wish to address is not whether subjects misuse diagnosticity in some quantitative way, but whether diagnosticity is a cross-situational, behaviorally significant task variable.

In the experiments reported below, the subjects were asked

to determine from which of two islands (corresponding to $H_1$ and $H_2$) an artifact had come. The artifact was described by a number of binary characteristics ($D_1$ and $D_2$ for each characteristic) and subjects sought information about the values of $P(D/H)$ of these characteristics from the two islands.

Base rates were manipulated to determine whether the phenomena described above would replicate in a concrete task with clearly defined statistical properties. The presence of confirmation bias in the task can be indirectly inferred by observing whether subjects sought information about a favored hypothesis rather than an alternative. Finally, and most importantly, the question of whether subjects comprehend diagnosticity can be addressed directly, by observing whether subjects choose data relevant to *both* hypotheses when given the opportunity to do so.

## The Experiment

METHOD

*Subjects.* The experiment was run in a large Introductory Psychology class. There were 152 participants.

*The task.* All instructions and manipulations were accomplished by varying the content of a 5-page booklet. The Ss' responses were made directly in the booklet. The first paragraph on the first page provided the context for the Ss:

> Imagine that you are an undersea explorer who has just found a pot on one of your dives. The pot looks very valuable and you would like to return it to its homeland. Although there are many islands in the surrounding area from which the pot could have come, you know that pottery-making exists on only two of these islands, Coral Island and Shell Island. You would like to determine from which of these two islands the pot came. One method you could use to try to determine this would be to compare the characteristics of your pot with what is known about pots and pottery on each of the two islands. Luckily, on each island there is a museum that contains information about the pottery made on that island. From your ship, you can place phone calls to the museums on each of the two islands. The

museum supervisors are very busy, however, so they will give
you only one piece of information about their collections during
each phone call.

You first make a careful examination of the pot you found,
and list its characteristics as follows:

Smooth clay (not rough)

Curved handles (not straight)

(Six more binary characteristics were then listed.)

The next paragraph then indicated that a phone call to each
island had produced information about the number of pots
that had been lost at sea by each. Differences in these numbers
provided the base rate manipulation.

Page 2 of the booklet instructed Ss that they could now make
two more "phone calls," which were simulated by peeling off
any two opaque stickers. One possible layout would be:

|                   | Coral Island | Shell Island |
| ----------------- | ------------ | ------------ |
| Curved handles    | 21%          | 87%          |
| straight handles  | 79%          | 13%          |
|                   |              |              |
| Smooth clay       | 19%          | 91%          |
| Rough clay        | 81%          | 9%           |

Each of the four pairs of percentages (e.g., 21% and 79%) was
covered by a circular sticker.

The Ss were then asked "which island do you think the pot
came from?" They checked either Coral or Shell, and then
wrote down the reasons for their choice.

On page 3 the Ss were faced with an array of 12 stickers
covering the proportions on each island of the other six binary
characteristics. They were permitted to peel off any six of the
12. They were asked to number each choice and state a decision after each. Page 4 asked for a final choice, and asked
two open ended questions about strategies. Page 5 provided
a check on whether Ss remembered the base rate, and asked
more open ended questions about their strategies.

There were two levels of base rate. An equal base rate treatment was provided by instructing the S that 1000 pots had
been lost at sea from each island. An unequal base rate treatment was provided by having 5000 pots lost by one island,
and only 500 by the other. The name of the island with the
high loss was Coral for some Ss, Shell for others. All unequal

base rate $Ss$ had data on page 2 which favored the opposite hypothesis than that favored by the base rate. However, since only two stickers could be peeled, $Ss$ could observe at most only one pair. In the example given above the pot was smooth and had curved handles. The data favored Shell Island whether the $Ss$ pulled the pair of stickers for texture or for handles. But the base rate was sufficiently extreme to outweigh either data set by a large margin. Suppose that an $S$ who favored Coral Island after seeing the base rate peeled the stickers describing the distributions of handle characteristics. If so, equation 1 tells us that the probability that the pot came from Coral Island is

$P$(Coral/curved handle)

$$= \frac{(0.909)(0.21)}{(0.909)(0.21) + (0.091)(0.87)} = 0.707. \quad (1)$$

The comparable value given information about the smooth–rough dichotomy is 0.676. Note also that an $S$ might select stickers so as to yield, from a formal point of view, no information. That is, they could select one sticker from each pot characteristic, or both from one island. Such choice behavior would not give the $S$ the values needed for the formally appropriate computations.

On page 3, the data associated with all six binary characteristics agreed in their implications for the hypotheses favored by the data on page 2. The ratios were of the same magnitudes as those shown above. For a control group ($N = 20$), the favored hypothesis was reversed.

There were six equal base rate conditions, which controlled for first order of mention of the island, the direction of the data on page 2 and the direction of the data on page 3. Since, as expected, no interesting differences involving these groups emerged, subsequent analyses are collapsed across them, and they will not be further described. A total of 66 $Ss$ had unequal base rates, while 86 had equal base rates.

*Procedure.* The students were previously informed that an in-class experiment was to be performed that day. Verbal instructions were brief, and consisted only of a request for silence and a comment that since there were many different forms of the booklet, the students should not look at their neighbors' booklets. The students were given 35 minutes. Any who completed the task early had the booklet collected and took out a book or notes to study. The remainder of the class period was given over to a discussion of the optimal responses, and why they were optimal.

RESULTS
The data will be discussed with respect to each of the three issues raised in the introduction.

*Base rates.* Of the 66 Ss who had unequal base rates, 60 stated that they thought the pot had come from the island favored by the data on page 2 of the booklet. Thus, base rate was ignored by these Ss in the face of individuating information. Of the 86 equal base rate Ss, 75 chose the pot favored by the data on page 2, one did not respond and one wrote out an explanation which indicated a reversal in the reading of the percentages. Nine Ss gave responses counter to the evidence, the reasons for which were uninterpretable.

*Confirmation bias.* Each S was categorized according to the number of pairs chosen on page 3 of the booklet and by the number of stickers peeled for each island. There were a maximum of three pairs per S, and six stickers for either island. The number of pairs chosen constrained the number of stickers for an island (for example an S who chose 3 pairs had to have the choices distributed equally). An S who chose no pairs could have any number of choices from 0 to 6 for a single island. Data were collapsed over island names. The instruction to record the choice sequence on page 3 was not sufficiently emphasized, and many Ss did not follow it, so only Ss who had the same final hypothesis on page 3 as they had on page 2 were counted. 11 Ss were dropped for switching hypotheses somewhere on page 3; 20 others were excluded since they were in the control condition which led them to switch on page 3. Thus, 121 Ss met the criterion of having the same initial hypothesis on page 2 as their final hypothesis. Their choice responses (i.e., stickers peeled) on page 3 can then be interpreted as being in the context of a favored hypothesis.

Table 19.1 presents these responses. These are the crucial data for this and the next section, and will be illustrated by example. Suppose an S hypothesized Coral Island on page 2, selected 6 stickers revealing data about pots from Coral Island, and then concluded that the pot was from Coral Island. That S would be one of the 7 in the upper right cell in table 19.1. So would an S who hypothesized Shell, picked 6 Shell stickers, and concluded Shell. The middle data column represents Ss who chose stickers equally from both islands. The right three columns indicate choices of data from the favored island (i.e., data to assess $P(D/H)$), while the left three columns are choices of data from the non-favored island ($P(D/\sim H)$).

The 61 Ss chose equally from the two islands. For the remaining 60 Ss, the data strongly suggest the operation of a bias to confirm. Of those 60, 51 chose more often from the favored island. A test of the significance of the obtained proportion for the asymmetric choices against a null hypothesis that $P = 0.5$ was highly significant ($z = 5.42, p < 0.01$).

*Pseudodiagnosticity.* The question of whether the Ss behave as though they have any working understanding of the concept of diagnosticity is also addressed by table 19.1. Very few of the Ss shown chose 3 pairs—the only rational strategy. The majority chose *no* pairs. Of the total number of 152 Ss, only 19 chose 3 pairs. Again the raw data are impressive. *Everyone* should have picked three pairs, had they even a rudimentary grasp of how to make sense out of the data that were available. A $\chi^2$ test for the data shown in table 19.1, testing the row totals

**Table 19.1** Number of subjects removing differing numbers of labels on page 3, categorized by number of pairs selected and number of labels selected relevant to favored island. A dash indicates that no entry in that cell was possible

| Number of pairs | Number of labels relevant to favored island | | | | | | | |
| | 0 | 1 | 2 | 3 | 4 | 5 | 6 | $\Sigma$ |
|---|---|---|---|---|---|---|---|---|
| 0 | 1 | 0 | 5 | 32 | 16 | 10 | 7 | 71 |
| 1 | – | 0 | 2 | 14 | 9 | 5 | – | 30 |
| 2 | – | – | 1 | 4 | 4 | – | – | 9 |
| 3 | – | – | – | 11 | – | – | – | 11 |
| $\Sigma$ | 1 | 0 | 8 | 61 | 29 | 15 | 7 | 121 |

**Table 19.2** Number of subjects[a] with given hypothesis on page 2, categorized by number of pairs chosen on page 2

| | Hypothesis | |
| --- | --- | --- |
| Number of pairs chosen | H favored by Bayes' rule | Incorrect H |
| 0 | 1 | 49 |
| 1 | 3 | 11 |

[a] Two subjects were unclassified since they peeled only one sticker on page 2.

against a null hypothesis that $Ss$ were selecting stickers by chance, is highly significant ($\chi^2 = 538.5$, $df = 3$, $p < 0.01$).

The data from page 2, especially from the unequal base rate $Ss$, are even more compelling. Recall that the data for either pair of characteristics were not sufficiently diagnostic to lead an optimal Bayesian to depart from the base rate. The $Ss'$ responses were tabulated by the number of pairs they chose (zero or one) and by the hypothesis they stated after removing the two stickers. Table 19.2 reports the results of this categorization for the unequal base rate $Ss$ only, collapsed across island names. Note that if all our $Ss$ had behaved optimally with respect to the selection and interpretation of evidence, all of them would have been in the lower left cell.

*Three* of the 64 $Ss$ were. If the limitation were only on the $Ss'$ ability to process the evidence presented, as is often assumed in experiments employing Bayes' theorem as a model, then all $Ss$ should have shown up in the bottom row, with the distribution over hypotheses determined by other factors. But only 14 of the 64 $Ss$ selected a pair of labels. Of the $Ss$ in the equal base rate conditions, only 20 percent selected a pair. It is clear that the diagnostic value of pairs of characteristics was not understood by $Ss$. Not only did they not select pairs, but they changed their hypotheses based on the pseudodiagnostic data they did select.

## Discussion

The hypothesis that people ignore base rates in the face of individuating information received very strong support from the present study. The tendency to ignore base rates is truly a robust phenomenon. Subjects abandoned the hypothesis strongly favored by the base rate, even though the data they selected were either insufficient to warrant such a change or, more often, worthless.

The hypothesis that people's inference behavior is limited by a strong bias to confirm is also supported by the data. That phenomenon, too, is robust. Choice responses on page 3 were clearly consistent with a confirmation bias hypothesis. There seems to be no reasonable alternative explanation for such an imbalance in the choice data.

The novel contribution of this study lies in the demonstration that people do not have a working knowledge of the concept of diagnosticity. When the subjects sought statistical data with which to evaluate the artifacts, the great majority did not ask for the data needed to form likelihood ratios. It is not a question of nonoptimal revision, given data. The subjects actively chose irrelevant information and ignored relevant information which was equally easily available. Furthermore, table 19.2 shows that the subjects who chose no pairs were almost uniformly (98 percent) swayed by these data. The term "pseudodiagnosticity" was chosen to highlight the fact that the data had a major impact on the decisions. In spite of the 10:1 prior probability (or base rate), the subjects revised their opinion, given worthless data.

Two other experiments dealing with the phenomenon were run, with a total of 66 subjects. These experiments did not permit the choice of pairs vs non-pairs, but called for a decision after only a single datum. The subjects revised their opinions based on that single, hence formally inadequate, $P(D/H)$. These two experiments serve primarily to achieve a modest cross-task generalization, since the operation used to demonstrate pseudodiagnosticity differed from the main study. The data of the two smaller studies were also consistent with the base rate and confirmation bias phenomena discussed above.

It might be instructive to speculate about how the subjects may have been attempting to solve the task. A theorist trying to rationalize the subjects' behavior would have to assume that the subjects were making explicit guesses about the distributional characteristics of the task. That is, one might posit that the subject who uncovered a high value of $P(D/H_1)$ might then guess that $P(D/H_2)$ is likely lower. Such a strategy would be rational in some circumstances, specifically those in which the subject has relevant experience to bring to bear on the estimates of unknown probabilities. However the present task was novel to the subjects, and there was no way in which they could reasonably have guessed the relative magnitude of $P(D/H_2)$, given $P(D/H_1)$. This is especially so since the pseudo-diagnosticity effect appeared on the first opportunity for it to do so, before any observations at all were made of $P(D/H_2)$. Even if some model which posits that subjects are making assumptions about hidden frequencies does describe the underlying process, the fundamental point of the research is nonetheless valid: the relevant data were there for the asking—and the subjects did not ask for them. While we do not yet have sufficient data to propose a formal process model, we believe that the basic phenomenon operating in this context, underlying both pseudodiagnosticity and confirmation bias, is a cognitive one: it is hard to think about one datum in relation to two hypotheses. It is much easier to think about the relevance of two data to one hypothesis.

It is worthwhile to contrast the picture of inference behavior which emerges from these experiments with that which emerges from traditional bookbag and poker chip studies. The bookbag tasks typically constrain subjects sharply with respect to the response which can be made, permitting subjects only to revise their subjective probabilities given the prior probabilities, data sequences, and diagnosticities which have been imposed by the experimenters. The picture which emerges is either that the subject is or is not a conservative Bayesian (see Pitz 1975, for a recent discussion). From the present task a more articulated picture of the inference process is possible. The subject, given a chance to ask for data, seems more likely to ask for data perceived as relevant to the favored hypothesis than to seek information about alternatives. Given any individuating information, base rates are readily abandoned. Once the datum is "in," the subject makes an immediate adjustment in the hypothesis state, even in the absence of the other data which form the necessary context in which to evaluate the obtained datum. This process, which we have labeled pseudodiagnosticity, can only abet premature closure with respect to the hypothesis, preclude consideration of the impact of the datum on other hypotheses, and inhibit the search for other evidence which may support alternative hypotheses.

There is an interesting parallel between our results and those observed in Wason's (1968b) four-card selection task. His subjects were required to seek hidden information to verify a general rule of the form "if P then Q." P referred to one side of a card, Q to the other side, and subjects were asked to choose between P, Q, not-P, and not-Q cards. Most chose only P and Q. In fact, however, only P and not-Q are informative, since these are the only choices that can falsify the rule. All other combinations are logically consistent with the rule, and so are uninformative. Wason's results have generally been explained as the results of reliance on intuitive matching processes, rather than on logical relations (Evans and Lynch 1973; Wason and Evans 1975). As in Wason's task, pseudodiagnosticity represents failure to seek the only potentially "falsifying" information available to our subjects. The choices actually made by subjects are consistent with the pot being from either island, and so are uninformative. This comparison may point toward understanding of the mechanisms which underlie pseudodiagnosticity. The phenomenon may be another example of what Wason and Evans (1975) have called "an 'intuitive' way of coping with conceptual difficulty." Our results may extend such processes to the domain of probabilistic reasoning.

Pseudodiagnosticity is clearly dysfunctional in the task presented our subjects. If it operates in more open problem spaces, as for example in scientific thinking, its consequences may be especially severe. Whereas practicing scientists often see confirmation bias as potentially valuable for its motivating effect and for its role in establishing "good hypotheses," it is hard to see what positive benefits could result from processing data in a pseudodiagnostic fashion. Such data cannot provide disconfirmatory evidence for a hypothesis. More seriously, they cannot even serve to confirm a hypothesis, and are, in the fullest sense of the word, worthless.

IAN I. MITROFF

# Scientists

# and Confirmation

# Bias

IN RESPONSE TO the opening interview questions on the relative plausibility of various scientific hypotheses associated with the moon, three scientists were overwhelmingly nominated as most attached to their own ideas.[1] The comments referring to these scientists were peppered with emotion. The following is typical:

> X is so committed to the idea that the moon is Q that you could literally take the moon apart piece by piece, ship it back to Earth, reassemble it in X's backyard and shove the whole thing . . . and X would still continue to believe that the moon is Q. X's belief in Q is unshakeable. He refuses to listen to reason or to evidence. I no longer regard him as a scientist. He's so hopped up on the idea of Q that I think he's unbalanced.

The three scientists most often perceived by their peers as most committed to their hypotheses and the object of such strong reaction were also judged to be among the most outstanding scientists in the program. They were simultaneously judged to be the most creative and the most resistant to change. The aggregate judgment was that they were "the most creative" for their continual creation of "bold, provocative, stimulating,

Excerpts from Norms and counter-norms in a select group of Apollo moon scientists: A case study of the ambivalence of scientists. *American Sociological Review* (1974), 39:579–95.

suggestive, speculative hypotheses," and "the most resistant to change" for "their pronounced ability to hang onto their ideas and defend them with all their might to theirs and everyone else's death." Because of the centrality of these scientists, the perception of them by their peers was studied over the course of the Apollo missions. The perceived intensity of commitment of these scientists to their pet ideas was systematically measured in terms of various attitude scales. Every scientist in the sample was asked to locate the scientific position of each of the three scientists with respect to a number of possible positions and to rate the intensity of their commitment to their position. *There was virtually no change in the perceived positions and the perceived intensity of their commitment to their ideas over the three and a half year period.*[2] . . .

*Every one of the scientists interviewed on the first round of interviews indicated that they thought the notion of the objective, emotionally disinterested scientist naïve.*

The vocal and facial expressions that accompanied the verbal responses were the most revealing of all. They ranged from mild humor and guffaws to extreme annoyance and anger. They indicated that the only people who took the idea of the purely objective, emotionally disinterested scientist literally and seriously were the general public or beginning science students. Certainly no working scientist, in the words of the overwhelming majority, "believed in that simple-minded nonsense." Because they actually did science and because they had to live with the day-to-day behavior of some of their more extreme colleagues, they knew better.

What was even more surprising was that the scientists rejected the notion of the "emotionally disinterested scientist" as a prescriptive ideal or standard. Strong reasons were evinced why a good scientist *ought* to be highly committed to a point of view. Ideally, they argued, scientists ought not to be without strong, prior commitments. Even though the general behavior and personality of their more extremely committed colleagues infuriated them, as a rule they still came out in favor of scientists having strong commitments. The following comments are typical:

Scientist A—Commitment, even extreme commitment such as bias, has a role to play in science and it can serve science well.

Part of the business [of science] is to sift the evidence and to come to the right conclusions, and to do this you must have people who argue for both sides of the evidence. This is the only way in which we can straighten the situation out. I wouldn't like scientists to be without bias since a lot of the sides of the argument would never be presented. We must be emotionally committed to the things we do energetically. No one is able to do anything with liberal energy if there is no emotion connected with it.

*Scientist B*—The uninvolved, unemotional scientist is just as much a fiction as the mad scientist who will destroy the world for knowledge. Most of the scientists I know have theories and are looking for data to support them; they're not sorting impersonally through the data looking for a theory to fit the data. You've got to make a clear distinction between not being objective and cheating. A good scientist will not be above changing his theory if he gets a preponderance of evidence that doesn't support it, but basically he's looking to defend it.

Without [emotional] commitment one wouldn't have the energy, the drive to press forward sometimes against extremely difficult odds.

You don't consciously falsify evidence in science but you put less priority on a piece of data that goes against you. No reputable scientist does this consciously but you do it subconsciously.

*Scientist C*—The [emotionally] disinterested scientist is a myth. Even if there were such a being, he probably wouldn't be worth much as a scientist. I still think you can be objective in spite of having strong interests and biases.

*Scientist D*—If you make neutral statements, nobody really listens to you. You have to stick your neck out. The statements you make in public are actually stronger than you believe in. You have to get people to remember that you represent a point of view even if for you it's just a possibility.

It takes commitment to be a scientist. One thing that spurs a scientist on is competition, warding off attacks against what you've published.

*Scientist E*—In order to be heard you have to overcommit yourself. There's so much stuff if you don't speak out you won't get heard but you can't be too outrageous or you'll get labeled as

a crackpot; you have to be just outrageous enough. If you have
an idea, you have to pursue it as hard as you can. You have to
ride a horse to the end of the road.

*Scientist F*—The notion of the disinterested scientist is really a
myth that deserves to be put to rest. Those scientists who are
committed to the myth have an intensity of commitment which
belies the myth.
Those scientists who are the movers are not indifferent. One
has to be deeply involved in order to do good work. There is
the danger that the bolder the scientist is with regard to the
nature of his ideas, the more likely he is to become strongly
committed to his ideas.
I don't think we have good science because we have adversaries
but that it is in the attempt to follow the creed and the ritual
of scientific method that the scientist finds himself unconsciously thrust in the role of an adversary.

*And finally, Scientist G*—You can't understand science in terms
of the simple-minded articles that appear in the journals. Science is an intensely personal enterprise. Every scientific idea
needs a personal representative who will defend and nourish
that idea so that it doesn't suffer a premature death. Most people
don't think of science in this way but that's because the image
they have of science only applies to the simplest, and for that
reason, almost non-existent, ideal cases where the evidence is
clear-cut and it's not a matter of scientists with different shades
of opinion. In every real scientific problem I've ever seen, the
evidence by itself never settled anything because two scientists
of different outlook could both take the same evidence, and
reach entirely different conclusions. You eventually settle the
differences, but not because of the evidence itself but because
you develop a preference for one set of assumptions over the
other. How you do this is not clear since there's not always a
good set of reasons for adopting one rather than the other.

Note that in this part of the discussion the scientists partly
reversed themselves and praised their more committed colleagues precisely for their extreme commitments:

The commitment of these guys to their ideas while absolutely
infuriating at times can be a very good thing too. One should
never give up an idea too soon in science—any idea, no matter
how outrageous it may be and no matter how beaten down it
seems by all the best evidence at the time. I've seen too many

totally disproven ideas come back to haunt us. I've learned by now that you never completely prove or disprove anything; you just make it more or less probable with the best of what means you've got at the time. It's true that these guys are a perpetual thorn in the side of the profession and for that reason a perpetual challenge to it too. Their value probably outweighs their disadvantages although I've wondered many times if we might not be better off without them. Each time I reluctantly conclude no. We need them around. They perpetually shake things up with their wild ideas although they drive you mad with the stick-to-itiveness that they have for their ideas.

The comments illustrate clearly the variety of reasons for the belief that scientists should be emotionally committed to their ideas. Above all, they reveal the psychological and sociological elements that permeate the structure of science. Psychologically, the comments indicate that commitment is a characteristic of scientists. The comments strongly support Merton's ideas on scientists' affective involvement with their ideas (1963:80). Sociologically, the comments reveal the social nature of science. Scientist E, for example, says there's so much "stuff in the system" that if one wants to be heard over the crowd, one must adopt a position more extreme than one believes in.

# NOTES

1. These same three scientists were nominated in open-ended conversation and also in response to the direct questions: "Which scientists are in your opinion most committed to their pet hypotheses?" and "Which scientists do you think will experience the most difficulty in parting with their ideas?" These questions were asked at each interview in the 3½ year period. No matter how they were asked or when, the responses are the same.

2. It is beyond the scope of this paper (cf. Mitroff 1974a) to report on this aspect of the study in detail. Measuring and assessing the differences in psychology between the scientists in the sample was a major focus of the study. Various typologies of different kinds of scientists were constructed from their comments. At the one extreme, are the three highly committed scientists who "wouldn't hesitate to build a whole theory of the solar system based on no tangible data at all; they're extreme speculative thinkers." At the other is the data-bound experimentalist who "wouldn't risk an extrapolation, a leap beyond the data if his life depended on it." Whereas the three highly committed scientists are perceived as biased, brilliant, theoretical, as extreme generalists, creative yet rigid, aggressive, vague, as theoreticians, and finally as extremely speculative in their thinking, the opposite extremes are seen as impartial,

dull, practical, as specialists, unimaginative yet flexible, retiring, precise, as experimentalists, and extremely analytical in their thinking. It is also beyond the scope of this paper to show (cf. Mitroff 1974a) that these psychological differences can be used, contrary to Merton (1957:638–40), to argue for a psychological explanation for the contentious behavior of scientists involved in priority disputes. This is not to say that such behavior must be explained purely psychologically or sociologically. Indeed, it is due to the interaction of both factors in that individual scientists react differently to the social institution of science. In other words, science probably does not attract contentious personalities more than other institutions. However, some kinds of scientists are more contentious than others and thus quicker to initiate and press their claims for priority.

HOWARD E. GRUBER

# The Rationalist

# Myth

# of Hypothesis

# Choice

IT HAS BEEN suggested that essentially the whole of Darwin's mature point of view is reflected in his earliest remarks on evolution, as though his ultimate theory sprang forth at once the moment he turned his thoughts to the matter. Nothing could be further from the truth. Darwin had to go through many stages of intellectual growth. To become the author of the *Origin of Species* he had to modify his principle of adaptive equilibrium so that it stressed relations among organisms rather than relations between each organism and its physical milieu; he had to eliminate completely the concept of monadism and monad life span. Working from his starting point, he had to develop the model of branching evolution first in a formal way, and then to transform it by suffusing it with the idea of struggle and selection. Although there are occasional moments of sharp insight, these moments are only nodal points in a slow growth process.

The picture of scientific thought is often painted as being

Excerpts from H. Gruber and P. H. Barrett, *Darwin on man*. New York: E. P. Dutton, 1974, pp. 146, 170–74. Copyright © 1974 by Howard E. Gruber and Paul H. Barrett. Reprinted by permission of the publisher E. P. Dutton.

carried forward by the construction of alternative hypotheses followed by the rational choice between them. Darwin's notebooks do not support this rationalist myth. Hypotheses are discovered with difficulty in the activity of a person holding *one* point of view, and they are the expression of that point of view. It is hard enough to have one reasonable hypothesis, and two at a time may be exceedingly rare. In Darwin's case, when he is forced to give up one hypothesis, he does not necessarily substitute another—he sometimes simply remains at a loss until his point of view matures sufficiently to permit the expression of a new hypothesis.

Finally, September 28, 1838, the great Malthusian moment of truth arrives. Does it strike Darwin with the sudden force of a thunderbolt? Is it an Archimedean "Eureka" experience? Does it transform his thinking from that moment forward and for all time to come? . . . Surely, this is not merely a great moment in the history of science but also a great culmination, a peak experience for the man Charles Darwin?

Not really. The insertions, the metaphors, the occasionally high-flown style, the underlining—these traits can be found all through the notebooks. As it happens, the crucial passage does not even contain a single exclamation point, although in other transported moments he used quite a few, sometimes in triplets.

More significantly, he does not drop all his other concerns and questions in a manner suggesting that he now feels he has the answer of answers to the "question of questions." The next day, September 29, we find a long entry on the behavior of various primates, much of it about their sexual curiosity.

Darwin had come to the summit. After a hard climb, the summit is not a simple achievement. It is no longer clear which way is up or down. Getting down is still a problem, and other peaks have become visible. Nor is the summit a sharp point, but rather a broad field with subtleties and ambiguities all its own.

The links between Darwin's concepts of the nature of variation, of the amount of variation, and of the superfecundity principle are intricate. So long as variation was conceived of as directly adaptive, there need be only one appropriate modification for each change of the environment. Only *after* he had given up the idea that variations are necessarily adaptive

would it become imperative to have many variations for selection to choose among. But the moment the number of variations required by the theory becomes large, the superfecundity principle comes to the foreground of attention. There
must be many individuals to serve as "carriers" of the many
variations. Variation and selection may be separate principles,
but they are both aspects of the same reality, a large and varying number of individuals.

In his *Autobiography* Darwin gives his recollection of the
Malthus insight:

> In October 1838, that is, fifteen months after I had begun my
> systematic enquiry, I happened to read for amusement Malthus
> on *Population*, and being well prepared to appreciate the strug-
> gle for existence which everywhere goes on from long-contin-
> ued observation of the habits of animals and plants, it at once
> struck me that under these circumstances favourable variations
> would tend to be preserved, and unfavourable ones to be de-
> stroyed. The result of this would be the formation of new spe-
> cies. Here, then, I had at last got a theory by which to work.
> . . . (Darwin 1958:120)

This paragraph follows shortly after a statement about his
method:

> My first note-book was opened in July 1837. I worked on true
> Baconian principles, and without any theory collected facts on
> a wholesale scale. . . . (Darwin 1958:119)

Taken together, these statements give an extremely misleading picture. Darwin certainly began the notebooks with
a definite theory, and when he gave it up it was for what he
thought was a better theory. True, when he gave up his second
theory he remained in a theoretical limbo for some months.
But even then he was always trying to solve special theoretical
problems, such as those related to hybridization, and he almost
never collected facts without some theoretical end in view. It
was not simply from observation but from hard theoretical
work that he was so well prepared to grasp the significance
of Malthus' essay.

A quick résumé of this effort will help to establish the point.
Darwin had worked his way through and then abandoned the
monad theory with its attendant premise of spontaneous generation; the ubiquitous transformation of inanimate into ani-

mate matter was a principle that competed with the superfecundity idea implicit in natural selection. Darwin had then
surrendered his theory of perpetual becoming—that is, he had
seen that variation by itself could not explain progressive evolution. He had begun at least to doubt that variations are
directly adaptive responses to environmental forces; if they
were, there would be no need for selection. He had seen that
the amount of variation in nature was very much greater than
previously realized. He had become aware of Ehrenberg's work
showing the dramatic superfecundity of micro-organisms. He
had searched valiantly for the causes of heritable variations,
and although he had not given up the search, he had become
better prepared to treat variation not as the conclusion of a
satisfying argument but as an unexplained premise. Although
he had attended to the subject of domestic breeding mainly
with the intent of contrasting artificial and natural selection,
it seems plausible that this effort may have prepared him to
see the useful analogy between them.

But the way was neither straight or narrow. We do come to
an apparent paradox. He still had to take the crucial step of
*combining* the premises of ubiquitous variation and Malthusian superfecundity in an argument that leads logically to the
conclusion of natural selection and progressive evolution.
This means that he must have already successfully disembed-
ded the idea of natural selection from the conservative matrix
in which, as we have seen, it had been previously utilized.
Are we saying that restructuring the argument depends on
seeing the principle of natural selection, while at the same
time seeing the principle depends on restructuring the argument?

Yes. This is a circular argument or, better, a helical process.
Understanding the structure of Darwin's argument as a growth
structure helps to explain why the path to the moment of
insight seems so tortuous, repetitive, cyclical. A path of such
shape is the only way in which you can construct an argument
out of parts which depend for their significance on the structure of the whole. The developed whole *must* be foreshadowed
or prefigured at earlier and more primitive steps in the process
of creative development.

# Part V

# Mathematics in Scientific Thinking

# Introduction

MATHEMATICS PLAYS A variety of roles in scientific thinking. It assists in defining and keeping track of observations and measurements. It provides rules to operate on measurements, and statistical rules for relating sets of measurement to the hypotheses in whose service the measurements were taken. Mathematics provides a "strong grammar for didactic discourse," and a form of reasoning which is "immensely more powerful than plain language when it comes to generating verifiable predictions, unpalatable conclusions, or unsuspected connections between known facts" (Ziman 1978:17–18). Many of the phenomena in which scientists are interested are conceptualized explicitly in terms of mathematics, and appear in mathematical models and mathematical theories. At the social level, it provides a language facilitating intersubjective agreement and communicability of observation and of ultimate agreement about the "truth status" of scientific statements.

Even at the level of fundamental presuppositions (Holton's themata), mathematics informs our thinking, as in the contrasting conceptions of the universe as causal and determinate on the one hand or probabilistic on the other. Kepler believed that the very soul of the universe was mathematical, and that the purpose of science was to discover the harmonies that reigned in the mind of the creator (Butterfield 1957). Einstein expressed a similar view:

> Our experience up to date justifies us in feeling sure that in nature is actualized the idea of mathematical simplicity.
>
> It is my conviction that pure mathematical construction enables us to discover the concepts and the laws connecting them, which give us the key to the understanding of the phenomena

of nature. Experience can, of course, guide us in our choice of serviceable mathematical concepts; it cannot possibly be the source from which they are derived.

In a certain sense, therefore, I hold it to be true that pure thought is competent to comprehend the real as the ancients dreamed. (Einstein, in Frank 1947:209)

Yet there remains a question: *Why* do so many of the relationships discovered by science find their best expressions in mathematical laws? To raise this as a question of metaphysics is not our purpose. Instead, we propose that it is usefully construed as a psychological question. Why is it so much easier to think about lawful relationships when they are expressed mathematically? To say that mathematics "helps" is merely to rephrase the question. *How* does it help? Under what conditions? Under what conditions does it hurt? Are there other ways to achieve the same facilitation? There is very little research on this topic, in spite of its centrality, and this section of our book merely raises some of the issues.

We begin with a selection from Descartes, in which the utility of mathematics as an aid to thought is defended on the grounds of its cognitive utility. Mathematics helps because it permits us to overcome the narrow limits of human memory and human attention. Like all of Descartes' "Rules for the Direction of the Mind" (which include diagrams, and the arranging of items in series—not unlike those advocated by Francis Bacon in selection 34), the point of mathematical representation is to permit focus upon those central characteristics of a problem which will permit "a distinct intuition of simple propositions" (1628:Rule XII).

That mathematics can play this role has been frequently attested in the history of science. As Joseph Fourier (1768–1830) said,

Mathematical analysis is as extensive as nature itself; it defines all perceptible relations, measures times, spaces, forces, temperatures; this difficult science is formed slowly, but it preserves every principle which it has once acquired; it grows and strengthens itself incessantly in the midst of the many variations and errors of the human mind (1822/1952:173).

But, since Descartes, few have tried to find out why it works this way.

In a short but insightful book the French mathematician Jacques Hadamard (1945) suggested part of the difficulty. He questioned mathematicians and surveyed what had been written on the problem of mathematical insight. He concluded that most of the work of insight was done at an unconscious level. This is not the sort of conclusion that rests easily among current psychologists, but it does, at the least, suggest some of the difficulties facing an introspective analysis of this issue. In concerning himself with the process of mathematical discovery, Hadamard (like Poincaré before him) found different *types* of thinkers. Before one can approach the question of *how* mathematics is used, one needs to determine how many kinds of users exist. Hadamard's point gives us an insight into a central problem for scientific reasoning. Thus, a nuclear physicist once described to us his conception of certain theoretical physicists as playing around with mathematical ideas until they had an interesting equation, then "crank-turning" to see what predictions ensued. He contrasted this with another (his own) approach to attacking physical problems, in which hypothetical entities were "as real to me as my own children." There was no implication in his discussion that one way was right and the other wrong. He was conveying the idea that *both* ways were right.

Because the issue is fundamental to broader psychological issues concerning aids to scientific thought, we have included a particularly lengthy second selection, by the psychologist Max Wertheimer. His account of the genesis of Albert Einstein's thinking about relativity theory touches on a range of issues, and especially upon those dealing with mathematical representation. That Wertheimer's account is an after-the-fact reconstruction is clear (cf. Miller 1975), but that should not obscure its value for us as an illustrative example.

Is mathematization, by itself, a worthy goal for scientific inquiry? Many have argued in the affirmative (e.g., Kuhn 1961), but it can also be argued that not all uses of mathematics are progressive. Thus, William James attacked Fechner's psychophysical laws as constituting a Pyrrhic victory: "Fechner's book (1860) was the starting point of a new department of literature, which it would be perhaps impossible to match for the quantities of thoroughness and subtlety, but of which, in the humble opinions of the present writer, the proper psychological outcome is just *nothing*" (James 1890 1:534). Psychologically, just what do we have when we have an equation that describes some data set, some functional relationship? Sometimes we have a sense of powerful explanation. Other times, our reaction, like James's, is "So what?" Surely these different reactions may have extraordinary consequences for future research.

But what governs the psychological reaction? What is it that gives a quantitative "explanation" its compellingness, when it is compelling? What gives a quantitative "explanation" the power to evoke the aesthetic response which so many scientists report? Perhaps it has something to do with the total lack of disanalogy—the lack of parts that don't fit—in contrast with other forms of analogical explanations, which necessarily have partial disanalogies. Mathematical explanations do not suffer from disanalogy since, paraphrasing the mad hatter, mathematical symbols mean precisely what we want them to mean! Like language itself, mathematics is a "language" whose terms and syntax we control. Unlike language, however, mathematical demonstrations have an "uncontrivable" character (Ziman 1978:31); given the presuppositions, everything follows inevitably.

Another possibility is that mathematicization might in fact be, under some conditions, dysfunctional. It may be that during some stages of scientific inquiry the attempt to develop questions in a mathematical way brings one to a premature closure, a failure to seek relations to other questions. This seems related to the possibility raised earlier that falsification may be dysfunctional in the early stages of theory construction. It is, for example, possible that the tendency of contemporary cognitive psychologists to develop mathematical models for phenomena is contributing to what some see as the non-cumulative nature of cognitive psychology (Newell 1973). To have a mathematical model is not enough. One would like to have a mathematical model which can generate some interesting, uncontrivable consequences. Physics has had many such, but psychology only very few.

Another question related to mathematical thought concerns the nature of anomalies. Anomalies in data are often ascribed to random variation, and an attempt is made to explain them away. But it is also the occasional failure to fit an equation perfectly that provides one sort of constructive anomaly. Mathematics provides the ability to describe regularity, even to *see* regularity in a mass of otherwise impenetrable, though highly regular, data. It is that power, the power to describe regularity, which provides the added possibility of knowing when variation or anomaly is something other than random, and hence perhaps something worth investigating.

The perception of regularity presupposes a fundamental psychological ability, the ability to perceive or conceptualize two or more events as similar (Tversky 1977). As a psychological phenomenon, this ability is not well understood, but it is one deserving of further study, and one which will have important implications for the study of the role of mathematics in science. This issue will be raised again since psychological similarity is also a key idea in understanding the role of analogies, metaphors, models, etc.

Certainly another set of questions which mathematical reasoning allows to be asked are questions about the efficiency of the human operator. We often compare human operators against mathematically optimal performance, for example in decision theoretic research, perceptual research, motor skills research, etc. Such studies are thereby analogous to research which uses the propositions of logic as a criterion against which to assess inference behavior, as exemplified in selections 18, 19 and 20. From work like this, we know that ordinary people do not have available certain key concepts commonly used by scientists, such as the concepts of a sampling distribution and regression to the mean (Tversky & Kahneman 1974).

Finally, we must raise the question of whether mathematics itself, pure mathematics, is different from scientific inference. The commonplace view is that science deals with empirical realities, while mathematics deals with constructed or invented realities, which may or may not correspond to the "real world." Whatever ontological meaning this distinction may possess, it may or may not reflect a psychological reality. The mathematician George Polya (1945; 1954) has argued that doing mathematics is a process of discovery, akin to scientific discovery, though dependent upon formal reasoning. Lakatos (1963) argued that mathematical discovery can be related to the same process of conjecture, test, and reformulation that

Popper has described for science. Lakatos wrote a dramatic reconstruction of the discovery and verification of Euler's theorem for polygons $(V - E + F = 2)$ to establish his point. In mathematics, as in science, discovered anomalies must be attended to, and the success or failure of inquiry is critically dependent upon how such anomalies are resolved. Whether or not Lakatos' reconstructed history is valid as history is of less concern than the suggestion that the processes of hypothesis generation and evaluation discussed earlier (part 4) may be observable in mathematical inquiry as well. It is interesting to note that Hadamard (1945:48–49) would take issue with this view. He argued that an excessive devotion to one's own ideas, which Claude Bernard saw as destructive for science, was in fact not a problem for mathematicians. Note that one feature of much of mathematics is relevant to the distinction drawn above in the discussion of what factors might determine whether falsification will be a fruitful strategy. In science, the relation between hypothesis and data is often far fuzzier than in mathematics. Further, the "data" of the mathematician are not laced with error of measurement in the same sense that empirical measurements are. Thus, confirmation and falsification may play rather different roles in science and mathematics. An ongoing study in our laboratory, using mathematicians who think aloud as they explore a number-theoretic conjecture, suggests that mathematical hypotheses are in fact abandoned rapidly in the face of counterexamples (L. Markowitz, personal communication).

We close this section with excerpts from two empirical studies carried out by experimental psychologists on the nature of one kind of mathematical reasoning: the induction of particular equations from sets of data points. In each study, subjects were observed to generate hypotheses and to test them against data in a recursive fashion. Bearing Lakatos in mind, the results suggest that psychological approaches to broader issues of mathematical thinking as well as the nature of mathematical thinking in science can be richly rewarding.

**Selection 22** _____

RENÉ DESCARTES

# The Cognitive

# Function

# of Mathematics

RULE XVI

*When we come across matters which do not require our present
attention, it is better, even though they are necessary to our
conclusion, to represent them by highly abbreviated symbols,
rather than by complete figures. This guards against error due
to defect of memory on the one hand, and, on the other, prevents that distraction of thought which an effort to keep those
matters in mind while attending to other inferences would
cause.*

But because our maxim is that not more than two different
dimensions out of the countless number that can be depicted
in our imagination ought to be the object either of our bodily
or of our mental vision, it is of importance so to retain all
those outside the range of present attention that they may
easily come up to mind as often as need requires. Now memory
seems to be a faculty created by nature for this very purpose.
But since it is liable to fail us, and in order to obviate the need
of expending any part of our attention in refreshing it, while
we are engaged with other thoughts, art has most opportunely
invented the device of writing. Relying on the help this gives

Excerpt from R. Descartes. *The Philosophical Works of Descartes Rendered into English by E.S. Haldane and G.R.T. Ross.* Vol. 1. Cambridge: Cambridge University Press,
1969 (1911), pp. 66–67; 70. Reprinted by permission of Cambridge University Press.
First published in 1628.

us, we leave nothing whatsoever to memory, but keep our imagination wholly free to receive the ideas which are immediately occupying us, and set down on paper whatever ought to be preserved. In doing so we employ the very briefest symbols, in order that, after distinctly examining each point in accordance with Rule IX, we may be able, as Rule XI bids us do, to traverse them all with an extremely rapid motion of our thought and include as many as possible in a single intuitive glance.

Everything, therefore, which is to be looked upon as single from the point of view of the solution of our problem, will be represented by a single symbol which can be constructed in any way we please. But to make things easier we shall employ the characters $a$, $b$, $c$, etc. for expressing magnitudes already known, and $A$, $B$, $C$, etc. for symbolising those that are unknown. To these we shall often prefix the numerical symbols, 1, 2, 3, 4, etc., for the purpose of making clear their number, and again we shall append those symbols to the former when we want to indicate the number of the relations which are to be remarked in them. Thus if I employ the formula $2a^3$ that will be the equivalent of the words 'the double of the magnitude which is symbolised by the letter $a$, and which contains three relations.' By this device not only shall we economize our words, but, which is the chief thing, display the terms of our problem in such a detached and unencumbered way that, even though it is so full as to omit nothing, there will nevertheless be nothing superfluous to be discovered in our symbols, or anything to exercise our mental powers to no purpose, by requiring the mind to grasp a number of things at the same time. . . .

Moreover, it must be observed that, as a general rule, nothing that does not require to be continuously borne in mind ought to be committed to memory, if we can set it down on paper. This is to prevent that waste of our powers which occurs if some part of our attention is taken up with the presence of an object in our thought which it is superfluous to bear in mind. What we ought to do is to make a reference-table and set down in it the terms of the problem as they are first stated. Then we should state the way in which the abstract formulation is to be made and the symbols to be employed, in order that, when the solution has been obtained in terms of these symbols, we may easily apply it, without calling in the aid of memory at all, to the particular case we are considering: for it is only in passing from a lesser to a greater degree of generality that abstraction has any *raison d'être*.

MAX WERTHEIMER

# Einstein:

# The Thinking

# That Led

# to the Theory

# of Relativity

THOSE WERE WONDERFUL days, beginning in 1916, when for hours and hours I was fortunate enough to sit with Einstein, alone in his study, and hear from him the story of the dramatic developments which culminated in the theory of relativity. During those long discussions I questioned Einstein in great detail about the concrete events in his thought. He described them to me, not in generalities, but in a discussion of the genesis of each question.

Einstein's original papers give his results. They do not tell the story of his thinking. In the course of one of his books he did report some steps in the process. I have quoted him in the proper places in this chapter.

The drama developed in a number of acts.

From M. Wertheimer, *Productive Thinking* (first published 1945). Enlarged edition. New York: Harper & Row, 1959, pp. 213–233. Reprinted by permission of the estate of Valentin J. T. Wertheimer.

*Act I. The Beginning of the Problem*

The problem started when Einstein was sixteen years old, a pupil in the Gymnasium (Aarau, Kantonschule). He was not an especially good student, unless he did productive work on his own account. This he did in physics and mathematics, and consequently he knew more about these subjects than his classmates. It was then that the great problem really started to trouble him. He was intensely concerned with it for seven years; from the moment, however, that he came to question the customary concept of time (see Act VII), it took him only five weeks to write his paper on relativity—although at this time he was doing a full day's work at the Patent Office.

The process started in a way that was not very clear, and is therefore difficult to describe—in a certain state of being puzzled. First came such questions as: What if one were to run after a ray of light? What if one were riding on the beam? If one were to run after a ray of light as it travels, would its velocity thereby be decreased? If one were to run fast enough, would it no longer move at all? . . . To young Einstein this seemed strange.

The same light ray, for another man, would have another velocity. What *is* "the velocity of light"? If I have it in relation to something, this value does not hold in relation to something else which is itself in motion. (Puzzling to think that under certain conditions light should go more quickly in one direction than another.) If this is correct, then consequences would also have to be drawn with reference to the earth, which is moving. There would then be a way of finding out by experiments with light whether one is in a moving system! Einstein's interest was captured by this; he tried to find methods by which it would be possible to establish or to measure the movement of the earth—and he learned only later that physicists had already made such experiments. His wish to design such experiments was always accompanied by some doubt that the thing was really so; in any case, he felt that he must try to decide.

He said to himself: "I know what the velocity of a light ray is in relation to a system. What the situation is if another system is taken into account seems to be clear, but the consequences are very puzzling."

*Act II.  Light Determines a State of Absolute Rest?*

Would operations with light lead to conclusions different in this respect from conclusions from mechanical operations? From the point of view of mechanics there seems to be no absolute rest; from the point of view of light there does seem to be. What of the velocity of light? One must relate it to something. Here the trouble starts. Light determines a state of absolute rest? However, one does not know whether or not one is in a moving system. Young Einstein had reached some kind of conviction that one cannot notice whether or not one is in a moving system; it seemed to him deeply founded in nature that there is no "absolute movement." The central point here became the conflict between the view that light velocity seems to presuppose a state of "absolute rest" and the absence of this possibility in the other physical processes.

Back of all this there had to be something that was not yet grasped, not yet understood. Uneasiness about this characterized young Einstein's state of mind at this time.

When I asked him whether, during this period, he had already had some idea of the constancy of light velocity, independent of the movement of the reference system, Einstein answered decidedly: "No, it was just curiosity. That the velocity of light could differ depending upon the movement of the observer was somehow characterized by doubt. Later developments increased that doubt." Light did not seem to answer when one put such questions. Also light, just as mechanical processes, seemed to know nothing of a state of absolute movement or of absolute rest. This was interesting, exciting.

Light was to Einstein something very fundamental. At the time of his studies at the Gymnasium, the ether was no longer being thought of as something mechanical, but as "the mere carrier of electrical phenomena."

*Act III.  Work on the One Alternative*

Serious work started. In the Maxwell equations of the electromagnetic field, the velocity of light plays an important role; and it is constant. If the Maxwell equations are valid with regard to one system, they are not valid in another. They would have to be changed. When one tries to do so in such a way that the velocity of light is not assumed to be constant, the matter becomes very complicated. For years Einstein tried to clarify the problem by studying and trying to change the Maxwell equations. He did not succeed in formulating these equations in such a way as to meet the difficulties satisfactorily. He tried hard to see clearly the relation between the velocity of light and the facts of movement in mechanics. But in whatever way he tried to unify the question of mechanical movement with the electromagnetic phenomena, he got into difficulties. One of his questions was: What would happen to the Maxwell equations and to their agreement with the facts if one were to assume that the velocity of light depends on the motion of the source of the light?

The conviction grew that in these respects the situation with regard to light could not be different from the situation with regard to mechanical processes (no absolute movement, no absolute rest). What took him so much time was this: he could not doubt that the velocity of light is constant and at the same time get a satisfactory theory of electromagnetic phenomena.

*Act IV.  Michelson's Result and Einstein*

The famous Michelson experiment confronted physicists with a disconcerting result. If you are running away from a body that is rushing toward you, you will expect it to hit you somewhat later than if you are standing still. If you run toward it, it will hit you earlier. Michelson did just this in measurements of the velocity of light. He compared the time light takes to travel in two pipes if these pipes meet at right angles to each other, and if one lies in the direction of the movement of the earth, while the other is vertical to it. Since the first pipe, in its lengthwise direction, is moving with the movement of the earth, the light traveling in it ought to reach the receding end of this pipe later than the light in the other pipe reaches *its* end. . . .

No difference was found. The experiment was repeated, and the negative result was clearly confirmed.

The result of the Michelson experiment in no way fitted the fundamental ideas of the physicists. In fact the result contradicted all their reasonable expectations.

For Einstein, Michelson's result was not a fact for itself. It had its place within his thoughts as they had thus far developed. Therefore, when Einstein read about these crucial ex-

periments made by physicists, and the finest ones made by
Michelson, their results were no surprise to him, although
very important and decisive. They seemed to confirm rather
than to undermine his ideas. But the matter was not yet entirely cleared up. Precisely how does this result come about?
The problem was an obsession with Einstein although he saw
no way to a positive solution.

*Act V. The Lorentz Solution*
   Not only Einstein was troubled; many physicists were. Lorentz, the famous Dutch physicist, had developed a theory
which formulated mathematically what had occurred in the
Michelson experiment. In order to explain this fact it seemed
necessary to him, as it had to Fitzgerald, to introduce an auxiliary hypothesis: he assumed that the entire apparatus used
in the measurement underwent a small contraction in the direction of the earth's motion. According to this theory, the
pipe in the direction of the movement of the earth was changed
in length, while the other pipe suffered only a change in width
and the length remained unaffected. The contraction had to
be assumed to be just the amount needed to compensate for
the effect of the earth's motion on the traveling of the light.
This was an ingenious hypothesis.
   There was now a fine, positive formula, determining the
Michelson results mathematically, and an auxiliary hypothesis, the contraction. The difficulty was "removed." But for
Einstein the situation was no less troublesome than before; he
felt the auxiliary hypothesis to be a hypothesis *ad hoc*, which
did not go to the heart of the matter.

*Act VI. Re-examination of the Theoretical Situation*
   Einstein said to himself: "Except for that result, the whole
situation in the Michelson experiment seems absolutely clear;
all the factors involved and their interplay seem clear. But *are*
they really clear? Do I really understand the structure of the
whole situation, especially in relation to the crucial result?"
During this time he was often depressed, sometimes in despair,
but driven by the strongest vectors.
   In his passionate desire to understand or, better, to see
whether the situation was really clear to him, he faced the
essentials in the Michelson situation again and again, espe-

cially the central point: the measurement of the speed of light under conditions of movement of the whole set in the crucial direction.

This simply would not become clear. He felt a gap somewhere without being able to clarify it, or even to formulate it. He felt that the trouble went deeper than the contradiction between Michelson's actual and the expected result.

He felt that a certain region in the structure of the whole situation was in reality not as clear to him as it should be, although it had hitherto been accepted without question by everyone, including himself. His proceeding was somewhat as follows: There is a time measurement while the crucial movement is taking place. "Do I see clearly," he asked himself, "the relation, the inner connection between the two, between the measurement of time and that of movement? Is it clear to me how the measurement of time works in such a situation?" And for him this was not a problem with regard to the Michelson experiment only, but a problem in which more basic principles were at stake.

*Act VII. Positive Steps toward Clarification*

It occurred to Einstein that time measurement involves simultaneity. What of simultaneity in such a movement as this? To begin with, what of simultaneity of events in different places?

He said to himself: "If two events occur in one place, I understand clearly what simultaneity means. For example, I see these two balls hit the identical goal at the same time. But . . . am I really clear about what simultaneity means when it refers to events in two different places? What does it mean to say that this event occurred in my room at the same time as another event in some distant place? Surely I can use the concept of simultaneity for different places in the same way as for one and the same place—but can I? Is it as clear to me in the former as it is in the latter case? . . . It is not!"

For what now followed in Einstein's thinking we can fortunately report paragraphs from his own writing. He wrote them in the form of a discussion with the reader. What Einstein here says to the reader is similar to the way his thinking proceeded: "Lightning strikes in two distant places. I assert that both bolts struck simultaneously. If now I ask you, dear reader, whether this assertion makes sense, you will answer, 'Yes, certainly'. But if I urge you to explain to me more clearly the meaning of this assertion, you will find after some deliberation that answering this question is not as simple as it at first appears.

"After a time you will perhaps think of the following answer: 'The meaning of the assertion is in itself clear and needs no further clarification. It would need some figuring out, to be sure, if you were to put me to the task of deciding by observation whether in a concrete case the two effects were actually simultaneous or not.'"

I now insert an illustration which Einstein offered in a discussion. Suppose somebody uses the word "hunchback." If this concept is to have any clear meaning, there must be some way of finding out whether or not a man has a hunched back. If I could conceive of no possibility of reaching such a decision, the word would have no real meaning for me.

"Similarly," Einstein continued, "with the concept of simultaneity. The concept really exists for the physicist only when in a concrete case there is some possibility of deciding whether the concept is or is not applicable. Such a definition of simultaneity is required, therefore, as would provide a method for deciding. As long as this requirement is not fulfilled, I am deluding myself as physicist (to be sure, as nonphysicist too!) if I believe that the assertion of simultaneity has a real meaning. (Until you have truly agreed to this, dear reader, do not read any further.)

"After some deliberation you may make the following proposal to prove whether the two shafts of lightning struck simultaneously. Put a set of two mirrors, at an angle of 90° to each other, at the exact halfway mark between the two light effects, station yourself in front of them, and observe whether or not the light effects strike the mirrors simultaneously."

Simultaneity in distant places here gets its meaning by being based on clear simultaneity in an identical place.

All these steps came not by way of isolated clarification of this special question, but as part of the attempt to understand the inner connection that was mentioned above, the problem of the measurement of speed during the crucial movement. In the mirror situation this means simply: What happens if, in the time during which the light rays approach my mirrors, I

move with them, away from one source of light and toward
the other? Obviously, if the two events appeared simultaneous
to a man at rest they would not then appear so to me, who am
moving with my mirrors. His statement and mine must differ.
We see then that our statements about simultaneity *involve
essentially reference to movement of the observer.* If simultaneity in distant places is to have real meaning, I must explicitly take into account the question of movement, and in
comparing my judgments with those of another observer, I
have to take into account the relative movement between him
and me. When dealing with "simultaneity in different places"
I must refer to the relative movement of the observer.

I repeat: suppose that I with my mirrors am traveling in a
train going in a straight line at a constant velocity. Two shafts
of lighting strike in the distance, one near the engine, the other
near the rear end of the train, my double mirror being right
in the middle between the two. As a passenger I use the train
as my frame of reference, I relate these events to the train. Let
us assume that just at the critical moment when the lightning
strikes, a man is standing beside the tracks, likewise with
double mirrors, and that his place at that moment coincides
with mine. What would my observations be and what would
his be?

"If we say that the bolts of lightning are simultaneous with
regard to the tracks, *this now means*: the rays of light coming
from two equidistant points meet simultaneously at the mirrors of the man on the track. But if the place of my moving
mirrors coincides with his mirrors at the moment the lightning
strikes, the rays will not meet exactly simultaneously in my
mirrors because of my movement.

"Events which are simultaneous in relation to the track are
not simultaneous in relation to the train, and vice versa. Each
frame of reference, each system of coordinates therefore has
*its special time;* a statement about a time has real meaning
only when the frame of reference is stated, to which the assertion of time refers."

It has always seemed simple and clear that a statement about
the "time difference" between two events is a "fact," independent of other factors, such as movement of the system. But,
in actual fact, is not the thesis that "the time difference between two events is independent of the movement of the sys-

tem" an arbitrary assumption? It did not hold, as we saw, for simultaneity in different places, and therefore it cannot hold even for the length of a second. To measure a time interval, we must use a clock or the equivalent of a clock, and look for certain coincidences at the beginning and at the end of the interval. Therefore the trouble with simultaneity is involved. We cannot dogmatically assume that the time which a certain event takes in relation to the train is the same as the time in relation to the track.

This applies also to the measurement of distances in space! If I try to measure exactly the length of a car by marking its end points on the roadbed, I must take care, when I have made my mark at one end, that the car does not move before I come to the other end! Unless I have explicitly given attention to this possibility, my measurements will be misleading.

I must therefore conclude that in every such measurement reference must be made to the movement of the system. For the observer within the moving system will get results which differ from those of an observer in another frame of reference. "Every system has its special time and space values. A time or space judgment has sense only if we know the system with reference to which the judgment was made." We must change the old view: the measurements of time intervals and of distances in space are *not* independent of the conditions of movement of the system in relation to the observer.

The old view had been a time-honored "truth." Einstein, seeing that it was questionable, came to the conclusion that space and time measurements depend on the movement of the system.

*Act VIII. Invariants and Transformation*

What followed was determined by two vectors which simultaneously tended toward the same question.

1. The system of reference may vary; it can be chosen arbitrarily. But in order to reach physical realities, I have to get rid of such arbitrariness. The basic laws must be independent of arbitrarily chosen coordinates. If one wants to get a description of physical events, the basic laws of physics must be invariant with regard to such changes.

Here it becomes clear that one might adequately call Einstein's theory of relativity just the opposite, an absolute theory.

2. Insight into the interdependence of time measurement and movement is certainly not enough in itself. What is now needed is a transformation formula that answers this question: "How does one find the place and time values of an event in relation to one moving system, if one knows the places and times as measured in another? Or better, how does one find the transformation from one system to another when they move in relation to each other?"

What would be the direct way? In order to proceed realistically, I would have to base the transformation on an assumption with regard to some physical realities which could be used as invariants.

The reader may think back to an old historical situation. Physicists in past ages tried to construct a *perpetuum mobile*. After many attempts which did not succeed, the question suddenly arose: how would physics look if nature were basically such as to make a *perpetuum mobile* impossible? This involved an enormous change, which recentered the whole field.

Similarly there arose in Einstein the following question, which was inspired by his early ideas mentioned in Acts II and III. How would physics look if, by nature, measurements of the velocity of light would under all conditions have to lead to the identical value? Here is the needed invariant! (Thesis of the basic constancy of the velocity of light.)

In terms of the desired transformation, this means: "Can a relation between the place and time of events in systems which move linearly to each other be so conceived that the velocity of light becomes a constant?"

Eventually Einstein reached the answer: "Yes!" The answer consisted of concrete and definite transformation formulas for distances in time and space, formulas that differed characteristically from the old Galilean transformation formulas.

3. In the discussions I had with Einstein in 1916 I put this question to him: "How did you come to choose just the velocity of light as a constant? Was this not arbitrary?"

Of course it was clear that one important consideration was the empirical experiments which showed no variation in the velocity of light. "But did you choose this arbitrarily," I asked, "simply to fit in with these experiments and with the Lorentz transformation?" Einstein's first reply was that we are entirely free in choosing axioms. "There is no such difference as you

just implied," he said, "between reasonable and arbitrary axioms. The only virtue of axioms is to furnish fundamental
propositions from which one can derive conclusions that fit
the facts." This is a formulation that plays a great role in
present theoretical discussions, and about which most theorists seem to be in agreement. But then Einstein himself smil-
ingly proceeded to give me a very nice example of an unreasonable axiom: "One could of course choose, say, the velocity
of sound instead of light. It would be reasonable, however, to
select not the velocity of just 'any' process, but of an 'outstanding' process. . . ." Questions like the following had occurred to Einstein: Is the speed of light perhaps the fastest
possible? Is it perhaps impossible to accelerate any movement
beyond the speed of light? As velocity increases, progressively
greater forces are required to increase it still further. Perhaps
the force required to increase a velocity beyond the velocity
of light is infinite?

It was marvelous to hear in Einstein's descriptions how
these bold questions and expectations had taken shape in him.
It was new, unthought of before, that the velocity of light might
be the greatest possible velocity, that an attempt to go beyond
that limit would require forces infinitely great.

If these assumptions brought clarity into the system, and if
they were proved by experiment, then it would make good
sense to take the velocity of light as the basic constant. (Cf. the
absolute zero of temperature which is reached when the molecular movements in an ideal gas approach zero.)

4. The derivations which Einstein reached from his transformation formulas showed mathematical coincidence with
the Lorentz transformation. The contraction hypothesis had
therefore been in the right direction, only now it was no longer
an arbitrary auxiliary hypothesis, but the outcome of improved
insight, a logically necessary derivation from the improved
view of fundamental physical entities. The contraction was
not an absolute event, but a result of the relativity of measurements. It was not determined by a "movement in itself
which possesses no real sense for us, but only by a movement
with reference to the chosen observation system."

*Act IX. On Movement, on Space, a Thought Experiment*
The last statement throws new light on the changes in thinking which were already involved in the earlier steps. "By the

motion of a body we always mean its change of position in relation to a second body," to a framework, or a system. If there is one body alone, it makes no sense to ask or to try to state whether it is moving or not. If there are two, we can state only whether they are approaching or moving away from each other, but, so long as there are only two, it makes no sense to ask, or to try to state, whether one is turning around the other; the essential in movement is change of position in relation to another object, a framework, or a system.

But is there not *one* outstanding system in regard to which there is *absolute* movement of a body, "the" space (Newtonian space, the space of the ether), the box in which all movement takes place?

Here I may mention something that happened not just at this point in the development of the process, but may illustrate what was really going on. It transcends the problems of the special theory of relativity: Is there no proof of the reality of such an outstanding system? A famous experiment of Newton's had been used as proof: When a sphere of oil rotates it becomes flattened. This is a real, physical, observable fact, apparently caused by an "absolute" movement.

But is this really a demonstration of such an absolute movement? It seems so certainly; but is it actually, if we think it through? In reality we have not a body moving alone in absolute space, but a body that moves within our fixed-star firmament. Is the flattening of that sphere perhaps an outcome of the movement of the sphere relative to the surrounding stars? What would happen if we took a very huge iron wheel, with a small hole at the center, if we suspended in this hole a little sphere of oil, and then rotated the wheel? Perhaps the little sphere would again become flattened. Then the flattening would have nothing to do with the rotation in an absolute space box; rather it would be determined by the systems moving in relation to each other, the big wheel or the firmament on the one hand and the little sphere of oil on the other.

Of course rotation already transcends the region of the so-called special relativity of Einstein. It became basic in the problem of the general theory of relativity.

*Act X.  Questions for Observation and Experiment*
Einstein was at heart a physicist. Thus all these developments aimed at real, concrete, experimental problems. As soon

as he reached clarification he concentrated on the point: "Is
it possible to find crucial physical questions to be answered
in experiments that will decide whether these new theses are
'true'; whether they fit facts better, give better predictions of
physical events than the old theses?"

He found a number of such crucial experiments, some of
which physicists could and later did carry out.


In actual fact, the problem leads on: it led in Einstein's mind
to the problems of the general theory of relativity. But let us
stop with the story here and ask ourselves: What were the
decisive characteristics of this thinking?

The physicist is interested in the relation of Einstein's theory
to established facts, in experimental proof, in the consequences for further development, in the mathematical formulas which follow from the theory of relativity in the various
parts of physics.

The theorist of knowledge is interested in the ideas of space,
time, and matter, in the "relativistic" character of the theory
(with all the wrong consequences in the direction of philosophical, sociological, or ethical relativism drawn by others),
in the problem of "testability" which played such an important role in Einstein's dealing with simultaneity (and later in
the developments of operationalism).

The psychologist, who is concerned with the problems of
thinking, wants to realize what went on psychologically.

If we were to describe the process in the way of traditional
logic, we would state numerous operations, like making ab-
stractions, stating syllogisms, formulating axioms and general
formulas, stating contradictions, deriving consequences by
combining axioms, confronting facts with these consequences,
and so forth.

Such a procedure is certainly good if one wishes to test each
of the steps with regard to its logical correctness. Einstein himself was passionately interested in logical correctness, logical
validity.

But what do we get if we follow such a procedure? We get
an aggregate, a concatenation of a large number of operations,
syllogisms, etc. Is this aggregate an adequate picture of what
has happened? What many logicians do, the way they think,

is somehow like this: A man facing a work of architecture, a fine building, focuses, in order to understand it, on the single bricks and also on the way in which they are cemented by the mortar. What he has at the end is not the building at all but a survey of the bricks and of their connections.[1]

In order to get at the real picture, we have to ask: How did the operations arise, how did they enter into the situation, what was their function within the actual process? Did they just drop in? Was the process a chain of happy accidents? Was the solution a consequence of trial and error, of mathematical guesswork? Why just these operations? No doubt there were other possibilities at some points. Why was Einstein moving in just this direction? How did it come about that after he made one step, he followed with just that other step?

I shall mention one specific point: How did the new axioms arise? Did Einstein just try any axioms of which certain ones then actually happened to work? Did he formulate some propositions, put them together, and observe what happened until eventually he was fortunate enough to find a proper set? Did such propositions leap into the picture accidentally, and did the changes in the role, place, and function of the items, did their new interrelatedness appear merely as derived consequences?

The technique of axioms is a very useful tool. It is one of the most efficient techniques so far invented in logic and mathematics; a few general propositions provide all that is needed in order to derive the details. One can deal with a gigantic sum of facts, with huge numbers of propositions, by substituting for them a few sentences which in a formal sense are equivalent to all that knowledge. Some great discoveries in modern mathematics became possible only because this extremely simplifying technique was at hand. Einstein, too, used this tool in the accounts which he gave of his theory of relativity.

But, to repeat, the question for the psychologist is: Were these axioms introduced before the structural requirements,[2] the structural changes of the situation were envisaged? Was it not the other way around? Surely, Einstein's thought did not put ready-made axioms or mathematical formulas together. The axioms were not the beginning but the outcome of what was going on. Before they came into the picture as formulated propositions, the situation as to the velocity of light and related topics had for a long time been structurally questionable to him, had in certain respects become inadequate, was in a state of transition. The axioms were only a matter of later formulation—after the real thing, the main discovery, had happened.[3]

When we proceed with an analysis in the sense of traditional logic, we easily forget that actually all the operations were parts of a unitary and beautifully consistent picture, that they developed as parts within one line of thinking; that they arose, functioned, and had their meaning within the whole process as the situation, its structure, its needs and demands were faced. In trying to grasp the structure of this great line of thinking, the reader may find himself at a loss in view of the wealth of events, of the breadth of the situation. What, then, were the decisive steps?

Let us recapitulate briefly.

First there was what we may call the foreperiod. Einstein was puzzled by the question, first, of the velocity of light when the observer is in motion. He considered, secondly, the consequences as to the question of "absolute rest." Thirdly, he then tried to make one alternative workable (is the velocity of light in Maxwell's equations a variable?), and obtained a negative result. There was, fourth, the Michelson experiment which confirmed the other alternative—and, fifth, the Lorentz-Fitzgerald hypothesis, which did not seem to go to the root of the trouble.

So far everything, including the meaning and structural role of time, space, measurement, light, etc., was understood in terms of traditional physics—structure I.

In this troubled situation the question arose: Is this structure itself, in which the Michelson result seems contradictory, really clear to me? This was the revolutionary moment. Einstein felt that the contradiction should be viewed without prejudice, that the time-honored structure should be requestioned. Was this structure I adequate? Was it clear just with regard to the critical point—the question of light in relation to the question of movement? Was it clear in the situation of the Michelson experiment? All these questions were asked in a passionate effort to understand. And then the procedure became more specific in one step after another.

How was the velocity of light to be measured in a moving system?

How was time to be measured under these circumstances? What does simultaneity mean in such a system?

But, then, what does simultaneity mean if the term is referred to different places?

The meaning of simultaneity was clear if two events occur in the same place. But Einstein was suddenly struck by the fact that it was *not* equally clear for events in distant places. Here was a gap in any real understanding. He saw: It is blind simply to apply the customary meaning of simultaneity to these other cases. If simultaneity is to have a real meaning, we must raise the question of its factual recognition so that in concrete cases we can tell whether or not the term applies. (Clearly, this was a fundamental logical problem.)

The meaning of simultaneity in general had to be based on the clear simultaneity in the case of spatial coincidence. But this required that in every case of different location of two events the relative movement be taken into account. Thus the meaning, the structural role of simultaneity in its relation to movement, underwent a radical change.

Immediately, corresponding requirements follow for the measurement of time in general, for the meaning, say, of a second, and for the measurement of space, since they must now depend upon relative movement. As a result, the concepts of time-flow, of space, and of the measurement both of time and space change their meaning radically.

At this point the introduction of the observer and his system of co-ordinates seemed to introduce a fundamentally arbitrary or subjective factor. "But the reality," Einstein felt, "cannot be so arbitrary and subjective." In his desire to get rid of this arbitrary element and, at the same time, to get a concrete transformation formula between various systems, he realized that a basic invariant was needed, some factor that remains unaffected by the transition from one system to another. Obviously, both demands went in the same direction.

This led to the decisive step—the introduction of the velocity of light as the invariant. How would physics look if recentered with this as a starting point? Bold consequences followed one after another, and a new structure of physics was the consequence.

When Einstein reached the concrete transformation formula on the basis of this invariant, the Lorentz transformation appeared as a derivation—but now it was understood in a deeper, entirely new way, as a necessary formulation within the new structure of physics. The Michelson result, too, was now seen in an entirely new light, as a necessary result when the interplay of all relative measurements within the moving system was taken into account. Not the result was troublesome—he had felt that from the very beginning—but the behavior of the various items in the situation before finding the solution. With the deeper understanding of these items the result was required.

The picture was now improved. Einstein could proceed to the question of experimental verification.

In the briefest formulation: In a passionate desire for clearness, Einstein squarely faced the relation between the velocity of light and the movement of a system, and confronted the theoretical structure of classicial physics and the Michelson result.

A part-region in this field became crucial and was subjected to a radical examination.

Under this scrutiny a great gap was discovered (in the classical treatment of time).

The necessary steps for dealing with this difficulty were realized.

As a result, the meaning of all the items involved underwent a change.

When a last arbitrariness in the situation had been eliminated, a new structure of physics crystallized.

Plans were made to subject the new system to experimental test.

Radical structural changes were involved in the process, changes with regard to separateness and inner relatedness, grouping, centering, etc.; thereby deepening, changing the meaning of the items involved, their structural role, place, and function in the transition from structure I to structure II. It may be advisable to explain once more in what sense Einstein's achievement meant a change of structure.

1) In the Michelson situation—as in classicial physics generally—time had been regarded as an independent variable and, therefore, as an independent tool in the business of measurement, entirely separate from, in no way functionally interdependent with the movements that were involved in that

observational situation. Accordingly, the nature of time had been of no interest with regard to the apparently paradoxical result.

In Einstein's thought there arose an intimate relationship between time-values and the physical events themselves. Thus the role of time within the structure of physics was fundamentally altered.

This radical change was first clearly envisaged in the consideration of simultaneity. In a way, simultaneity split in two: the clear simultaneity of events in a given place and, related to it, but related by means of specific physical events, the simultaneity of events in different places, particularly under conditions of movement of the system.

2) As a consequence, space-values also changed their meaning and their role in the structure of physics. In the traditional view they, too, had been entirely separated from, independent of time and of physical events. Now an intimate relation was established. Space was no longer an empty and wholly indifferent container of physical facts. Space geometry became integrated with the dimension of time in a four-dimensional system, which in turn formed a new unitary structure with actual physical occurrences.

3) The velocity of light had so far been one velocity among many. Although the highest velocity known to the physicist, it had played the same role as other velocities. It had been fundamentally unrelated to the way in which time and space are measured. Now it was considered as closely bound up with time- and space-values, and as a fundamental fact in physics as a whole. Its role changed from that of a particular fact among many to that of a central issue in the system.

Many more items could be mentioned which changed their meaning in the process, such as mass and energy, which now proved to be closely related. But it will not be necessary to discuss further particulars.

In appraising these transformations we must not forget that they took place in view of a gigantic given system. Every step had to be taken against a very strong gestalt—the traditional structure of physics, which fitted an enormous number of facts, apparently so flawless, so clear that any local change was bound to meet with the resistance of the whole strong and well-articulated structure. This was probably the reason why it took so long a time—seven years—until the crucial advance was made.

One could imagine that some of the necessary changes occurred to Einstein by chance, in a prodecure of trial and error.[4] Scrutiny of Einstein's thought always showed that when a step was taken this happened because it was required. Quite generally, if one knows how Einstein thought, one knows that any blind and fortuitous procedure was foreign to his mind.

The only point at which there could have been some doubt in this respect was the introduction of the constancy of light velocity in Einstein's general transformation formulas. In a thinker of lesser stature this could have happened through mere tentative generalization of the Lorentz formula. But actually the essential step was not reached in this fashion; there was no mathematical guesswork in it.

In late years Einstein often told me about the problems on which he was working at the time. There was never a blind step. When he dropped any direction, it was only because he realized that it would introduce ununderstandable, arbitrary factors. Sometimes it happened that Einstein was faced with the difficulty that the mathematical tools were not far enough developed to allow a real clarification; nonetheless he would not lose sight of his problem and would often succeed in finding a way eventually, in which the seemingly insuperable difficulties could be surmounted.

## Notes

1. "I am not sure," Einstein said once in this context, "whether there can be a way of really understanding the miracle of thinking. Certainly you are right in trying to get at a deeper understanding of what really goes on in a thinking process. . . ."

2. In our discussions Einstein focused on the material content of the steps. He did not use the terms of the preceding sentences of the text, terms which follow from the structural approach of this book.

3. In this respect I wish to report some characteristic remarks of Einstein himself. Before the discovery that the crucial point, the solution, lay in the concept of time, more particularly in that of simultaneity, axioms played no role in the thought process—of this Einstein is sure. (The very moment he saw the gap, and realized the relevance of simultaneity, he knew this to be the crucial point for the solution.) But even afterward, in the final five weeks, it was not the axioms that came first. "No really productive man thinks in such a paper fashion," said Einstein. "The way the two

triple sets of axioms are contrasted in the Einstein-Infeld book is not at all the way things happened in the process of actual thinking. This was merely a later formulation of the subject matter, just a question of how the thing could afterwards best be written. The axioms express essentials in a condensed form. Once one has found such things one enjoys formulating them in that way; but in this process they did not grow out of any manipulation of axioms."

He added, "These thoughts did not come in any verbal formulation. I very rarely think in words at all. A thought comes, and I may try to express it in words afterward." When I remarked that many report that their thinking is always in words, he only laughed. I once told Einstein of my impression that "direction" is an important factor in thought processes. To this he said, "Such things were very strongly present. During all those years there was a feeling of direction, of going straight toward something concrete. It is, of course, very hard to express that feeling in words; but it was decidedly the case, and clearly to be distinguished from later considerations about the rational form of the solution. Of course, behind such a direction there is always something logical; but I have it in a kind of survey, in a way visually."

4. In Act III, when Einstein examined whether a particular alternative would work, he actually did try several procedures. But although these attempts did not lead to a solution, they were by no means blind. At that stage it was wholly reasonable to test such possibilities.

L. ROWELL HUESMANN
and CHAO-MING CHENG

# Mathematical

# Induction:

# An Experimental

# Study

AN INDIVIDUAL ENGAGED in solving a difficult problem is faced with two tasks: the surface task is simply to solve the problem, but the deeper coincident task is to derive general heuristics, algorithms, and methods of representation that can be applied to new problems. Obviously, this process of extracting information out of specific examples to form general rules is induction. On the surface, induction in this situation seems quite different from most concept-learning situations constructed in the laboratory. The problem solver knows only that algorithms, heuristics, and representations exist which should be employed when indicated by certain ill-defined quantitative features from a large set of features related to the problem. His goal is to learn which features are relevant, and what the functions are that map the features into the methods.

Most strategies that a problem solver derives may be stated in many different equivalent forms. One such form would be

Excerpts from L. R. Huesmann and C. Cheng, A theory for the induction of mathematical functions. *Psychological Review* (1973), 80:126–38. Copyright © 1973 by the American Psychological Association. Reprinted by permission.

a mathematical function that maps observed values of relevant stimulus variables into observed utilities of solution methods. Representing strategies as mathematical functions, however, is particularly useful if one follows the conception of Newell and Simon (1961) that problem solvers form tables of connections from characteristics of the task environment to solution methods.

Our view is that a problem solver begins the learning process by constructing part of a table of connections between states of the task environment and solution methods. When enough data have been collected in such a table, he induces rules that specify which solution methods are good in each situation. The induction of these rules based upon the data in the table of connections could be accomplished by an induction process that maps classes of stimulus values into utilities for solution methods. More formally, we can say that to induce a rule for solving a problem one must find a function $f(x_1, x_2, \ldots x_p)$, such that if $x_1, x_2, \ldots x_p$ are stimulus variables and $y$ is a variable representing the utility of a particular solution method, then $f(x_1, x_2, \ldots x_p) = y$ for all observed instances of $(x_1, x_2, \ldots x_p, y)$.

By modeling the derivation of problem-solving strategies as a case of inducing a mathematical function, we have not intended to imply that humans necessarily see strategy formation in such a light. Rather, we wished to show that the induction of a mathematical function is typical of the type of induction task that problem solvers face. Thus, an understanding of how mathematical functions are derived should contribute to an understanding of how problem-solving strategies are learned.

## Induction

## of Mathematical Functions

## and Concept Formation

Clearly, the problem of inducing a mathematical function is a concept formation task. Yet the induction of a general mathematical function to explain observed numerical data is different enough from the concept-learning tasks studied in the laboratory that little can be inferred with certainty. Traditional concept formation experiments have investigated the attainment of Boolean functions rather than arithmetic or other types of functions. Hunt, Marin, and Stone (1966) have suggested a general model for induction based on concept-learning studies, but the assumptions underlying that model are violated when the function being induced is not Boolean. In particular, the induction of arithmetic functions differs from learning Boolean rules for pattern classification in that (a) the number of output classes may be infinite in the arithmetical task (e.g., the real numbers), (b) the input attributes may assume an infinite set of values in the arithmetical task, and (c) there may be no limit of possible arithmetical rules for mapping the input variables into the output variable. However, many of the processes employed in Boolean concept learning are suggestive of processes that could be employed in deriving general mathematical functions. The hypotheses-testing theories of Restle (1962), Bower and Trabasso (1964), Gregg and Simon (1967), and their supporting experimentation suggest that induction is primarily a heuristic generate-and-test process.

Studies of serial-pattern processing (Feldman, Tonge, and Kanter 1963; Gregg 1967; Leeuwenberg 1969; Restle 1970; Simon and Kotovsky 1963; and Simon and Sumner 1968) also suggest that subjects construct formal rules for specifying sequences by means of heuristic generate-and-test processes. For example, the Simon and Kotovsky model for series-completion tasks asserts that subjects first search for a regularity that defines a cycle and then generate and test hypotheses until one is found that explains the series. Recently, Simon (1972) has shown that these various theories proposed to explain serial pattern processing are essentially the same, and all include the generation of a formal rule by the subject and his production of new sequences with the rule.

Other evidence for this view of induction as a heuristic search process stems from artificial intelligence research. As Plotkin (1971) and others have pointed out, problems in induction are theoretically insoluble in that no *general* method can exist for finding laws to explain particular phenomena. One can only prove that a particular must follow from a generality, not vice versa. It should not be surprising, then, that many computer programs which have been built to perform induction employ heuristic generate-and-test techniques. Pivar and Finkelstein's (1964) program for inducing the serial pattern in a sequence of letters or numbers uses a variety of clever devices to generate plausible pattern descriptions for the given series. Each is a heuristic technique not guaranteed to succeed, but the combination of them constitutes an impressive program. Similarly, Williams (1972) and Buchanan, Sutherland, and Feigenbaum (1969) employed heuristic generate-and-test processes to solve induction problems typical of intelligence tests and organic chemistry, respectively. . . .

Bongard (1970) has used these devices in a program called ARITHMETIC that "learns" discriminant functions for recognizing the mathematical functions relating one variable to several others. While such "perceptrons" may have the potential for being models of perceptual processes, their limited applicability to induction in problem solving is well known (Minsky and Papert 1969). Finally, one should recognize that the problem of inducing a mathematical function to explain observed data is quite similar to the problem of inducing a grammar to explain an observed language. Hence, it is worthwhile to note Feldman's (1972) proof that a program that generates grammars *in order of increasing complexity* can find the best grammar for a language.

Studies of problem solving have also uncovered information suggestive of how people proceed with induction problems. Wason (1968a) asked subjects to supply the rule that governed what the next element of a numerical series would be. While the rule actually used was that the next number had to be greater, subjects generated a wide variety of complicated hypotheses. Interestingly, when a subject was told to generate the next instance in the series, he tended to try to generate an instance confirming his current hypothesis rather than one that might disconfirm.

Other evidence of (apparently) inefficient search behavior in hypothesis testing has been reported by Smedslund (1967). He found that humans do poorly at detecting correlations between dichotomous variables. Smedslund suggests that the overreliance of subjects on positive instances supporting incorrect hypotheses contributed to this deficit. At the same time, evidence exists that some subjects, when given the opportunity, use very systematic search procedures to induce rules rapidly. Duncan (1967) found that when subjects in an induction experiment were allowed to control the variation of the stimulus variables, those who varied them one at a time derived the principle most rapidly. To summarize, induction seems to be characterized by a heuristic generate-and-test process during which positive confirmations add to the credibility attached to an hypothesis. Subjects who use systematic search techniques seem to learn more rapidly.

# Analyzing

# Inductive

# Behaviors

While the preceding evidence has been suggestive of how people might induce mathematical functions from observed data, the differences between most concept formation tasks studied and the task of inducing a mathematical function are not trivial. Hence, before proposing a theory, one should obtain more specific information about the processes that humans employ in solving such induction problems. To do this, we designed a problem-solving experiment in which our objective was to obtain verbal protocols and solution times for subjects who were solving problems of inducing arithmetic functions from data.

EXPERIMENT I

*Problems.* Nine problems requiring the induction of an arithmetic function were used in this experiment. Each problem consisted of six instances of the form $a \$ b = c$. The two operands on the left, $a$ and $b$, could be mapped into the operand on the right, $c$, by some arithmetic expression constructed from the two operands: the operator's addition, subtraction, multiplication, division, and exponentiation; and integer constants between 1 and 4. No more than three operators were used in any expression. The nine problems and their solutions are shown in table 24.1.

*Subjects.* Eighteen undergraduates were employed as subjects. Only those subjects were analyzed who solved at least two out of three problems. In order to obtain 18 valid subjects, 22 subjects had to be tested.

*Procedure.* The stimuli for one induction problem consisted of six simple mathematical equations of the form $a \$ b = c$, for example, $2 \$ 3 = 8$. Each of the six equations represented a positive instance of the correct concept (function). No negative instances were provided. The problem was printed on a large white sheet about 2 meters from the subject. The six instances were listed in one column beginning at the top of the page. All six instances were exposed simultaneously.

**Table 24.1** Induction Problems Used in Experiment I

| Number of operators in correct function | Number of alternative functions fitting at least half the instances of data and having no more operators than the correct function | | | Solution $f(A, B)$ |
|---|---|---|---|---|
| | 0 | 1–2 | ≥2 | |
| 1 | $f(1, 0) = 1$ | $f(2, 2) = 4$ | $f(2, 2) = 4$ | $A^B$ |
| | $f(-2, 2) = 4$ | $f(1, 0) = 1$ | $f(1, 0) = 1$ | |
| | $f(3, 0) = 1$ | $f(-1, 2) = 1$ | $f(-1, 2) = 1$ | |
| | $f(1, 4) = 1$ | $f(2, 3) = 8$ | $f(2, 1) = 2$ | |
| | $f(-3, 2) = 9$ | $f(-3, 2) = 9$ | $f(3, 1) = 3$ | |
| | $f(2, 1) = 2$ | $f(-1, 3) = -1$ | $f(-3, 2) = 9$ | |
| 2 | $f(2, 2) = 5$ | $f(2, 1) = 3$ | $f(2, 1) = 3$ | $AB + 1$ |
| | $f(2, -1) = -1$ | $f(0, -1) = 1$ | $f(1, 0) = 1$ | |
| | $f(2, 3) = 7$ | $f(3, 2) = 7$ | $f(3, 2) = 7$ | |
| | $f(2, 1) = 3$ | $f(2, 2) = 5$ | $f(-1, 3) = -2$ | |
| | $f(-3, 2) = -5$ | $f(2, -1) = -1$ | $f(1, 2) = 3$ | |
| | $f(3, -1) = -2$ | $f(2, 0) = 1$ | $f(1, 4) = 5$ | |
| 3 | $f(3, 2) = 5$ | $f(3, 2) = 5$ | $f(3, 2) = 5$ | $A^2 - 2B$ |
| | $f(2, 1) = 2$ | $f(2, 2) = 0$ | $f(-2, 2) = 0$ | |
| | $f(3, 3) = 3$ | $f(1, 2) = -3$ | $f(4, 2) = 12$ | |
| | $f(4, 3) = 10$ | $f(2, 1) = 2$ | $f(1, 2) = -3$ | |
| | $f(2, 4) = -4$ | $f(3, 3) = 3$ | $f(4, 4) = 8$ | |
| | $f(5, 5) = 15$ | $f(2, 4) = -4$ | $f(-3, 4) = 1$ | |

The subject was told before the experiment that the symbol "$" could represent one of the simple mathematical operators ($+$, $-$, $*$, $/$, $\uparrow$), or some combination of them, but not more than a four-level combination. He was also told that the function might include a constant, for example, $3a - b^2$, but that any constant would be an integer between one and four inclusive. The subject was reminded of these requirements if he violated them. The subject was told in written instructions to say whatever came to his mind and to specify whatever instance he was looking at at any moment during the experiment.

While no time restriction was announced, the time for solving any one problem was limited to 25 minutes. If the subject failed to solve a problem, he was told the correct answer and then was asked to perform the next one. Before beginning work on the test problems, the subject performed three examples of different difficulties. Then, if he had no questions, he was given the first test problem. During the problem-solving period, his verbalizations were recorded, and his solution time was measured. Whenever the subject was silent for long periods, the experimenter asked him to speak up. . . .

*Results.* Two types of data were available for analysis: solution times and verbal protocols. The solution times (see table 24.2) support the theory that subjects begin generating simple hypotheses and then systematically increase the complexity of the hypotheses they try. Solution times increased significantly with the number of operations in the correct hypothesis ($F = 7.88$, $df = 2/30$). In addition, the greater the number of

**Table 24.2**  Median Solution Times in Experiment I[a]

| Number of operators in correct function | Number of alternative functions fitting at least half the instances of data and having no more operators than the correct function | | | |
|---|---|---|---|---|
| | 0 | 1–2 | ≥2 | Ṁ |
| 1 | 52.5 | 97.5 | 137.5 | 95.8 |
| 2 | 180.0 | 180.0 | 382.5 | 247.5 |
| 3 | 502.5 | 675.0 | 690.0 | 622.3 |
| Ṁ | 245.0 | 317.5 | 403.3 | 321.9 |

[a] All times are in seconds.

alternative hypotheses fitting the majority of instances, the longer were the solution times ($F = 3.60$, $df = 2/30$). This latter result indicates that subjects had difficulty abandoning a wrong hypothesis that fit a major portion of the data.

*Protocols.* The protocol data, while less easily quantitized, contain more specific information. One way to test the hypothesis that functions with $i$ operators are generated before functions with $i + 1$ operators is by extracting the order in which the hypotheses were generated from the protocols. Having found the order for the initial problem every subject solved, we computed a correlation between the number of operators in a hypothesis and its rank in the sequence of hypotheses. The correlations ranged from .30 to .95 for the 17 subjects with usable protocols. The average correlation was .65. Such correlations are imperfect and conservative measures since even a perfect ordering would not yield a correlation of unity.

Furthermore, many of the later occurrences of simple hypotheses were repetitions of earlier occurrences. Hence, it seems fair to say that every subject displayed a tendency to generate hypotheses with $i$ operators before those with $i + 1$ operators. One can perceive some regularities in the discrepancies from this procedure. For example, the majority of the discrepancies on the six more difficult problems were characterized by the subject skipping ahead to try a few more complex functions and then backing up to see if he missed a simpler function.

The number of operators was not the only factor that could be seen to influence the order of generation. Within a set of hypotheses having the same number of operators, subjects appeared to have fairly rigid orderings based on a variety of rules: (a) Addition, subtraction, and multiplication were usually tried before exponentiation or division. For example, in only 2 of 51 problem protocols was a hypothesis with division attempted before at least two operators from the set $(+, -, *)$ were tried. Similarly, in only 21 of 51 problems was exponentiation used before two of the others even though exponentiation occurred in the correct solutions of six of the nine problems. (b) Subjects displayed a clear left-to-right bias in generating noncommutative hypotheses. For example, while virtually every subject generated $a/b$, $a - b$, and $a^b$ at some time, only two subjects generated $b/a$ and none generated $b - a$ or $b^a$. (c) If an additive constant was appended to a function, it was usually done right after the function had been negated by an instance.

The frequencies with which particular hypotheses were used adds some other evidence in support of a primary generate-and-test process. If the same clues in the problem were used extensively in the same way by subjects to control what hypotheses were attempted next, one would expect to find many hypotheses used by a large number of subjects even when the hypotheses were quite complex. On the other hand, a less directed generate-and-test process with more random variation between subjects would yield extensive common use among subjects of the simple hypotheses which everyone generates and little agreement on the more complex hypotheses at the bottom of the generation tree. The distribution of noncorrect alternatives supports this latter view. While every subject tried each of five one-operator hypotheses, only two of the 23 three-operator and four-operator hypotheses used were tried by more than one subject.

Two facts about the testing of hypotheses on the data were also apparent from the protocols. In the majority of the cases a subject clearly began testing an hypothesis on either the first (top) instance or on the instance that negated the previous hypothesis. In addition, subjects more frequently retested an hypothesis that had been negated if that hypothesis had been supported by other instances previously. For example, the incorrect alternative solutions to the problems in the rightmost column of table 24.1, each of which was supported by at least three instances of data, were repeatedly tested an average of 1.1 times per subject. On the other hand, other alternatives were repeated only .2 times per subject.

*Discussion.* The analysis of the subjects' solution times and protocols has yielded evidence consonant with the results of other researchers and specific enough to be the basis for a theory of inductive processing.

The central finding was that subjects appear to generate and test hypotheses in an order that is independent of the data, but in accord with several well-specified heuristics. The most important of these was that functions with $i$ operators were tried before functions with more than $i$ operators. The evidence from protocols and solution times also supported the theory that subjects persevere on those wrong hypotheses for which some confirming evidence exists.

## Formal

## Model

## for Induction

We will now present a formal theory for how humans induce mathematical functions from data. This model is based upon the results of our induction experiment and previous concept-learning experiments. Of course, the validity of the theory must be confirmed by demonstrating its predictive utility. In order to specify the details of the model as precisely and unambiguously as possible, it was formulated as a computer simulation model called INDUCT-1. The program was designed so that its performance on a particular problem would vary from run to run in accord with the between-subject variations found. Its average performance should predict the average performance of the group.

The essence of the induction model is that subjects generate and test hypotheses in an order that is mostly independent of the data and partially determined by heuristics. To represent this process for problems of the type studied in Experiment I, the hypotheses that subjects had used in Experiment I were divided into pools. Within each pool the model selects any hypothesis for testing at random *without replacement*. However, the order in which the pools are generated is determined quite precisely in accord with heuristics discovered in Experiment I. Pools of hypotheses having $i$ operators are tried (with a few exceptions to be noted below) before pools with $i + 1$ operators. Within hypotheses having the same number of operators, order is partially determined. In particular, (a) noncommutative operators are used in the order "$a$ op $b$" before they are used as "$b$ op $a$": (b) additive constants are tried before any addition of one of the parameters is attempted; (c) addition, subtraction, and multiplication are tried before

division and exponentiation; and (d) multiplication and exponentiation by a constant are attempted before the same operation by a parameter.

Within a pool the search procedure is to generate a hypothesis at random (without replacement) and test it on either the
instance of data that negated the previous hypothesis or the
first instance in the sequence of data. The hypothesis is tested
successively on every instance in the sequence of data until
it is negated or all instances are fit. If all instances are fit, the
solution has been found; but if a negation is found, the current
hypothesis is immediately rejected and a new one from the
same pool is generated. When the current pool is exhausted,
generation of hypotheses from the next pool begins. One
should recognize that under this procedure the order in which
hypotheses are generated is independent of the problem being
solved. . . .

The program deviates from this basic search algorithm in
only two ways. Some subjects were observed skipping certain
pools and only trying them much later after more complex
hypotheses had been found wanting. To model this phenomena, we associated a probability with each pool that approx-
imated the observed probability that a subject would leave
that pool on any generation trial and skip ahead to the next
pool. At the same time, a probabilistic decision branch was
added so that a subject would (after a minimal time had
elapsed) go back to recheck simpler pools for untested hypotheses. The probabilities were estimated from the protocol
data and then left unchanged during the simulations.

The second variation in the search algorithm was derived
from our discovery that subjects fixated on incorrect hypotheses that fitted a fair portion of the data. Hence, whenever
a hypothesis was discovered that worked for half or more of
the instances, the simulated subjects would retest that hypothesis on all the instances before abandoning the current
pool. These two variations are incorporated in the algorithm.
. . .

For predicting solution times the INDUCT-1 model has four
free parameters representing times for generating and testing
hypotheses. Given that a hypothesis has $n$ operators, the time
for generating it once would be $nT_g$; the time for regenerating
it, after it had been tried at least once, would be $nT_r$; and the

time for testing it on one instance of data would be $T_{t1} + nT_{t2}$. While $T_g$, $T_r$, $T_{t1}$, and $T_{t2}$ are free parameters, they represent times and were constrained to be positive and to obey the relation $T_g > T_r$. In other words, the time to generate a hypothesis initially must be greater than the time needed to regenerate it. . . .

## Implications

## of the Model

The major implication of this study is that humans induce simple mathematical functions by a process that is neither mysterious nor complicated. They generate hypotheses from predetermined pools without replacement. The pools are ordered in accord with a few heuristics independently of the data. Among the most important heuristics is that hypotheses with $i$ operators are generated before those with more than $i$ operators. Once generated, an hypothesis is tested until it is negated. However, negated hypotheses that have been supported by some data may be periodically retested.

While a few readers may have difficulty accepting the central thesis of this model—that induction, the major process in scientific inference, is primarily a generate-and-test process— the conclusion is consistent with most of the other research on induction in humans. One should not be surprised by this assertion unless one believes that men are above the laws of logic that govern every automaton's behavior. Nor should one conclude from this finding that men are blind and unintelligent induction machines. Intelligence in induction can only mean that the problem solver heuristically directs his generate-and-test process to find a solution rapidly.

While few data-dependent heuristics were employed on the problems studied, good data-independent heuristics were used extensively. What evidence is there that more complex heuristics would have yielded quicker solutions to these problems? One cannot conclude that data-dependent heuristics would not be used on more difficult problems because they were not used on the studied problems. On other tree-searching problems, for example, games, the findings have been that humans only employ complex heuristics when the tree becomes too massive for simpler techniques. When that point is reached, however, humans wield quite sophisticated heuristic generate-and-test techniques as has been amply demonstrated in studies of chess and other games (Newell and Simon 1972).

With the appropriate heuristic techniques, a problem solver need not enumerate all or even a major portion of the tree or hypotheses before finding the solution. To one familiar with the recent progress in artificial intelligence, it seems quite reasonable that the major inductive feats of man could be accomplished by an automaton using a heuristically directed generate-and-test process.

Perhaps a more important criticism of man's inductive ability raised by this model is its assertion that he holds on to a theory disconfirmed by negative evidence if it has been confirmed by positive evidence. This tendency has been observed before (Wason 1968a) and appears to pervade even the most sophisticated types of inductive reasoning. As Kuhn (1970) has pointed out, the history of science is rife with cases where anomalies have not led to the rejection of a theory. Old theories are usually rejected only when a new theory is devised.

Why should man possess such inefficient persistence? Part of the reason may lie in man's recognition that he is not an errorless processor and should not trust a single disconfirming computation. The role of noise in the world may add to man's skepticism. Scientists certainly are aware that any single test can be erroneous. One might also argue that it is better to maintain a theory that brings some order to the data than to have no theory at all. Yet anyone who studies the protocols showing how subjects time and again fixate on incorrect functions that fit part of the data must question the adequacy of the above explanations.

Without rejecting any of the above theories we would like to suggest that a subject's persistence in maintaining a false hypothesis may also be due to his misperception of the probability that a false hypothesis could be confirmed. In particular, we propose that the a priori probability of a confirmation of a false hypothesis is perceived by most subjects to be much lower than it is. As a result, the decreases in subjective probability (of a hypothesis being correct) caused by a disconfirmation are erroneously offset by the increases that prior confirmations produced. For example, assume that the a priori probability that some hypothesis is correct is perceived to be .50, and the a priori probability of a confirmation of a correct hypothesis is presumed to be .90: then, if the a priori probability of a confirmation of an incorrect hypothesis were viewed as only .10, the computed probability that some hypothesis is correct following two confirmations and one disconfirmation would be (according to Bayes' law) .90. On the other hand, if the a priori probability of a confirmation of an incorrect hypothesis were perceived (more realistically) to be .50, the computed probability of correctness after two confirmations and one disconfirmation would be only .39. Thus, too low an a priori estimate of the likelihood that a false hypothesis could be confirmed can lead a subject to believe that a negated hypothesis has a higher probability of being correct than one that has not been tried.

# Selection 25

DONALD GERWIN
and PETER NEWSTED

## Strategies

## in Mathematical

## Induction

AT LEAST TWO general thinking strategies are employed in function finding: generate and test, and heuristic search. Their appearance in this newly explored task environment further testifies to their position as fundamental cognitive processes. The Heusmann and Cheng (HC) model's version of generate and test, (1) generating an alternative at random given certain ordering restrictions and (2) testing the alternative, is purely logical in nature, is associated with an avoidance of the data, and does not involve much elimination of alternatives. The Heuristic Dendral (HD: Buchanan, Sutherland, and Feigenbaum 1969; Buchanan and Lederberg, 1971) and DG (Gerwin 1974) models, based on heuristic search, consist of (1) determination of the data's significant features, (2) inferring a general class of hypotheses, (3) generation of specific hypotheses consistent with the general class, (4) testing predictions in order to choose a specific hypothesis, and (5) possible refinement on the basis of comparing actual and predicted values. This strategy is characterized by pattern finding and logical

Excerpt from D. Gerwin and P. Newsted, A comparison of some inductive inference models. *Behavioral Science* (1977), 22:1–11. Reprinted by permission of James Grier Miller, M.D., Ph.D., Editor.

elements, extensive data utilization and concentration on relatively few alternatives. It implies that many cognitive models, based on heuristic search but not incorporating pattern finding aspects, present too narrow a view of thinking. This conclusion is reinforced by recent results from chess studies (Chase and Simon 1973). Further, the expanded view is more compatible with accounts of scientific inference offered by Hanson (1958) and others.

The specific brand of heuristic search in the second strategy appears to have generality beyond function finding. Consider, for example, sequence extrapolation which depends upon inferring the rule from which a letter sequence is developed. Research conducted by Kotovsky and Simon (1973) and Simon and Kotovsky (1963) indicates that subjects first scan the sequence for a relation that repeats itself at regular intervals or that is interrupted at regular intervals (Step 1 above).[1] This facilitates inferring the sequence's period which acts as a constraint on the rules subsequently considered (Step 2). The details of a rule's composition are determined by finding relations among successive letters in the period or between letters in corresponding positions of successive periods (Step 3). Next, the rule is tested against the data (Step 4). The second strategy also should be useful in understanding scientific inference as nontrivial aspects appear in various accounts of discovery (Gerwin 1974), and the first four steps are employed in heuristic DENDRAL which solves scientific problems.

Three features of the function finding environment and two characteristics of inference makers conceivably account for the use of the HC or DG strategy. Random error, a graphical representation, sophisticated technology, experience with similar problems, and a geometrical cognitive style should produce a heuristic search. Deterministic data, a list representation, unsophisticated instrumentation, lack of familiarity with the problem and a numerical cognitive style should lead to generate and test. Since only two of these factors have been studied here, the effects of the other three remain conjecture. Of particular concern in our attempts to understand scientific inference is study of the way in which instrumentation influences thinking processes. Sophisticated technology such as computers may facilitate heuristic search by augmenting external memory or speeding up calculations. Telescopes may accomplish the same purpose by enhancing pattern-finding capacities.

Our empirical efforts to relate experience and representation to type of thinking strategy was based more on end result measures than upon process oriented measures. Number of hypotheses and number of hypotheses per unit time fall into the former category; the relational measure is somewhat more process oriented. While these measures should distinguish between generate and test or heuristic search they cannot reveal much about the sequence of detailed steps in each strategy. This requires in-depth protocol analysis and perhaps model development which is the aim of subsequent research.

High experience and a graphical representation appear to be associated with heuristic search, and low experience and a list representation with generate and test. Our conclusion is based on the former combination's association with fewer hypotheses, fewer hypotheses per unit time, and higher relatedness of hypotheses but not on all problems. Further, a better case was made for experience than representation. While the observed effects were not pervasive, we believe this is due in part to the nature of Heusmann and Cheng's problem environment. Experience effects may have been mitigated by the ease with which problems could be solved. It took much less time to solve them than the ones in the DG study. Representation effects may have been lessened because only six coordinates were given for functions of two variables.

What is the process by which high experience and a graphical representation lead to heuristic thinking? We can offer some tentative answers. More experienced inference makers are better able to perceive meaningful patterns in the data. They are also more likely to have mathematical functions stored in long-term memory which are evoked by the pattern information. Evidence from the chess task environment (Chase and Simon 1973) reinforces this view. Chess masters are able to perceive the current board position as a few familiar patterns of pieces. The patterns are associated in long-term memory with a number of plausible moves. Selection of a move occurs after examining the consequences of these alternatives. Less experienced players retain smaller and fewer patterns in long-term memory and therefore have more difficulty deciding on a good move.

A graphical representation encourages heuristic thinking because it facilitates the organization of information into useful patterns and because it makes the pattern information more apparent. The organizing function is quite close to the role seen by Hadamard (1945) for geometrical representations in mathematical thinking. In his view, they insure that useful hookups of ideas remain apparent. These observations are a modest beginning to our understanding the significance of Einstein's thought experiments involving speeding railroad cars with mirrors on ceiling and floor, Watson and Crick's Tinker-Toy-type model, Darwin's irregularly branching tree, and countless other examples.

Finally, our study underscores the growing realization that process oriented cognitive research has devoted too little time to the effects of task environment characteristics and virtually no time to the influence of individual differences. With respect to the latter category prior knowledge and experience has been suggested as one of the most potentially significant factors (Norman 1976). Our empirical results provide some support for this contention. We have also demonstrated that the professional background of faculty members is a variable worthy of further consideration. Science professors behaved more heuristically than engineering professors. However, at this point it is not clear whether professional background is related to experience or is an independent factor. With respect to scientific inference, the potential influence of individual differences offers a guideline. In the psychology of discovery, it is best to speak not of the way in which a discovery is made, but of the way in which a given scientist makes a discovery.

# Note

1. Gerwin and Newsted's elaboration of these steps is not provided in this selection. [Eds.]

Part VI ─────────────────────────

# Statistics
# in Scientific
# Thinking

# Introduction

IN 1889 FRANCIS Galton reverentially extolled the power of statistics to yield regularity from apparent chaos (selection 26). Influenced by the power of Charles Darwin's theory of evolution, Galton felt that analysis of the variation of human traits would produce psychological laws of great utility and generality (Galton 1883; Pearson 1914–1930). At about the same time statistical models emerged in physics and chemistry and led to the triumphs of statistical mechanics (Merz 1896). Probabilistic concepts were also of central importance to the emergence of quantum mechanics in the twentieth century, though their use as fundamental constructs (rather than as analytic tools) was problematic for many scientists. Albert Einstein, for one, refused to accept such models, asserting that in some respects, the quantum theory " . . . is so very contrary to my scientific instinct that I cannot forego the search for a more complete conception" (Einstein 1977:91).

It should be clear, then, that statistics can be used in science in one of two ways: as a conceptual model of a natural process, or as a tool to unravel a complex natural process. Our concern in this section is only with statistics as a tool for analyzing data, especially as it is used in the so-called "soft sciences." There is an extensive literature on the use of statistical methods for making inferences from research data, and controversy over, for example, the use of classical statistical inference, or null hypothesis testing, vs. Bayesian methods. However, our major interest is in the cognitive operations that may be elicited by situations in which there is considerable error. Are there fundamental differences between scientific activity in "soft" areas like psychology and in more highly developed sciences, differences related to the relative amount of error in the data? That such differences exist has been suggested by Paul Meehl (selection 27), who has argued that the heavy use of statistics as an analytical tool in psychology results in a unique relationship between data and theory, a relationship very different than that in physics. This is a claim that deserves serious attention.

What, exactly, is meant by the term "error"? Under what circumstances are statistics appropriate to deal with it? When variation itself is the object of study, as for example, when zoologists examine individual characteristics within a species, or physicists examine the distribution of velocities within a gas, then it is appropriate to use statistical indices to describe directly the phenomena of interest. When, however, variation is due to imprecision of measurement or to intrinsic properties which are *not* under study, the situation is different.

In the typical report of a psychological investigation, we find graphs of mean performance, with accompanying text describing the level of significance at which the null hypothesis was rejected. The significance test constitutes a "decision rule" which governs whether or not the experimental outcome is worthy of further consideration. To be sure, the way in which the investigator formally *reports* his or her use of a significance test may or may not correspond to the informal way in which an experimental outcome is considered (Medawar 1964). Even so, it is clear that statistical tests do constitute a decision rule (sometimes a very rigid one!) for journal editors (Mahoney 1976).

In most instances, significance tests involve computation of a ratio, the denominator of which includes "error," that is, variation which cannot be accounted for by treatment effects. What are the components of this error? There is always some theoretically irreducible amount of measurement error. No matter how elegant the measuring instrument, no matter how well understood the object of measurement, there is and will always be the $\pm$ appended to the number that results from the measurement operation. In sciences where the $\pm$ is small, it is often numerically specified. In sciences where the $\pm$ is embarrassingly large, it is usually left to the imagination of the reader. Many problems in "soft" science rests on the difficulty of reducing this sort of error to reasonable levels.

Another source of difficulty stems from uncertainty about what is being measured. There is considerable disagreement and public debate over the construct validity of even the very best psychological tests, tests of general intelligence. On the other hand, experimenters often worry too little about what their "dependent variables" measure. Different investigators may select indicator responses for the same construct which have unknown relations to one another. There is often, then, great imprecision in the relationship between the conceptual status of a term in a hypothesis and the operational definition of that term; too often, in behavioral research, little attention is paid to the coordinating definitions.

The importance of this issue has been humorously captured in an unpublished paper by "Isidore Nabi" (no date), which described a statistical approach to mechanics, based upon statistically derived regularities. For example, "Drowning men moved upward 3/7 of the time, and downward 4/7" and "Apples did indeed drop. A stochastic model showed that the probability of apple drop increases through the summer" (p. 3). The resultant "laws" make an effective point: "Bodies at rest remain at rest with a probability of .96 per hour, and objects in motion tend to continue in motion with a probability of .86. . . . For 95% ± .06 of all actions, there is a corresponding reaction at an angle of 175 ± 6° from the first. . . " (p. 4). In a more serious vein, a recent statistics textbook (Freedman, Pisani, and Purves 1978) showed that, for a given sample of rectangles, the correlation between perimeter and area was +0.87.[1] The point of both examples should not be restricted to classrooms or jokes: No scientist can expect statistical analysis to substitute for a powerful conceptualization.

How can one know that the "right" variables have been chosen for study? Through "tacit knowing" (Polanyi 1966)? Or, recalling Austin's (1978) account of his enzyme research, through sheer luck? Garner, Hake, and Eriksen (1956) proposed that psychologists use a procedure called "converging operations," in which each suspected prime variable must be related to a multiplicity of converging measures on both the stimulus and the response side. The procedure has proved extremely powerful in perceptual research (e.g., Schuck 1973), but has been slow to find acceptance elsewhere (Garner 1974). Converging operations is a hard doctrine, but the modern psychologist needs to bear in mind that no science will progress beyond "mere empiricism" until it can guarantee the measurement of fundamental units.

One source of interesting variation in psychological research lies in individual differences. It is likely that one randomly selected hydrogen atom is more similar to another randomly selected hydrogen atom than some randomly selected human being is to another randomly selected human being. Thus, when physicists or chemists talk about hydrogen atoms, they are communicating more clearly—with less error—than when philosophers, psychologists, or sociologists talk about people. Individual differences may be endemic to all science, but they are a massive source of "error" in the behavioral sciences. Cronbach's paper (1957; selection 28) deals precisely with the pervasive and serious problem of what to do with the large amounts of variance due to individual differences. The answers given by correlational psychologists and experimentalists are very different from one another. Correlationalists focus upon individual differences, whereas experimentalists almost always relegate such differences between people to an error term—to an everlasting netherworld of unanalyzed, unexplained anonymity.

The difference in conceptualization of error is related to differences in the kinds of statistical analyses used. Between 1930 and 1950, analysis of variance emerged as the dominant analytic tool of experimental psychology (Rucci and Tweney 1980) whereas correlational psychologists have relied heavily on such statistical techniques as factor analysis and multiple regression. The latter approach has been justified on the grounds of practical utility. Indeed, if a paper-and-pencil test can be shown to correlate with, say, job performance at the 0.30 level, then it may not be hard to justify use of the test on economic grounds. Use of the test as a selection device will reduce the costs of a personnel program. It is unfortunate that the practical utility of such devices is often conflated with scientific utility. In fact, a test with validity of 0.30 leaves over 90% of the variation unaccounted for.

Psychologically, error may be like a Rorschach inkblot. Given sufficiently ambiguous data one may be able to "perceive" all sorts of patterns. This "Rorschach effect" shows up in other ways, too. The confluence of substantial error (not just error of measurement, remember, but "unexplained variance") with the use of null hypothesis testing makes the confirmation of favored substantive hypotheses relatively easy, and relatively uninformative. Conversely, if one sets out to falsify someone else's hypothesis, sufficient doses of error make it relatively easy to set up a "straw man" test of that hypothesis, and to disconfirm it in some uninteresting way. But *sharp* disconfirmations of scientific, psychological hypotheses are virtually precluded by the presence of large amounts of error. Selection 30, which we have written for this volume, deals further with the relation of strong inference and null hypothesis testing.

The sources of variation in behavior are legion, and the tendency to lump them together as error must lead to a loss of many interesting phenomena, and to an inappropriate focus upon many uninteresting ones. The loss of information that stems from averaging over heterogeneous function forms, as in growth curves and learning curves, is well known. But the potential role of error as a contributor to the noncumulativeness of psychology has not been widely explored. Surely it is one of the contributing factors to what Lakatos (1970) would call the degenerateness of the research programs in many areas of psychology. The result is a succession of theories, each of which deals not with new phenomena, but with the failures of earlier theories. Such degeneration may be due in part to the fact that many psychological effects are not "robust"—that is, they are small in magnitude relative to other unknown and uncontrolled causal factors. Thus, apparently trivial changes in independent variables may cause phenomena to change in unpredicted ways. Theoretically important effects too often turn out to be due to experimenter demands (Chapanis and Chapanis 1964), or to procedural artifacts (Cartwright 1973). Meehl (selection 27) has written an incisive critique of soft psychology, in which he explores such problems.

We do not wish to paint too bleak a picture. We certainly do not agree with one widely held belief that a science of psychology is impossible since human beings have free will (the choices of which must certainly translate into unexplained variance). One can reconcile an appreciation of individual differences, even to the point of acknowledgment of the uniqueness of the individual, with the idea that a science of psychology is possible and desirable. Every hydrogen atom is unique in some sense. Every oak tree is more obviously unique. But physics and biology progress nevertheless. There is at present no resolution to the problem raised by individual differences for the psychologist. In any case the consequences of extreme individual differences and of error in general provide an intriguing scientific problem for the psychology of scientific thinking.

Probably the most extensive series of systematic investigations into people's ability to make inferences from statistical data is the work of Brehmer (1979) and his associates. Brehmer presents data to subjects, usually in the form of bar graphs. The data are based on a number of "cues" and can be used to predict a "criterion." The subject's task is to infer the function forms and composition rules, and to use the information to predict the criterion values based on the cue sets. The research resembles that of Huesmann and Cheng (selection 24) and Gerwin and Newsted (selection 25), in that subjects must determine a function form. However, Brehmer's subjects must make inferences in a statistical context. He has found that performance in such tasks can improve with experience, and that such improvement appears to be based on the construction of new hypothesis sets relevant to the cue-criterion combinations. Hoffman and Slovic (personal communication) are currently conducting experiments in which such tasks are presented with either numerical cues and criteria, or with tones—nonnumerical transforms of the cues and criteria. They are finding substantial differences in the cognitive processes used in the two task situations. The research is clearly relevant to the question of how scientists use numerical information.

We are, of course, merely introducing questions about the influence of error on scientific thinking. In some domains the question is moot—if the data are sloppy, the experimental physicist cleans up the experiment to get better data. This is often done by inventing new methods, or new instruments. In other domains, including much of psychology, the problem is more serious. In psychology the answer has too often been to run more subjects, thereby increasing the statistical power of an analysis of variance design. But that begs the question that we have been reiterating; what are the long-term effects for psychological science of treating individual differences as error?

Where the use of mathematics aids one in overcoming the limits of the human information-processing system, perhaps some uses of statistics magnify those limitations. The final selection, by Tversky and Kahnemann, is an elegant, simple empirical study which probes scientists' fallible intuitions about error. Having opened with Galton's paean to the law of large numbers, we close with Tversky and Kahnemann's demonstration that too often scientists treat small numbers as though they are large ones.

# Note

1. It is interesting to note here that five-year-old children judge the area of rectangles using a height plus width rule, rather than the correct height times width rule (Anderson & Cuneo 1978).

# Selection 26 _____

## FRANCIS GALTON

# The Charms

# of Statistics

### The Charms of Statistics

IT IS DIFFICULT to understand why statisticians commonly limit their inquiries to Averages, and do not revel in more comprehensive views. Their souls seems as dull to the charm of variety as that of the native of one of our flat English counties, whose retrospect of Switzerland was that, if its mountains could be thrown into its lakes, two nuisances would be got rid of at once. An Average is but a solitary fact, whereas if a single other fact be added to it, an entire Normal Scheme, which nearly corresponds to the observed one, starts potentially into existence.

Some people hate the very name of statistics, but I find them full of beauty and interest. Whenever they are not brutalized, but delicately handled by the higher methods, and are warily interpreted, their power of dealing with complicated phenomena is extraordinary. They are the only tools by which an opening can be cut through the formidable thicket of difficulties that bars the path of those who pursue the science of man. . . .

### Order in Apparent Chaos

I know of scarcely anything so apt to impress the imagination as the wonderful form of cosmic order expressed by the "Law

Excerpts from F. Galton, *Natural Inheritance*. London: Macmillan & Co., 1889, pp. 62; 66.

of Frequency of Error." The law would have been personified by the Greeks and deified, if they had known of it. It reigns with serenity and in complete self-effacement amidst the wildest confusion. The huger the mob, and the greater the apparent anarchy, the more perfect is its sway. It is the supreme law of Unreason. Whenever a large sample of chaotic elements are taken in hand and marshalled in the order of their magnitude, an unsuspected and most beautiful form of regularity proves to have been latent all along. The tops of the marshalled row form a flowing curve of invariable proportion; and each element, as it is sorted into place, finds, as it were, a preordained niche, accurately adapted to fit it. If the measurement at any two specified Grades in the row are known, those that will be found at every other Grade, except towards the extreme ends, can be predicted in the way already explained, and with much precision.

# Selection 27

PAUL E. MEEHL

# How Psychology

# Differs

# from Physics

THE PURPOSE OF the present paper is not so much to pro-
pound a doctrine or defend a thesis (especially as I should be
surprised if either psychologists or statisticians were to disagree with whatever in the nature of a "thesis" it advances),
but to call the attention of logicians and philosophers of science to a puzzling state of affairs in the currently accepted
methodology of the behavior sciences which I, a psychologist,
have been unable to resolve to my satisfaction. The puzzle,
sufficiently striking (when clearly discerned) to be entitled to
the designation "paradox," is the following: *In the physical
sciences, the usual result of an improvement in experimental
design, instrumentation, or numerical mass of data, is to increase the difficulty of the "observational hurdle" which the
physical theory of interest must successfully surmount; whereas,
in psychology and some of the allied behavior sciences, the
usual effect of such improvement in experimental precision
is to provide an easier hurdle for the theory to surmount.*
Hence what we would normally think of as improvements in
our experimental method tend (when predictions materialize)
to yield *stronger* corroboration of the theory in physics, since
to remain unrefuted the theory must have survived a more

Excerpts from P. E. Meehl, Theory-testing in psychology and physics: A methodological paradox. *Philosophy of Science* (1967), 34:103–115. Reprinted by permission.

difficult test; by contrast, such experimental improvement in psychology typically results in a *weaker* corroboration of the theory, since it has now been required to survive a more lenient test (Popper 1959, 1962; Bunge 1964). . . .

But this is not the worst of the story. Inadequate appreciation of the extreme weakness of the test to which a substantive theory $T$ is subjected by merely predicting a directional statistical difference $d > 0$ is then compounded by a truly remarkable failure to recognize the logical asymmetry between, on the one hand, (formally invalid) "confirmation" of a theory via affirming the consequent in an argument of form: $[T \supset H_1, H_1,$ infer $T]$, and on the other hand the deductively tight *refutation* of the theory *modus tollens* by a falsified prediction, the logical form being: $[T \supset H_1, \sim H_1,$ infer $\sim T]$.

While my own philosophical predilections are somewhat Popperian, I daresay any reader will agree that no full-fledged Popperian philosophy of science is presupposed in what I have just said. The destruction of a theory *modus tollens* is, after all, a matter of deductive logic; whereas the "confirmation" of a theory by its making successful predictions involves a much weaker kind of inference. This much would be conceded by even the most anti-Popperian "inductivist." The writing of behavior scientists often reads as though they assumed—what it is hard to believe anyone would explicitly assert if challenged—that successful and unsuccessful predictions are practically on all fours in arguing for and against a substantive theory. Many experimental articles in the behavioral sciences, and, even more strangely, review articles which purport to survey the current status of a particular theory in the light of all available evidence, treat the confirming instances and the disconfirming instances with equal methodological respect, as if one could, so to speak, "Count noses," so that if a theory has somewhat more confirming than disconfirming instances, it is in pretty good shape evidentially. Since we know that this is already grossly incorrect on purely formal grounds, it is a mistake *a fortiori* when the so-called "confirming instances" have themselves a prior probability, as argued above, somewhere in the neighborhood of ½, quite apart from any theoretical considerations.

Contrast this bizarre state of affairs with the state of affairs in physics. While there are of course a few exceptions, the usual situation in the experimental testing of a physical theory at least involves the prediction of a *form* of function (with parameters to be fitted); or, more commonly, the prediction of a quantitative magnitude (point-value). Improvements in the accuracy of determining this experimental function-form or point-value, whether by better instrumentation for control and making observations, or by the gathering of a larger number of measurements, has the effect of *narrowing* the band of tolerance about the theoretically predicted value. What does this mean in terms of the significance-testing model? It means: *In physics, that which corresponds, in the logical structure of statistical inference, to the old-fashioned point-null hypothesis $H_0$ is the value which flows as a consequence of the substantive theory $T$*; so that an increase in what the statistician would call "power" or "precision" has the methodological effect of stiffening the experimental test, of setting up a more difficult observational hurdle for the theory $T$ to surmount. Hence, in physics the effect of improving precision or power is that of *decreasing* the prior probability of a successful experimental outcome if the theory lacks verisimilitude, that is, precisely the reverse of the situation obtaining in the social sciences.

As techniques of control and measurement improve or the number of observations increases, the methodological effect in physics is that a successful passing of the hurdle will mean a greater increment in corroboration of the substantive theory; whereas in psychology, comparable improvements at the experimental level result in an empirical test which can provide only a progressively weaker corroboration of the substantive theory. . . .

This methodological paradox would exist for the psychologist even if he played his own statistical game fairly. The reason for its existence is obvious, namely, that most psychological theories, especially in the so-called "soft" fields such as social and personality psychology, are not quantitatively developed to the extent of being able to generate point-predictions. In this respect, then, although this state of affairs is surely unsatisfactory from the methodological point of view, and stands in great need of clarification (and, hopefully, of constructive suggestions for improving it) from logicians and philosophers of science, one might say that it is "nobody's fault," it being difficult to see just how the behavior scientist could extricate himself from this dilemma without making unrealistic attempts at the premature construction of theories which are sufficiently quantified to generate point-predictions for refutation.

However, there are five social forces and intellectual traditions at work in the behavior sciences which make the research consequences of this situation even worse than they may have to be, considering the state of our knowledge. In addition to (a) failure to recognize the marked evidential asymmetry between confirmation and *modus tollens* refutation of theories, and (b) inadequate appreciation of the extreme weakness of the hurdle provided by the mere directional significance test, there exists among psychologists (c) a fairly widespread tendency to report experimental findings with a liberal use of *ad hoc* explanations for those that didn't "pan out." This last methodological sin is especially tempting in the "soft" fields of (personality and social) psychology, where the profession highly rewards a kind of "cuteness" or "cleverness" in experimental design, such as a hitherto untried method for inducing a desired emotional state, or a particularly "subtle" gimmick for detecting its influence upon behavioral output. The methodological price paid for this highly-valued "cuteness" is, of course, (d) an unusual ease of escape from *modus tollens* refutation. For, the logical structure of the "cute" component typically involves use of complex and rather dubious auxiliary assumptions, which are required to mediate the original prediction and are therefore readily available as (genuinely) plausible "outs" when the prediction fails. It is not unusual that (e) this *ad hoc* challenging of auxiliary hypotheses is repeated in the course of a series of related experiments, in which the auxiliary hypothesis involved in Experiment 1 (and challenged *ad hoc* in order to avoid the latter's *modus tollens* impact on the theory) becomes the focus of interest in Experiment 2, which in turn utilizes further plausible but easily challenged auxiliary hypotheses, and so forth. In this fashion a zealous and clever investigator can slowly wend his way through a tenuous nomological network, performing a long series of related experiments which appear to the uncritical reader as a fine example of "an integrated research program," *without ever once refuting or corroborating so much as a single strand of the network.* Some of the more horrible examples of this process would require the combined analytic and reconstructive efforts of Carnap, Hempel, and Popper to unscramble the logical relationships of theories and hypotheses to evidence. Meanwhile our eager-beaver researcher, undismayed by logic-of-science considerations and relying blissfully on the "exactitude" of modern statistical hypothesis-testing, has produced a long publication list and been promoted to a full professorship. In terms of his contribution to the enduring body of psychological knowledge, he has done hardly anything. His true position is that of a potent-but-sterile intellectual rake, who leaves in his merry path a long train of ravished maidens but no viable scientific offspring.

Detailed elaboration of the intellectual vices (a)–(c) and their scientific consequences must be left for another place, as must constructive suggestions for how the behavior scientist can improve his situation. My main aim here has been to call the attention of logicians and philosophers of science to what, as I think, is an important and difficult problem for psychology, or for any science which is largely in a primitive stage of development such that its theories do not give rise to point-predictions.

LEE J. CRONBACH

# Experimental

# and Correlational

# Psychology

IN THE BEGINNING, experimental psychology was a substitute for purely naturalistic observation of man-in-habitat. The experimenter placed man in an artificial, simplified environment and made quantitative observations of his performance. The initial problem was one of describing accurately what man felt, thought, or did in a defined situation. Standardization of tasks and conditions was required to get reproducible descriptions. All experimental procedures were tests, all tests were experiments. Kraepelin's continuous-work procedure served equally the general study of fatigue and the diagnosis of individuals. Reaction time was important equally to Wundt and to Cattell.

The distinctive characteristic of modern experimentation, the statistical comparison of treatments, appeared only around 1900 in such studies as that of Thorndike and Woodworth on transfer. The experimenter, following the path of Ebbinghaus, shifted from measurement of the average mind to measuring the effect of environmental change upon success in a task (Woodworth 1918). Inference replaced estimation: the mean and its probable error gave way to the critical ratio. The stand-

Excerpt from L. J. Cronbach, The two disciplines of scientific psychology. *American Psychologist* (1957), 12:671–84. Copyright © 1957 by the American Psychological Association. Reprinted by permission.

ardized conditions and the standardized instruments remained, but the focus shifted to the single manipulated variable, and later, following Fisher, to multivariate manipulation. The experiment thus came to be concerned with between-treatments variance. I use the word "treatment" in a general sense; educational and therapeutic treatments are but one type. Treatment differences are equally involved in comparing rats given different schedules of reinforcement, chicks who have worn different distorting lenses, or social groups arranged with different communication networks.

The second great development in American experimental psychology has been its concern with formal theory. At the turn of the century, theory ranged far ahead of experiment and made no demand that propositions be testable. Experiment for its part, was willing to observe any phenomenon whether or not the data bore on theoretical issues. Today, the majority of experimenters derive their hypotheses explicitly from theoretical premises and try to nail their results into a theoretical structure. This deductive style has its undeniable defects, but one can not question the net gains from the accompanying theoretical sophistication. Discussions of the logic of operationism, intervening variables, and mathematical models have sharpened both the formulation of hypotheses and the interpretation of results.

Individual differences have been an annoyance rather than a challenge to the experimenter. His goal is to control behavior, and variation within treatments is proof that he has not succeeded. Individual variation is cast into that outer darkness known as "error variance." For reasons both statistical and philosophical, error variance is to be reduced by any possible device. You turn to animals of a cheap and short-lived species, so that you can use subjects with controlled heredity and controlled experience. You select human subjects from a narrow subculture. You decorticate your subject by cutting neurons or by giving him an environment so meaningless that his unique responses disappear (cf. Harlow 1953). You increase the number of cases to obtain stable averages, or you reduce N to 1, as Skinner does. But whatever your device, your goal in the experimental tradition is to get those embarrassing differential variables out of sight.

The correlational psychologist is in love with just those

variables the experimenter left home to forget. He regards individual and group variations as important effects of biological
and social causes. All organisms adapt to their environments,
but not equally well. His question is: what present characteristics of the organism determine its mode and degree of
adaptation?

Just as individual variation is a source of embarrassment to
the experimenter, so treatment variation attenuates the results
of the correlator. His goal is to predict variation within a treatment. His experimental designs demand uniform treatment for
every case contributing to a correlation, and treatment variance
means only error variance to him.

Differential psychology, like experimental, began with a
purely descriptive phase. Cattell at Hopkins, Galton at South
Kensington, were simply asking how much people varied.
They were, we might say, estimating the standard deviation
while the general psychologists were estimating the central
tendency.

The correlation coefficient, invented for the study of hereditary resemblance, transformed descriptive differential research into the study of mental organization. What began as
a mere summary statistic quickly became the center of a whole
theory of data analysis. Murphy's words, written in 1928, recall the excitement that attended this development:

> The relation between two variables has actually been found
> to be stable in other terms than those of experiment. . . . [More-
> over,] Yule's method of "partial correlation" has made possible
> the mathematical "isolation" of variables which cannot be iso-
> lated experimentally. . . .[Despite the limitations of correla-
> tional methods,] what they have already yielded to psychology
> . . . is nevertheless of such major importance as to lead the
> writer to the opinion that the only twentieth-century discovery
> comparable in importance to the conditioned-response method
> is the method of partial correlations. (Murphy 1930:410)

Today's students who meet partial correlation only as a mo-
mentary digression from their main work in statistics may find
this excitement hard to comprehend. But partial correlation
is the starting place for all of factor analysis.

Factor analysis is rapidly being perfected into a rigorous
method of clarifying multivariate relationships. Fisher made
the experimentalist an expert puppeteer, able to keep untan-

gled the strands to half-a-dozen independent variables. The correlational psychologist is a mere observer of a play where Nature pulls a thousand strings; but his multivariate methods make him equally an expert, an expert in figuring out where to look for the hidden strings.

His sophistication in data analysis has not been matched by sophistication in theory. The correlational psychologist was led into temptation by his own success, losing himself first in practical prediction, then in a narcissistic program of studying his tests as an end in themselves. A naive operationism enthroned theory of test performance in the place of theory of mental processes. And premature enthusiasm exalted a few measurements chosen almost by accident from the tester's stock as the ruling forces of the mental universe.

In former days, it was the experimentalist who wrote essay after anxious essay defining his discipline and differentiating it from competing ways of studying mind. No doubts plagued correlationists like Hall, Galton, and Cattell. They came in on the wave of evolutionary thought and were buoyed up by every successive crest of social progress or crisis. The demand for universal education, the development of a technical society, the appeals from the distraught twentieth-century parent, and finally the clinical movement assured the correlational psychologist of his great destiny. Contemporary experimentalists, however, voice with ever-increasing assurance their program and social function; and the fact that tonight you have a correlational psychologist discussing disciplinary identities implies that anxiety is now perched on *his* windowledge.

PAUL E. MEEHL

# Sir Karl,

# Sir Ronald,

# and Soft Psychology

I HAD SUPPOSED that the title gave an easy tipoff to my topic, but some puzzled reactions by my Minnesota colleagues show otherwise, which heartens me because it suggests that what I am about to say is not trivial and universally known. The two knights are Sir Karl Raimund Popper (1959, 1962, 1972; Schilpp 1974) and Sir Ronald Aylmer Fisher (1956, 1966, 1967), whose respective emphases on subjecting scientific theories to grave danger of refutation (that's Sir Karl) and major reliance on tests of statistical significance (that's Sir Ronald) are, at least in current practice, not well integrated—perhaps even incompatible. If you have not been accustomed to thinking about this incoherency, and my remarks lead you to do so (whether or not you end up agreeing with me), this article will have served its scholarly function.

I consider it unnecessary to persuade you that most so-called "theories" in the soft areas of psychology (clinical, counseling, social, personality, community, and school psychology) are scientifically unimpressive and technologically worthless. Documenting that statement would of course require a considerable amount of time, but you can quickly get the flavor

Excerpts from P. E. Meehl, Theoretical risks and tabular asterisks: Sir Karl, Sir Ronald, and the slow progress of soft psychology. *Journal of Consulting and Clinical Psychology* (1978), 46:806–34. Copyright © 1978 by the American Psychological Association. Reprinted by permission.

by having a look at Braun (1966); Fiske (1974); Gergen (1973); Hogan, DeSoto, and Solano (1977); McGuire (1973); Meehl (1960/1973a, 1959/1973b); Mischel (1977); Schlenker (1974); Smith (1973); and Wiggins (1973). These are merely some high visible and forceful samples; I make no claim to bibliographic completeness on the large theme of "What's wrong with 'soft' psychology." A beautiful hatchet job, which in my opinion should be required reading for all PhD candidates, is by the sociologist Andreski (1972). Perhaps the easiest way to convince yourself is by scanning the literature of soft psychology over the last 30 years and noticing what happens to theories. Most of them suffer the fate that General MacArthur ascribed to old generals—they never die, they just slowly fade away. In the developed sciences, theories tend either to become widely accepted and built into the larger edifice of well-tested human knowledge or else they suffer destruction in the face of recalcitrant facts and are abandoned, perhaps regretfully as a "nice try." But in fields like personology and social psychology, this seems not to happen. There is a period of enthusiasm about a new theory, a period of attempted application to several fact domains, a period of disillusionment as the negative data come in, a growing bafflement about inconsistent and unreplicable empirical results, multiple resort to ad hoc excuses, and then finally people just sort of lose interest in the thing and pursue other endeavors. . . .

I do not think that there is any dispute about this matter among psychologists familiar with the history of the other sciences. It is simply a sad fact that in soft psychology theories rise and decline, come and go, more as a function of baffled boredom than anything else; and the enterprise shows a disturbing absence of that *cumulative* character that is so impressive in disciplines like astronomy, molecular biology, and genetics. . . .

The word *nomological* is in soft psychology at best an extension of meaning and at worst a misleading corruption of the logician's terminology. Originally it designated strict laws as in W. E. Johnson's (1921/1964) earlier use of "nomic necessity" (p. 61). The lawlike relationships we have to work with in soft psychology are rarely (never?) of this strict kind, errors of measurement aside. Instead, they are correlations, tendencies, statistical clusterings, increments of probabilities, and altered stochastic dispositions. The ugly neologism *sto-chastological* (as analogue to *nomological*) is at least shorter than the usual "probabilistic relation" or "statistical depend-ence," so I shall adopt it. We are so accustomed to our im-mersion in a sea of stochastologicals that we may fail to notice what a terrible disadvantage this sort of probabilistic law net-work puts us under, both as to the clarity of our concepts and, more importantly, the testability of our theories. . . .

When the observational corroborators of the theory consist wholly of percentages, crude curve fits, correlations, signifi-cance tests, and distribution overlaps, it is difficult or impos-sible to see clearly when a given batch of empirical data refutes a theory or even when two batches of data are (in any inter-esting sense) "inconsistent." All we can usually say with quasi-certainty is that context-dependent statistics should *not* be numerically identical in different studies of the same prob-lem. (A dramatic recent example of this was the discovery that some of Sir Cyril Burt's correlation coefficients were *too con-sistent* to have been derived from the different tests and pop-ulations that he reported!)

In heading [not shown—Eds.] this section "Context-Depend-ent Stochastologicals," I mean to emphasize the aspect of this problem that seems to me most frustrating to our theoretical interests, namely, that the statistical dependencies we observe are always somewhat, and often strongly, dependent on the institution-cum-population setting in which the measure-ments were obtained. Lacking a "complete (causal) theory" of what influences what, *and how much*, we simply cannot com-pute expected numerical changes in stochastic dependencies when moving from one population or setting to another. Some-times we cannot even rationally predict the direction of such changes. If the difference between two Pearson correlations were safely attributable to random sampling fluctuation alone, we could use the statistician's standard tools to decide whether Jones's study "fails to replicate" Smith's. But the usual situation is not one of simple cross-validation shrinkage (or "boostage")—rather, it involves the validity generalization problem. For this, there are no standard statistical procedures. We may be able, relying on strong theorems in general statistics plus a backlog of previous experience and a smattering of theory, to say some fairly safe things about restriction of range and the like. However, thoughtful theorists realize how little *quantitatively* we can say with sufficient confidence to warrant counting an unexpected shift in a stochastic quantity as a strong "discorroborator." This being so, we cannot fairly count an "in the ball park" predicted value as a strong corroborator. For example, Meehl's Mental Measure correlates .50 with SES in Duluth junior high school students, as predicted from Fisbee's theory of sociability. When Jones tries to replicate the finding on Chicano seniors in Tucson, he gets r = .34. Who can say anything theoretically cogent about this difference? Does any sane psychologist believe that one can do much more than shrug?

Although probability concepts (in the theory) and statistical distributions (in the data) sometimes appear in both classical and quantum physics, their usual role differs from that of context-dependent stochastologicals in social science. Without exceeding space limitations or my competence, let me briefly suggest some differences. When probabilities appear in physics and chemistry, they often drop out in the course of the derivation chain, yielding a quasi-nomological at its termination (e.g., derivation of gas laws or Graham's diffusion law from the kinetic theory of heat in which the postulates are nomological, the "conditions" are probability distributions, and the resulting theorems are again nomological). Second, when the predicted observational result still contains statistical notions, their numerical values are either not context dependent or the context dependencies permit precise experimental manipulation. A statistical scatter function for photons or electrons can be finely tuned by altering a very limited number of experimental variables (e.g., wavelength, slit width, screen distance), and the law of large numbers assures that the expected "probabilistic" values of, say, photon incidence in a specified band will be indiscernibly different from the observed (finite but huge) numbers.

All this is very unlike the stochastologicals of soft psychology, in which strong context dependence prevails, but we do not know (a) the complete list of contextual influences, (b) the function form of context dependency for those influences that we can list, (c) the numerical values of parameters in those function forms that we know or guess, or (d) the values of the context variables if we are so fortunate as to get past Ignorances a–c. Finally, unlike physics, our sample sizes are usually such that the Bernoulli theorem does not guarantee a close fit between theoretical and observed frequencies—perhaps one of the few good uses for significance tests? . . .

Not to be overly pessimistic, let me mention (without proof) five noble traditions in clinical psychology that I believe have permanent merit and will still be with us 50 or 100 years from now, despite the usual changes. Some of these are currently unpopular among those addicted to one of the contemporary fly-by-night theories, but that does not bother me. These five noble traditions are (a) descriptive clinical psychiatry, (b) psychometric assessment, (c) behavior genetics, (d) behavior modification (I lump under this rubric positive contingency management, aversion therapy, and desensitization), and (e) psychodynamics. . . .

These five noble traditions differ greatly in the methods they use and their central concepts, and I am hard put to say what is common among them. Some of them, such as behavior modification, are not conceptually exciting to those of us who are interested in ideas like Freud's, but they more than make up for that by their remarkable technological power. I shall focus the remainder of my remarks on one feature that they have in common with the developed sciences (physical or biological); to wit, they were originally developed with negligible reliance on *statistical significance testing*. Even the psychometric assessment tradition in its early stages paid little attention to significance testing except (sometimes) for finding good items. Binet did not know anything about *t* tests, but he drew graphs of the developmental change of items. I suggest to you that Sir Ronald has befuddled us, mesmerized us, and led us down the primrose path. I believe that the almost universal reliance on merely refuting the null hypothesis as the standard method for corroborating substantive theories in the soft areas is a terrible mistake, is basically unsound, poor scientific strategy, and one of the worst things that ever happened in the history of psychology. . . .

In the typical *Psychological Bulletin* article reviewing research on some theory, we see a table showing with asterisks (hence, my title) whether this or that experimenter found a difference in the expected direction at the .05 (one asterisk), .01 (two asterisks!), or .001 (three asterisks!!) levels of significance. Typically, of course, some of them come out favorable and some of them come out unfavorable. What does the reviewer usually do? He goes through what is from the standpoint of the logician an almost meaningless exercise; to wit, he *counts noses*. If, say, Fisbee's theory of the mind has a batting average of 7:3 on 10 significance tests in the table, he concludes that Fisbee's theory seems to be rather well supported, "although further research is needed to explain the discrepancies." This is scientifically a preposterous way to reason. It completely neglects the crucial asymmetry between confirmation, which involves an inference in the formally invalid third figure of the implicative syllogism (this is why inductive inferences are ampliative and dangerous and why we can be objectively wrong even though we proceed correctly), and refutation, which is in the valid fourth figure, and which gives the *modus tollens* its privileged position in inductive inference. Thus the adverse *t* tests, seen properly, do Fisbee's theory far more damage than the favorable ones do it good.

I am not making some nit-picking statistician's correction. I am saying that the whole business is so radically defective as to be scientifically almost pointless. This is not a technical hassle about whether Fisbee should have used the varimax rotation, or how he estimated the communalities, or that perhaps some of the higher order interactions that are marginally significant should have been lumped together as a part of the error term, or that the covariance matrices were not quite homogeneous. I am not a statistician, and I am not making a statistical complaint. I am making a philosophical complaint or, if you prefer, a complaint in the domain of scientific method. I suggest that when a reviewer tries to "make theoretical sense" out of such a table of favorable and adverse significance test results, what the reviewer is actually engaged in, willy-nilly or unwittingly, is meaningless substantive constructions on the properties of the statistical power function, and almost nothing else.

This feckless activity is made worse by the almost universal practice of what I call *stepwise low validation*. By this I mean that we rely on one investigation to "validate" a particular instrument and some other study to validate another instrument, and then we correlate the two instruments and claim to have validated the substantive theory. I do not argue that this is a scientific nothing, but it is about as close to a nothing as you can get without intending to. Consider that I first show that Meehl's Mental Measure has a validity coefficient (against the criterion I shall here for simplicity take to be quasi-infallible or definitive) of, say, .40—somewhat higher than we usually get in personology and social psychology! Then I show that Glotz's Global Gauge has a validity for its alleged variable of the same amount. Relying on these results, having stated the coefficient and gleefully recorded the asterisks showing that these coefficients are not zero (!), I now try to corroborate the Glotz-Meehl theory of personality by showing that the two instruments, each having been duly "validated," correlate .40, providing, happily, some more asterisks in the table. Now just what kind of a business is this? Let us suppose that each instrument has a reliability of .90 to make it easy. That means that the portion of construct-valid variance for each of the devices is around one fifth of the reliable variance and the same for their overlap when correlated with each other. I do not want to push the discredited (although recently revived) principle of indifference, but without other knowledge, it is easily possible, and one could perhaps say rather likely, that the correlation between the two occurs in a region of each one's components that has literally nothing to do with either of the two criterion variables used in the validity studies relied on. This is, of course, especially dangerous in light of the research that we have on the contribution of methods variance.

I seem to have trouble conveying to my students and colleagues just how dreadful a mess of flabby inferences this kind of thing involves. It is as if we were interested in the effect of sunlight on the mating behavior of birds, but not being able to get directly at either of these two things, we settle for correlating a proxy variable like field-mice density (because the birds tend to destroy the field mice) with, say, incidence of human skin cancer (since you can get that by spending too much time in the sun!) You may think this analogy dreadfully unfair: but I think it is a good one. Of course, the whole idea of simply counting noses is wrong, because a theory that has seven facts for it and three facts against it is *not* in good shape, and it would not be considered so in any developed science.

You may say, "But, Meehl, R. A. Fisher was a genius, and

we all know how valuable his stuff has been in agronomy. Why shouldn't it work for soft psychology?" Well, I am not intimidated by Fisher's genius, because my complaint is not in the field of mathematical statistics; and as regards inductive logic and philosophy of science, it is well-known that Sir Ronald permitted himself a great deal of dogmatism. I remember my amazement when the late Rudolf Carnap said to me, the first time I met him, "But, of course, on this subject Fisher is just mistaken; surely you must know that." My statistician friends tell me that it is not clear just how useful the significance test has been in biological science either, but I set that aside as beyond my competence to discuss. . . .

Sometimes in the other sciences it has been possible to concoct a middling weak theory that, while incapable of generating numerical point values, entails a certain *function form*, such as a graph should be an ogive or that it should have three peaks and that these peaks should be increasingly high, and that the distance on the abscissa between the first two peaks should be less than the distance between the second two. In the early history of quantum theory, physicists relied on Wien's law, which related "some (unknown) function" of wavelength to energy multiplied by the fifth power of wavelength. In the cavity radiation experiment, the empirical points were simply plotted at varying temperatures, and it was evident by inspection that they fell on the same curve, even though a formal expression for that curve was beyond the theory's capabilities (Eisberg 1961:50–51).

Talking of Wien's law is a good time for me to recommend to psychologists who disagree with my position to have a look at any textbook of theoretical chemistry or physics, where one searches in vain for a statistical significance test (and finds few confidence intervals). The power of the physicist does not come from exact assessment of probabilities that a difference exists (which physicists would view as a ludicrous thing to show), nor by the verbal precision of so-called "operational definitions" in the embedding text. The physicist's scientific power comes from two other sources, namely, the immense deductive fertility of the formalism and the accuracy of the measuring instruments. The scientific trick lies in conjoining rich mathematics and experimental precision, a sort of "invisible hand wielding fine calipers." The embedding text is

sometimes surprisingly loose, free-wheeling, even metaphor-
ical—as viewers of television's *Nova* are aware, seeing Nobel
laureates discourse whimsically about the charm, strangeness,
and gluons of nuclear particles (see, e.g., Nambu, 1976). One
gets the impression that when you have a good science going,
with potent mathematics and accurate instruments, you can
be relaxed and easygoing about the words. Nothing is as stuffy
and pretentious as the verbal "pseudorigor" of the soft
branches of social science. In my modern physics text, I am
unable to find one single test of statistical significance. What
happens instead is that the physicist has a sufficiently powerful invisible hand theory that enables him to generate an
expected curve for his experimental results. He plots the observed points, looks at the agreement, and comments that "the
results are in reasonably good accord with theory." Moral: *It
is always more valuable to show approximate agreement of
observations with a theoretically predicted numerical point
value, rank order, or function form, than it is to compute a
"precise probability" that something merely differs from
something else. . . .*

I want now to state as strongly as I can a prescription that
we should adopt in soft psychology to help get away from the
feeble practice of significance testing: *Wherever possible, two
or more nonredundant estimates of the same theoretical quantity should be made, because multiple approximations to a
theoretical number are always more valuable, provided that
methods of setting permissible tolerances exist, than a so-
called exact test of significance, or even an exact setting of
confidence intervals.* This is a special case of what my philosopher colleague Herbert Feigl refers to as "triangulation in
logical space." It is, as you know, standard procedure in the
developed sciences. We have, for instance, something like a
dozen independent ways of estimating Avogadro's number,
and since they all come out "reasonably close" (again, I have
never seen a physicist do a *t* test on such a thing!), we are
confident that we know how many molecules there are in a
mole of chlorine.

This last point may lead you to ask, "If consistency tests are
as important as Meehl makes them out to be, why we don't
hear about them in chemistry and physics?" I have a perfect
answer to that query. It goes like this: *Consistency tests are*

*so much a part of standard scientific method in the developed disciplines, taken so much for granted by everybody who researches in chemistry or physics or astronomy or molecular biology or genetics, that these scientists do not even bother having a special name for them!* It shows the sad state of soft psychology when a fellow like me has to cook up a special metatheory expression to call attention to something that in respectable science is taken as a matter of course.

Having presented what seems to me some encouraging data, I must nevertheless close with a melancholy reflection. The possibility of deriving consistency tests in the taxonic situation rests on the substantive problems presented by fields like medicine and behavior genetics, and it is not obvious how we would go about doing this in soft areas that are nontaxonic. It may be that the nature of the subject matter in most of personology and social psychology is inherently incapable of permitting theories with sufficient conceptual power (especially mathematical development) to yield the kinds of strong refuters expected by Popperians, Bayesians, and unphilosophical scientists in developed fields like chemistry. This might mean that we could most profitably confine ourselves to low-order inductions, a (to me, depressing) conjecture that is somewhat corroborated by the fact that the two most powerful forms of clinical psychology are atheroretical psychometrics of prediction on the one hand and behavior modification on the other. Neither of these approaches has the kind of conceptual richness that attracts the theory-oriented mind, but I think we ought to acknowledge the possibility that there is never going to be a really impressive theory in personality or social psychology. I dislike to think that, but it might just be true.

MICHAEL E. DOHERTY,
RYAN D. TWENEY,
and CLIFFORD R. MYNATT

# Null Hypothesis Testing,

# Confirmation Bias

# and Strong Inference

NO DOMAIN HAS been more dependent upon statistical analysis than has recent American experimental psychology. This dependence dates from the introduction of Fisher's (1925/ 1967) conception of statistical decision making in experimental research. For a variety of reasons, the incorporation of null hypothesis testing, and its most powerful tool, the analysis of variance, was rapid and thorough in America (Rucci & Tweney 1980). In fact, Cronbach (selection 28) characterized two domains of psychology, the correlational and the experimental, solely on the basis of the latter's use of analysis of variance. While this state of affairs is not without its critics (e.g., Bakan 1967), no scientific psychologist can today avoid the necessity of a close acquaintance with analysis of variance and of null hypothesis testing.

There is an abundant literature on the powers and pitfalls of null hypothesis testing. This paper will not attempt to review that literature, for which the reader is referred to the complete text of Meehl's (1978, selection 29) recent paper. In spite of its imperfections, classical statistical inference is an

Written especially for this volume.

intellectual achievement of the first order, a decidely imperfect, often badly abused, yet powerful, partial solution to the problem of induction.

The power of the null hypothesis test lies in its ability to detect slight but theoretically important effects which would otherwise be obscured by large individual differences and other sources of "error." It allows the scientist to detect the "signal" in the sea of "noise." We know from several lines of research that people are good at detecting weak signals in perceptual tasks, but not very good at making comparable judgments about numerical data (Tversky & Kahneman 1974). Statistical techniques help overcome these limitations.

The problem, however, lies deeper than a mere difficulty with numerical data. It is related to a more fundamental problem that plagues all induction. That inductive inference is inadequate to the task of proving universal generalizations has been acknowledged since the penetrating insights of David Hume (1748). Popper's falsificationist prescription (1959) is one attempted solution to this problem. In psychological research, particularly in research conducted to test a theory, there is a practice which seems superficially to reflect adherence to this prescription. Almost without exception, psychologists adduce support for substantive hypotheses by "falsifying" a statistical null hypothesis. It is textbook wisdom that one goes about research by setting up null hypotheses which one *wants* to reject in order to confirm the hypothesis one *wants* to support. Further, it is standard practice to discount as uninteresting and uninformative those data that fail to "achieve significance." Failure to reject the null hypothesis may be ascribed to insufficient sample sizes, measurement error, or to other sources of design deficiency, but conventional wisdom holds that it is *not* necessary to conclude that the theory which predicted a difference is incorrect. Thus, rejection of the null hypothesis is regarded as highly informative and reflective of powerful research and good theory: failure to reject is regarded as uninformative and perhaps reflective of bad research, but not necessarily of bad theory. Failure to reject means that the experiment "didn't come out" and is "unpublishable."

What consequence does this duality of attitude have for the

experimenter's *substantive* hypothesis? The answer is simple. The experimenter is in a "no-lose" situation. No empirical outcome can be taken as evidence against the truth of the substantive hypothesis. The only methodologically admissible outcome is confirmation of the experimenter's belief. There is, built in at the methodological-statistical core of our enterprise, a practice which institutionalizes and legitimizes confirmation bias of the most unproductive sort.

Is the practice, in fact, unproductive? There are cases where confirmation bias appears to be facilitative, cases ranging from the first attempts of children to extend their knowledge about physical relationships (Karmiloff-Smith and Inhelder 1975), to adults solving complex problem solving tasks (Mynatt, Doherty, and Tweney 1978), to strongly committed research scientists engaged in major research projects (Mitroff 1974b). Yet the particular *kind* of confirmation bias that results from null hypothesis testing is not facilitative, because it insulates the substantive hypothesis from falsification by the experimenter *or by anyone else*. Failure to reject a null hypothesis is not acceptable as disproof of a hypothesis. In fact, such outcomes are difficult or impossible to publish (Mahoney 1976). Recent psychological journals offer many examples of relationships which are difficult or impossible to replicate, "unclothed emporers" which cannot be exposed. Into this category fall the "risky shift" phenomena in social psychology (Cartwright 1973), a plethora of relationships between personality and behavior in psychometric test situations (see, e.g., Lykken's 1968 review of one such case, and Meehl 1973b), and a host of findings in cognitive psychology (see Allport's 1975 review for examples).

We do not quarrel with the logical and statistical reasons which underlie the duality of attitude to which we referred. Nor do we imply that confirmation bias is necessarily unproductive, as our arguments in the introduction to part 4 should make clear. We do believe that the particular mechanism of confirmation bias described above ought to be a matter of great concern to those committed to scientific psychology. The problems are exacerbated when other substantive theories also predict the same outcomes, or can be made to do so once outcomes are known.

## Strong Inference

The major purpose of this paper is to explore the possibility that a modification of Chamberlin's method of multiple hypotheses (selection 13), modified in important ways, may provide a way out of the seemingly paradoxical role played by null hypothesis testing in psychology. The procedure, which is similar to Platt's strong inference (1964), consists of devising alternative hypotheses, devising experiments which will exclude one or more of the hypotheses, carrying out the experiments, and recycling the procedure.

In sciences in which the quality of hypothesis and data are high, and the relation between them is clear, investigators can, given good experiments, obtain falsification in a very strong sense. In much psychological research, however, theoretically important effects are slight relative to the error associated with individual differences. As a consequence, statistical techniques such as null hypothesis testing are necessary.

Under certain circumstances, null hypothesis tests can escape the criticisms leveled against them. Suppose we stopped evaluating *the* experimental (substantive) hypothesis by testing the experimental outcome against *the* statistical null hypothesis. Suppose instead that we conceptualized phenomena in terms of sets of alternative substantive hypotheses. Suppose further that we designed at least some experimental tests such that at least one of the interesting substantive hypotheses entailed a prediction of no difference. Using null hypothesis testing in the context of multiple alternative hypotheses would mean that one or more *substantive* hypotheses would be directly assessed by the null hypothesis test. In this way, null hypothesis testing could serve the goal of falsification in the strong sense.

An example of this prescription can be formulated. Similarity is a basic psychological concept, implicated in one way or another in theories of perception, learning, social psychology, creativity (as we shall argue in part 7) and much else. Similarity has historically been considered to be symmetric and to be complementary to dissimilarity (Gregson 1975). Symmetry implies that the similarity relation of *a* to *b* is the same as the similarity relation of *b* to *a*. Complementarity implies that similarity, properly scaled, equals 1 − dissimilarity. These assumptions are explicitly embodied in many approaches to multidimensional scaling (Shepard, Romney, and Nerlove 1972). Tversky (1977) recently presented a feature-matching theory of similarity, in which he postulated that similarity is not necessarily either symmetric or complementary. Thus, on the one hand, we have a position which implies an identity: the complementariness assumption entails the prediction that a set of objects would be ordered the same way whether judged for similarity or for dissimilarity. On the other hand, Tversky's theory explicitly predicts different orderings under at least some circumstances. Neither theory is a "straw man," set up only to be rejected.

Given two such substantive hypotheses, an experiment can be designed in which the null hypothesis corresponds precisely to one of the substantive hypotheses. Rejection of the null hypothesis in this case would provide not only the usual indirect evidence for a substantive hypothesis (Tversky's) but be a literal and strong falsification of the complementariness hypothesis inherent in multidimensional scaling. Similarly, it should also be possible to design experiments in which Tversky's hypothesis is represented by the null hypothesis, since the theory is both explicit and quantitative.

Suppose that one follows the strong inference model and generates a set of hypotheses, one of which entails a prediction which is not only unexpected, but is not entailed by any other hypothesis. Most scientists would, with Lakatos (1978), regard confirmation of that prediction as highly informative, irrespective of its parlous logical status. Most would regard it as such even if the limitations of the science were such that confirmation came via rejection of the null hypothesis! As we have argued elsewhere (Mynatt, Doherty, & Tweney 1978; this volume, part 4), confirmation is not necessarily a hindrance in science, in spite of Popper's claims. The situation described is clearly one such example.

Thus, in psychology the null hypothesis test does not have to play the unproductive role that Meehl and others have ascribed to it. It can function as a needed decision aid to the scientist, and it may be an especially productive decision aid in the context of multiple hypothesis testing, that is, in an expanded version of strong inference. But one of two conditions must be met: 1) either a substantive hypothesis must be tested by a prediction of no difference, *or* 2) a substantive hypothesis must have consequences which are unexpected, given alternative substantive hypotheses.

AMOS TVERSKY
and DANIEL KAHNEMAN

# The Law

# of Small

# Numbers

"SUPPOSE YOU HAVE run an experiment on 20 subjects, and have obtained a significant result which confirms your theory ($z = 2.23$, $p < .05$, two-tailed). You now have cause to run an additional group of 10 subjects. What do you think the probability is that the results will be significant, by a one-tailed test, separately for this group?"

If you feel that the probability is somewhere around .85, you may be pleased to know that you belong to a majority group. Indeed, that was the median answer of two small groups who were kind enough to respond to a questionnaire distributed at meetings of the Mathematical Psychology Group and of the American Psychological Association.

On the other hand, if you feel that the probability is around .48, you belong to a minority. Only 9 of our 84 respondents gave answers between .40 and .60. However, .48 happens to be a much more reasonable estimate than .85.[1]

Apparently, most psychologists have an exaggerated belief in the likelihood of successfully replicating an obtained find-

Excerpts from A. Tversky & D. Kahneman, Belief in the law of small numbers. *Psychological Bulletin*, (1971), 76:105–10. Copyright © 1971 by the American Psychological Association. Reprinted by permission.

ing. The sources of such beliefs, and their consequences for the conduct of scientific inquiry, are what this paper is about. Our thesis is that people have strong intuitions about random sampling; that these intuitions are wrong in fundamental respects; that these intuitions are shared by naive subjects and by trained scientists; and that they are applied with unfortunate consequences in the course of scientific inquiry.

We submit that people view a sample randomly drawn from a population as highly representative, that is, similar to the population in all essential characteristics. Consequently, they expect any two samples drawn from a particular population to be more similar to one another and to the population than sampling theory predicts, at least for small samples.

The tendency to regard a sample as a representation is manifest in a wide variety of situations. When subjects are instructed to generate a random sequence of hypothetical tosses of a fair coin, for example, they produce sequences where the proportion of heads in any short segment stays far closer to .50 than the laws of chance would predict (Tune 1964). Thus, each segment of the response sequence is highly representative of the "fairness" of the coin. Similar effects are observed when subjects successively predict events in a randomly generated series, as in probability learning experiments (Estes 1964) or in other sequential games of chance. Subjects act as if *every* segment of the random sequence must reflect the true proportion: if the sequence has strayed from the population proportion, a corrective bias in the other direction is expected. This has been called the gambler's fallacy.

The heart of the gambler's fallacy is a misconception of the fairness of the laws of chance. The gambler feels that the fairness of the coin entitles him to expect that any deviation in one direction will soon be cancelled by a corresponding deviation in the other. Even the fairest of coins, however, given the limitations of its memory and moral sense, cannot be as fair as the gambler expects it to be. This fallacy is not unique to gamblers. Consider the following example:

> The mean IQ of the population of eighth graders in a city is known to be 100. You have selected a random sample of 50 children for a study of educational achievements. The first child tested has an IQ of 150. What do you expect the mean IQ to be for the whole sample?

The correct answer is 101. A surprisingly large number of people believe that the expected IQ for the sample is still 100. This expectation can be justified only by the belief that a random process is self-correcting. Idioms such as "errors cancel each other out" reflect the image of an active self-correcting process. Some familiar processes in nature obey such laws: a deviation from a stable equilibrium produces a force that restores the equilibrium. The laws of chance, in contrast, do not work that way: deviations are not canceled as sampling proceeds, they are merely diluted.

Thus far, we have attempted to describe two related intuitions about chance. We proposed a representation hypothesis according to which people believe samples to be very similar to one another and to the population from which they are drawn. We also suggested that people believe sampling to be a self-correcting process. The two beliefs lead to the same consequences. Both generate expectations about characteristics of samples, and the variability of these expectations is less than the true variability, at least for small samples.

The law of large numbers guarantees that very large samples will indeed be highly representative of the population from which they are drawn. If, in addition, a self-corrective tendency is at work, then small samples should also be highly representative and similar to one another. People's intuitions about random sampling appear to satisfy the law of small numbers, which asserts that the law of large numbers applies to small numbers as well.

Consider a hypothetical scientist who lives by the law of small numbers. How would his belief affect his scientific work? Assume our scientist studies phenomena whose magnitude is small relative to uncontrolled variability, that is, the signal-to-noise ratio in the messages he receives from nature is low. Our scientist could be a meteorologist, a pharmacologist, or perhaps a psychologist.

If he believes in the law of small numbers, the scientist will have exaggerated confidence in the validity of conclusions based on small samples. To illustrate, suppose he is engaged in studying which of two toys infants will prefer to play with. Of the first five infants studied, four have shown a preference for the same toy. Many a psychologist will feel some confidence at this point, that the null hypothesis of no preference is false. Fortunately, such a conviction is not a sufficient condition for journal publication, although it may do for a book. By a quick computation, our psychologist will discover that the probability of a result as extreme as the one obtained is as high as 3/8 under the null hypothesis.

To be sure, the application of statistical hypothesis testing to scientific inference is beset with serious difficulties. Nevertheless, the computation of significance levels (or likelihood ratios, as a Bayesian might prefer) forces the scientist to evaluate the obtained effect in terms of a *valid* estimate of sampling variance rather than in terms of his subjective-biased estimate. Statistical tests, therefore, protect the scientific community against overly hasty rejections of the null hypothesis (i.e., Type I error) by policing its many members who would rather live by the law of small numbers. On the other hand, there are no comparable safeguards against the risk of failing to confirm a valid research hypothesis (i.e., Type II error).

Imagine a psychologist who studies the correlation between need for Achievement and grades. When deciding on sample size, he may reason as follows: "What correlation do I expect? $r = .35$. What $N$ do I need to make the result significant? (Looks at table.) $N = 33$. Fine, that's my sample." The only flaw in this reasoning is that our psychologist has forgotten about sampling variation, possibly because he believes that any sample must be highly representative of its population. However, if his guess about the correlation in the population is correct, the correlation in the sample is about as likely to lie below or above .35. Hence, the likelihood of obtaining a significant result (i.e., the power of the test) for $N = 33$ is about .50.

In a detailed investigation of statistical power, Cohen (1962, 1969) has provided plausible definitions of large, medium, and small effects and an extensive set of computational aids to the estimation of power for a variety of statistical tests. In the normal test for a difference between two means, for example, a difference of $.25\sigma$ is small, a difference of $.50\sigma$ is medium, and a difference of $1\sigma$ is large, according to the proposed definitions. The mean IQ difference between clerical and semiskilled workers is a medium effect. In an ingenious study of research practice, Cohen (1962) reviewed all the statistical analyses published in one volume of the *Journal of*

*Abnormal and Social Psychology,* and computed the likelihood of detecting each of the three sizes of effect. The average power was .18 for the detection of small effects, .48 for medium effects, and .83 for large effects. If psychologists typically expect medium effects and select sample size as in the above example, the power of their studies should indeed be about .50.

Cohen's analysis shows that the statistical power of many psychological studies is ridiculously low. This is a self-defeating practice: it makes for frustrated scientists and inefficient research. The investigator who tests a valid hypothesis but fails to obtain significant results cannot help but regard nature as untrustworthy or even hostile. Furthermore, as Overall (1969) has shown, the prevalence of studies deficient in statistical power is not only wasteful but actually pernicious: it results in a large proportion of invalid rejections of the null hypothesis among published results.

Because considerations of statistical power are of particular importance in the design of replication studies, we probed attitudes concerning replication in our questionnaire.

> Suppose one of your doctoral students has completed a difficult and time-consuming experiment on 40 animals. He has scored and analyzed a large number of variables. His results are generally inconclusive, but one before-after comparison yields a highly significant $t = 2.70$, which is surprising and could be of major theoretical significance.
>
> Considering the importance of the result, its surprisal value, and the number of analyses that your student has performed—
>
> Would you recommend that he replicate the study before publishing? If you recommend replication, how many animals would you urge him to run?

Among the psychologists to whom we put these questions there was overwhelming sentiment favoring replication: it was recommended by 66 out of 75 respondents, probably because they suspected that the single significant result was due to chance. The median recommendation was for the doctoral student to run 20 subjects in a replication study. It is instructive to consider the likely consequences of this advice. If the mean and the variance in the second sample are actually identical to those in the first sample, then the resulting value of $t$ will be 1.88. Following the reasoning of note 1, the student's chance of obtaining a significant result in the replication is only slightly above one-half (for $p = .05$, one-tail test). Since we had anticipated that a replication sample of 20 would appear reasonable to our respondents, we added the following question:

Assume that your unhappy student has in fact repeated the initial study with 20 additional animals, and has obtained an insignificant result in the same direction, $t = 1.24$. What would you recommend now? Check one: [the numbers in parentheses refer to the number of respondents who checked each answer]

(a) He should pool the results and publish his conclusion as fact. (0)
(b) He should report the results as a tentative finding. (26)
(c) He should run another group of [median = 20] animals. (21)
(d) He should try to find an explanation for the difference between the two groups. (30)

Note that regardless of one's confidence in the original finding, its credibility is surely enhanced by the replication. Not only is the experimental effect in the same direction in the two samples but the magnitude of the effect in the replication is fully two-thirds of that in the original study. In view of the sample size (20), which our respondents recommended, the replication was about as successful as one is entitled to expect. The distribution of responses, however, reflects continued skepticism concerning the student's finding following the recommended replication. This unhappy state of affairs is a typical consequence of insufficient statistical power.

In contrast to Responses $b$ and $c$, which can be justified on some grounds, the most popular response, Response $d$, is indefensible. We doubt that the same answer would have been obtained if the respondents had realized that the difference between the two studies does not even approach significance. (If the variances of the two samples are equal, $t$ for the difference is .53.) In the absence of a statistical test, our respondents followed the representation hypothesis: as the difference between the two samples was larger than they expected, they viewed it as worthy of explanation. However, the attempt to "find an explanation for the difference between the two groups" is in all probability an exercise in explaining noise.

Altogether our respondents evaluated the replication rather

harshly. This follows from the representation hypothesis: if we expect all samples to be very similar to one another, then almost all replications of a valid hypothesis should be statistically significant. The harshness of the criterion for successful replication is manifest in the responses to the following question:

> An investigator has reported a result that you consider implausible. He ran 15 subjects, and reported a significant value, $t = 2.46$. Another investigator has attempted to duplicate his procedure, and he obtained a nonsignificant value of $t$ with the same number of subjects. The direction was the same in both sets of data.
>
> You are reviewing the literature. What is the highest value of $t$ in the second set of data that you would describe as a failure to replicate?

The majority of our respondents regarded $t = 1.70$ as a failure to replicate. If the data of two such studies ($t = 2.46$ and $t = 1.70$) are pooled, the value of $t$ for the combined data is about 3.00 (assuming equal variances). Thus, we are faced with a paradoxical state of affairs, in which the same data that would increase our confidence in the finding when viewed as part of the original study, shake our confidence when viewed as an independent study. This double standard is particularly disturbing since, for many reasons, replications are usually considered as independent studies, and hypotheses are often evaluated by listing confirming and disconfirming reports.

Contrary to a widespread belief, a case can be made that a replication sample should often be larger than the original. The decision to replicate a once obtained finding often expresses a great fondness for that finding and a desire to see it accepted by a skeptical community. Since that community unreasonably demands that the replication be independently significant, or at least that it approach significance, one must run a large sample. To illustrate, if the unfortunate doctoral student whose thesis was discussed earlier assumes the validity of his initial result ($t = 2.70$, $N = 40$), and if he is willing to accept a risk of only .10 of obtaining a $t$ lower than 1.70, he should run approximately 50 animals in his replication study. With a somewhat weaker initial result ($t = 2.20$, $N$

= 40), the size of the replication sample required for the same power rises to about 75.

That the effects discussed thus far are not limited to hypotheses about means and variances is demonstrated by the responses to the following question:

> You have run a correlational study, scoring 20 variables on 100 subjects. Twenty-seven of the 190 correlation coefficients are significant at the .05 level; and 9 of these are significant beyond the .01 level. The mean absolute level of the significant correlations is .31, and the pattern of results is very reasonable on theoretical grounds. How many of the 27 significant correlations would you expect to be significant again, in an exact replication of the study, with $N = 40$?

With $N = 40$, a correlation of about .31 is required for significance at the .05 level. This is the mean of the significant correlations in the original study. Thus, only about half of the originally significant correlations (i.e., 13 or 14) would remain significant with $N = 40$. In addition, of course, the correlations in the replication are bound to differ from those in the original study. Hence, by regression effects, the initially significant coefficients are most likely to be reduced. Thus, 8 to 10 repeated significant correlations from the original 27 is probably a generous estimate of what one is entitled to expect. The median estimate of our respondents is 18. This is more than the number of repeated significant correlations that will be found if the correlations are recomputed for 40 subjects randomly selected from the original 100! Apparently, people expect more than a mere duplication of the original statistics in the replication sample; they expect a duplication of the significance of results, with little regard for sample size. This expectation requires a ludicrous extension of the representation hypothesis; even the law of small numbers is incapable of generating such a result.

The expectation that patterns of results are replicable almost in their entirety provides the rationale for a common, though much deplored practice. The investigator who computes all correlations between three indexes of anxiety and three indexes of dependency will often report and interpret with great confidence the single significant correlation obtained. His confidence in the shaky finding stems from his belief that the

obtained correlation matrix is highly representative and readily replicable.

In review, we have seen that the believer in the law of small
numbers practices science as follows:

• He gambles his research hypotheses on small samples
without realizing that the odds against him are unreasonably
high. He overestimates power.

• He has undue confidence in early trends (e.g., the data of
the first few subjects) and in the stability of observed patterns
(e.g., the number and identity of significant results). He overestimates significance.

• In evaluating replications, his or others' he has unreasonably high expectations about the replicability of significant
results. He underestimates the breadth of confidence intervals.

• He rarely attributes a deviation of results from expectations to sampling variability, because he finds a causal "explanation" for any discrepancy. Thus, he has little opportunity
to recognize sampling variation in action. His belief in the law
of small numbers, therefore, will forever remain intact.

Our questionnaire elicited considerable evidence for the
prevalence of the belief in the law of small numbers. Our
typical respondent is a believer, regardless of the group to
which he belongs. There were practically no differences between the median responses of audiences at a mathematical
psychology meeting and at a general session of the American
Psychological Association convention, although we make no
claims for the representativeness of either sample. Apparently,
acquaintance with formal logic and with probability theory
does not extinguish erroneous intuitions. What, then, can be
done? Can the belief in the law of small numbers be abolished
or at least controlled?

Research experience is unlikely to help much, because sampling variation is all too easily "explained." Corrective experiences are those that provide neither motive nor opportunity for spurious explanation. Thus, a student in a statistics
course may draw repeated samples of given size from a population, and learn the effect of sample size on sampling variability from personal observation. We are far from certain,
however, that expectations can be corrected in this manner,
since related biases, such as the gambler's fallacy, survive considerable contradictory evidence.

Even if the bias cannot be unlearned, students can learn to

recognize its existence and take the necessary precautions. Since the teaching of statistics is not short on admonitions, a warning about biased statistical intuitions may not be out of place. The obvious precaution is computation. The believer in the law of small numbers has incorrect intuitions about significance level, power, and confidence intervals. Significance levels are usually computed and reported, but power and confidence limits are not. Perhaps they should be.

Explicit computation of power, relative to some reasonable hypothesis, for instance, Cohen's (1962, 1969) small, large, and medium effects, should surely be carried out before any study is done. Such computations will often lead to the realization that there is simply no point in running the study unless, for example, sample size is multiplied by four. We refuse to believe that a serious investigator will knowingly accept a .50 risk of failing to confirm a valid research hypothesis. In addition, computations of power are essential to the interpretation of negative results, that is, failures to reject the null hypothesis. Because readers' intuitive estimates of power are likely to be wrong, the publication of computed values does not appear to be a waste of either readers' time or journal space.

In the early psychological literature, the convention prevailed of reporting, for example, a sample mean as $\bar{X} \pm PE$, where $PE$ is the probable error (i.e., the 50% confidence interval around the mean). This convention was later abandoned in favor of the hypothesis-testing formulation. A confidence interval, however, provides a useful index of sampling variability, and it is precisely this variability that we tend to underestimate. The emphasis on significance levels tends to obscure a fundamental distinction between the size of an effect and its statistical significance. Regardless of sample size, the size of an effect in one study is a reasonable estimate of the size of the effect in replication. In contrast, the estimated significance level in a replication depends critically on sample size. Unrealistic expectations concerning the replicability of significance levels may be corrected if the distinction between size and significance is clarified, and if the computed size of observed effects is routinely reported. From this point of view, at least, the acceptance of the hypothesis-testing model has not been an unmixed blessing for psychology.

The true believer in the law of small numbers commits his

multitude of sins against the logic of statistical inference in good faith. The representation hypothesis describes a cognitive or perceptual bias, which operates regardless of motivational factors. Thus, while the hasty rejection of the null hypothesis is gratifying, the rejection of a cherished hypothesis is aggravating, yet the true believer is subject to both. His intuitive expectations are governed by a consistent misperception of the world rather than by opportunistic wishful thinking. Given some editorial prodding, he may be willing to regard his statistical intuitions with proper suspicion and replace impression formation by computation whenever possible.

## Note

1. The required estimate can be interpreted in several ways. One possible approach is to follow common research practice, where a value obtained in one study is taken to define a plausible alternative to the null hypothesis. The probability requested in the question can then be interpreted as the power of the second test (i.e., the probability of obtaining a significant result in the second sample) against the alternative hypothesis defined by the result of the first sample. In the special case of a test of a mean with known variance, one would compute the power of the test against the hypothesis that the population mean equals the mean of the first sample. Since the size of the second sample is half that of the first, the computed probability of obtaining $z \geq 1.645$ is only .473. A theoretically more justifiable approach is to interpret the requested probability within a Bayesian framework and compute it relative to some appropriately selected prior distribution. Assuming a uniform prior, the desired posterior probability is .478. Clearly, if the prior distribution favors the null hypothesis, as is often the case, the posterior probability will be even smaller.

# Part VII

# Scientific Creativity (1): Observation and Classification

# Introduction

EARLIER SECTIONS OF the book have raised questions concerning what scientists do with hypotheses once they get them. The remaining sections are concerned with scientific creativity, with the perhaps less tractable question of the origin of hypotheses. The creative process is central to science, and is implicated at all levels: in the choice of observations, in the generation of hypotheses, in the design of tests of hypotheses, even (as we shall see) in the evaluation of the adequacy of theories.

Existing literature on creativity in science has been unsatisfactory in our view. Most of it has been based upon case studies of single individuals, and a good deal of uncritical reliance has been placed on the statements of scientists themselves—their diaries, their notebooks, their memoirs. Like case studies in clinical psychology and psychiatry, these works are extraordinarily interesting and a rich source of scientific hypotheses. Many of these hypotheses may be amenable to the methods of modern cognitive psychology, but the literature as a whole has not come to grips with a central problem. Some mysterious creative force is too often invoked to explain any and all findings. In this respect, case studies provide the existing literature with great power but with great limitations as well. Generalizations *from* the unique case are risky and generalizations *about* the unique case are tenuous.

There is widespread belief that Einstein used vivid visual images in grappling with the abstract and difficult problems he set himself. But the evidence for such a claim comes entirely from Einstein's self-reports. Even if one places special trust in *Einstein's* self-reports, in spite of psychologists' great distrust of this method as a general source of valid data concerning cognitive processes, one must still confront the possibility that this vivid visual imagery was *not* instrumental in the development of the theory of relativity. There is nothing to rule out the possibility that the imagery may even have impeded Einstein's intellectual progress, or, perhaps less fantastic, that it may have been epiphenomenal, or merely a means of facilitating communication. Whether or not such questions are inherently intractable is not clear at the present time, but we feel certain of one thing; they *will* remain intractable unless and until more sophisticated techniques are brought to bear on their study. Whether techniques derived from cognitive psychology will lead to answers to those questions is by no means certain, but we believe they show promise. Further, what is true of the role of imagery in science is, we feel, equally true for all questions involving scientific creativity: case studies are necessary but not sufficient.

To point the way toward a more powerful cognitive account of these processes, we have chosen selections that manifest potential critical psychological variables in scientific creativity. They are based upon retrospective case studies, but most point to broader issues, issues that appear to be testable. In a sense, the remainder of this book consists of a single section. We have divided it into three parts, but the reader should bear in mind the extraordinary interrelatedness of the issues addressed by the selections. In part 7 we address the issue of creativity in observation, in part 8 of creativity in experiment and theory, and in part 9 we consider the role of imagery in science.

## Observation

## and Similarity

The question of scientific creativity actually begins with a basic psychological question, and its obverse: What psychological mechanisms render two entities *similar* in the mind of the beholder? What mechanisms render two entities *different*? That the perception of sameness and difference is a fundamental psychological capability has been long recognized (cf. Locke 1690), but attention to its role in science is more recent. Similarity holds one key to understanding scientific creativity because it is by the construction of classes of things that are **alike** that the scientist extends knowledge. When Kekulé's dream revealed the benzene molecule as **similar to** a ring structure, then Kekulé had achieved the creative breakthrough. When Galileo saw that the dynamic laws of free-falling bodies were **like** those of balls rolling on an inclined plane, he had achieved a key insight. When Newton saw the formal similarity between terrestrial and celestial mechanics, he transformed our view of the universe.

Cognitive psychology has approached the problem of similarity in a number of ways. Bruner, Goodnow, and Austin's classic study of concept formation was motivated in part by a desire to explain the seeming paradox of "the existence of discrimination capacities which, if fully used, would make us slaves to the particular" (1956:1). Garner's (1974) work on information integration has been directed at understanding how a stimulus is processed when it constitutes one of a set of related (hence, similar) stimuli (whether those are actually present or only inferred). Recent studies of comprehension have focused upon "how meaning arises from perceiving an entity's participation in various events" (Bransford and McCarrell 1974: 379). Shepard and Cermak (1973) explored the circumstances under which mathematically generated "free-form" stimuli were grouped into regions that reflected **perceived** similarity, rather than the parametrically defined similarity used by the experimenter to generate stimuli.

The construction of classes of similar objects is a central activity in science. In a sense, no data can be collected until publicly verifiable rules are set for the regulation of the observational process. This point has long been recognized, and there is a literature which concerns itself with some of the underpinnings of this process. It will not surprise the reader that we regard the literature as pointing to the need for psychological approaches to this issue in science. When Niels Bohr said (in effect) "The hydrogen atom is **like** a planetary system," he was utilizing a similarity which then allowed him to construct a powerful formal model of the emission of spectra. The formal model was a different sort of representation of the real phenomena than was the original metaphoric similarity, but it was a similarity nonetheless.

Similarities may, of course, be seriously misleading as well. Thus, Kenelm Digby's "Powder of Sympathy" (1669) was based on the notion that a wound would heal more quickly if touched by the weapon that caused the wound. The perhaps more sophisticated correspondence between the radii of planetary orbits and the Euclidean regular solids led Kepler to believe there was a transcendent harmony in the "music of the spheres" (Kepler 1619). Such misleading use of similarity is not absent in modern science; it can be seen in the "discovery" of "N-Rays" by Blondlot in 1903 (Rostand 1960), and in any number of the coincidences that fuel popular pseudoscience (e.g., von Daniken's prehistoric aliens).

When appropriate taxonomy is created, the sense of unity and the recognition of diversity are both provided for. An accepted taxonomic system can serve as a general reference system for scientific communication but it can subserve creative functions as well. Thus, at one level, Mendeleev's achievement was to devise a classification of the chemical elements that systematically displayed their chemical and physical properties. The taxonomy, in turn, facilitated the task of deriving a relation between atomic number and physical properties, and was instrumental in the prediction of new elements to plug the "holes" in the table.

In spite of the promising beginning made by Bruner, Goodnow, and Austin in 1956, little relevant empirical work exists in the literature of experimental cognitive psychology. While concept formation research bears some relevance to the development of taxonomies (cf. Bourne 1966) we believe that the relationship is more apparent than real. The number of features potentially defining the concepts studied have been very small, and the features themselves extremely simple. Subjects in laboratory taxonomy problems are often told when they are right after each instance is presented and classified. Subjects have generally been precluded from seeking data; that is, the selection paradigm has been used relatively infrequently compared to the reception paradigm (in the former, the subject can select the instances to judge, in the latter the order of presentation of stimulus objects is fixed, with the subject reduced to a passive observer).[1] Such research can provide at best a background for research related to scientific taxonomy, but does not directly address the central issues. Thus, we do not feel that research such as that of Levine (1975) is relevant to the understanding of scientific theories.

Some recent research on taxonomies has explored the structure underlying everyday conceptual systems. Rosch, et al. (1976) found evidence for the existence of "basic" categories which are defined by certain key examples around which less paradigmatic instances are found, a conception similar to Whewell's Method of Gradation (Whewell 1858). Thus, a robin better exemplifies the bird category than does a penguin. Kuhn (1974) has speculated on the relevance of a similar theory of conceptualization to the emergence of scientific concepts. We see this as a promising domain for investigation.

We shall take the creation of taxonomies as a starting point, as our first example of the function of similarity in science. Selection 32 by Darwin comments briefly on the making of taxonomies by the "young naturalist." John Herschel, in selection 33, speculates about the psychological processes (including induction) that enter into taxonomic work, the roles of similarity ("resemblances") and labeling, and the place of taxonomy in bringing a sense of order to a complex world. An extensive selection (34) from Bacon's *Novum Organum* lists the positive and negative instances of a phenomenon, in a self-conscious effort to arrive at an inductive understanding. Bacon's Tables of Presentation prescribe a sort of taxonomic approach to scientific discovery, in this case, the discovery of the nature of heat. Another selection (35) by Herschel follows, in which he discusses the role of analogy in scientific explanation. It should thus become apparent that the problem of induction is psychologically related to the taxonomic problem and to the problem of the role of analogy and metaphor in science. All of these operations involve the creative use of similarity.

It is a commonplace to observe that no scientist works inductively, as Bacon advocated.[2] In selection 36, Hanson sums up an interesting position on this question. Referring to the inductionist view, he says "There is something wrong with the older view: it is false." Hanson explores Kepler's discovery that planetary orbits were elliptical and, in a distinctively psychological analysis, concludes that Kepler's approach is not properly considered either induction or deduction, but is what Peirce called *abduction* or *retroduction*.

What the scientist makes out of the sensory world—whether abstract theory or concrete classification—is not solely a function of that sensory world itself. Holton (1973, 1978; selection 37) makes the point in a compelling way. What he has called themata, the ontological and methodological presuppositions held by the scientist, play a major role in guiding the scientist in the search for truth. Thematic concepts include the concept of a causal, mechanistic universe, the absoluteness of time, space, and simultaneity, and the belief in absolutes in nature. Methodological themata include, for example, a preference for parsimony, and the assumption that controlled experiments are preferable to any alternative method of investigation. Examples of thematic hypotheses include the constancy of the speed of light, and, in psychology, the assumption that the search for pleasure is the fundamental human motive. Certainly a major methodological thema is expressed in the reliance on mathematical formulations as a means of producing valid knowledge about the world.

Holton conceptualizes themata operating primarily as opposing couples, or even as triads, and as being subject to change. An example of a thema-antithema couple would be the need to see the world as both constant and changing. One fundamental thema is the assumption of order—an assumption without which science would not make much sense. But there is also the antithema of the assumption of intrinsic differentiation, of variability in species and in individuals.

Poincaré (selection 36) closes this section of the book with a consideration of where the scientist ought to look for relevant facts. His answer, that the best facts are simple, regular, and beautiful, reveals a thema without which science would not progress.

## Notes

1. An interesting exception to this generalization, and a study of great relevance, is based on the learning of artificial grammatical systems (Shipstone, 1960). In spite of promising initial results, there have been relatively few studies following up the initial efforts.

2. It is interesting to note that, historically, this was not always believed. Paeans to Baconian induction were once commonplace, and any suggestion that it was *not* the principal tool of science was dismissed. This remained true at least until the time of Whewell (Laudan 1969).

Selection 32 _____

CHARLES DARWIN

# On the

# Making

# of Taxonomies

WHEN A YOUNG naturalist commences the study of a group
of organisms quite unknown to him, he is at first much per-
plexed in determining what differences to consider as specific,
and what as varietal; for he knows nothing of the amount and
kind of variation to which the group is subject; and this shows,
at least, how very generally there is some variation. But if he
confine his attention to one class within one country, he will
soon make up his mind how to rank most of the doubtful
forms. His general tendency will be to make many species, for
he will become impressed, just like the pigeon or poultry
fancier, . . . with the amount of difference in the forms which
he is continually studying; and he has little general knowledge
of analogical variation in other groups and in other countries,
by which to correct his first impressions. As he extends the
range of his observations, he will meet with more cases of
difficulty; for he will encounter a greater number of closely-
allied forms. But if his observations be widely extended, he
will in the end generally be able to make up his own mind:
but he will succeed in this at the expense of admitting much
variation,—and the truth of this admission will often be dis-

Excerpt from C. Darwin, *The Origin of Species by means of Natural Selection* (originally published 1859). New York: The Modern Library, no date, pp. 44–45

puted by other naturalists. When he comes to study allied forms brought from countries not now continuous, in which case he cannot hope to find intermediate links, he will be compelled to trust almost entirely to analogy, and his difficulties will rise to a climax.

JOHN F. W. HERSCHEL

# The Classification

# of Natural

# Objects

(129.) THE NUMBER AND variety of objects and relations which the observation of nature brings before us are so great as to distract the attention, unless assisted and methodized by such judicious distribution of them in classes as shall limit our view to a few at a time, or to groups so bound together by general resemblances that, for the immediate purpose for which we consider them, they may be regarded as individuals. Before we can enter into any thing which deserves to be called a general and systematic view of nature, it is necessary that we should possess an enumeration, if not complete, at least of considerable extent, of her materials and combinations; and that those which appear in any degree important should be distinguished by names which may not only tend to fix them in our recollection, but may constitute, as it were, nuclei or centres, about which information may collect into masses. The imposition of a name on any subject of contemplation, be it a material object, a phenomenon of nature, or a group of facts and relations, looked upon in a peculiar point of view, is an epoch in its history of great importance. It not only enables us readily to refer to it in conversation or writing, without

Excerpts from J. F. W. Herschel, *A Preliminary Discourse on the Study of Natural Philosophy*. London: for Longman, Rees, Orme, Brown, & Green, 1830, pp. 135–38; 140–41.

circumlocution, but, what is of more consequence, it gives it a recognized existence in our own minds, as a matter for separate and peculiar consideration; places it on a list for examination; and renders it a head or title, under which information of various descriptions may be arranged; and, in consequence, fits it to perform the office of a connecting link between all the subjects to which such information may refer.

(130.) For these purposes, however, a temporary or provisional name, or one adapted for common parlance, may suffice. But when a very great multitude of objects come to be referred to one class, especially of such as do not offer very obvious and remarkable distinctions, a more systematic and regular nomenclature becomes necessary, in which the names shall recall the differences as well as the resemblances between the individuals of a class, and in which the direct relation between the name and the object shall materially assist the solution of the problem, "given the one, to determine the other." How necessary this may become, will be at once seen, when we consider the immense number of individual objects, or rather species, presented by almost every branch of science of any extent; which absolutely require to be distinguished by names. Thus, the botanist is conversant with from 80,000 to 100,000 species of plants; the entomologist with, perhaps, as many, of insects: the chemist has to register the properties of combinations, by twos, threes, fours, and upwards, in various doses, of upwards of fifty different elements, all distinguished from each other by essential differences; and of which though a great many thousands are known, by far the greater part have never yet been formed, although hundreds of new ones are coming to light, in perpetual succession, as the science advances; all of which are to be named as they arise. The objects of astronomy are, literally, as numerous as the stars of heaven; and although not more than one or two thousand require to be expressed by distinct names, yet the number, respecting which particular information is required, is not less than a hundred times that amount; and all these must be registered in lists, (so as to be at once referred to, and so that none shall escape,) if not by actual names, at least by some equivalent means.

(131.) Nomenclature, then, is, in itself, undoubtedly an important part of science, as it prevents our being lost in a wilderness of particulars, and involved in inextricable confusion. Happily, in those great branches of science where the objects of classification are most numerous, and the necessity for a clear and convenient nomenclature most pressing, no very great difficulty in its establishment is felt. The very multitude of the objects themselves affords the power of grouping them in subordinate classes, sufficiently well defined to admit of names, and these again into others, whose names may become attached to, or compounded with, the former, till at length the particular species is identified. The facility with which the botanist, the entomologist, or the chemist, refers by name to any individual object in his science shows what may be accomplished in this way when characters are themselves distinct. In other branches, however, considerable difficulty is experienced. This arises mostly where the species to be distinguished are separated from each other chiefly by difference in degree, of certain qualities common to all, and where the degrees shade into each other insensibly. Perhaps such subjects can hardly be considered ripe for systematic nomenclature; and that the attempt to apply it ought only to be partial, embracing such groups and parcels of individuals as agree in characters evidently natural and generic, and leaving the remainder under trivial or provisional denominations, till they shall be better known, and capable of being scientifically grouped. . . .

(134.) The classifications by which science is advanced, however, are widely different from those which serve as bases for artificial systems of nomenclature. They cross and intersect one another, as it were, in every possible way, and have for their very aim to interweave all the objects of nature in a close and compact web of mutual relations and dependence. As soon, then, as any resemblance or analogy, any point of agreement whatever, is perceived between any two or more things,— be they what they will, whether objects, or phenomena, or laws,—they immediately and *ipso facto* constitute themselves into a group or class, which may become enlarged to any extent by the accession of such new objects, phenomena, or laws, agreeing in the same point, as may come to be subsequently ascertained. It is thus that the materials of the world become grouped in natural families, such as chemistry furnishes examples of, in its various groups of acids, alkalies, sulphurets,

&c.; or botany, in its euphorbiaceae, umbelliferae, &c. It is thus, too, that phenomena assume their places under general points of resemblance; as, in optics, those which refer themselves to the class of periodic colours, double refraction, &c.; and that resemblances themselves become traced, which it is the business of induction to generalize and include in abstract propositions.

**Selection 34** _____

FRANCIS BACON

# The Basis

# of Induction

X.

HAVING THUS SET up the mark of knowledge, we must go on to precepts, and that in the most direct and obvious order. Now my directions for the interpretation of nature embrace two generic divisions; the one how to educe and form axioms from experience; the other how to deduce and derive new experiments from axioms. The former again is divided into three ministrations; a ministration to the sense, a ministration to the memory, and a ministration to the mind or reason.

For first of all we must prepare a _Natural and Experimental History_, sufficient and good; and this is the foundation of all; for we are not to imagine or suppose, but to discover, what nature does or may be made to do.

But natural and experimental history is so various and diffuse, that it confounds and distracts the understanding, unless it be ranged and presented to view in a suitable order. We must therefore form _Tables and Arrangements of Instances_, in such a method and order that the understanding may be able to deal with them.

And even when this is done, still the understanding, if left to itself and its own spontaneous movements, is incompetent and unfit to form axioms, unless it be directed and guarded.

Excerpts from Francis Bacon, _Novum Organum_ (originally published 1620). In J. Spedding, R. L. Ellis, and D. D. Heath, eds., _The Works of Francis Bacon._ Popular edition. New York: Hurd & Houghton, 1878, pp. 178–82; 184; 193–94; 203–05; 210–11. This translation first published 1857–1859.

Therefore in the third place we must use *Induction*, true and legitimate induction, which is the very key of interpretation. But of this, which is the last, I must speak first, and then go back to the other ministrations.

XI.

The investigation of Forms proceeds thus: a nature being given, we must first of all have a muster or presentation before the understanding of all known instances which agree in the same nature, though in substances the most unlike. And such collection must be made in the manner of a history, without premature speculation, or any great amount of subtlety. For example, let the investigation be into the Form of Heat.

*Instances Agreeing in the Nature of Heat*

1. The rays of the sun, especially in summer and at noon.
2. The rays of the sun reflected and condensed, as between mountains, or on walls, and most of all in burning-glasses and mirrors.
3. Fiery meteors.
4. Burning thunderbolts.
5. Eruptions of flame from the cavities of mountains.
6. All flame.
7. Ignited solids.
8. Natural warm-baths.
9. Liquids boiling or heated.
10. Hot vapours and fumes, and the air itself, which conceives the most powerful and glowing heat, if confined; as in reverbatory furnaces.
11. Certain seasons that are fine and cloudless by the constitution of the air itself, without regard to the time of year. . . .
27. Even keen and intense cold produces a kind of sensation of burning;

Nec Boreae penetrabile frigus adurit.[1]

28. Other instances.

This table I call the *Table of Essence and Presence*.

XII.

Secondly, we must make a presentation to the understanding of instances in which the given nature is wanting; because the Form, as stated above, ought no less to be absent when the given nature is absent, than present when it is present. But to note all these would be endless.

The negatives should therefore be subjoined to the affirmatives, and the absence of the given nature inquired of in those subjects only that are most akin to the others in which it is present and forthcoming. This I call the *Table of Deviation, or of Absence in Proximity*

*Instances in Proximity where the Nature of Heat is Absent.*

1. The rays of the moon and of stars and comets are not found to be hot to the touch; indeed the severest colds are observed to be at the full moons.

The larger fixed stars however, when passed or approached by the sun, are supposed to increase and give intensity to the heat of the sun; as is the case when the sun is in the sign Leo, and in the Dog-days.

2. The rays of the sun in what is called the middle region of the air do not give heat; for which there is commonly assigned not a bad reason, viz. that that region is neither near enough to the body of the sun from which the rays emanate, nor to the earth from which they are reflected. And this appears from the fact that on the tops of mountains, unless they are very high, there is perpetual snow. . . .

4. Try the following experiment. Take a glass fashioned in a contrary manner to a common burning-glass, and placing it between your hand and the rays of the sun, observe whether it diminishes the heat of the sun, as a burning-glass increases and strengthens it. For it is evident in the case of optical rays that according as the glass is made thicker or thinner in the middle as compared with the sides, so do the objects seen through it appear more spread or more contracted. Observe therefore whether the same is the case with heat.

5. Let the experiment be carefully tried, whether by means of the most powerful and best constructed burning glasses, the rays of the moon can be so caught and collected as to produce even the least degree of warmth. But should this degree of warmth prove too subtle and weak to be perceived and apprehended by the touch, recourse must be had to those glasses which indicate the state of the atmosphere in respect of heat and cold. Thus, let the rays of the moon fall through a burning-

glass on the top of a glass of this kind, and then observe
whether there ensues a sinking of the water through warmth. . . .

XIII.

Thirdly, we must make a presentation to the understanding
of instances in which the nature under inquiry is found in
different degrees, more or less; which must be done by making
a comparison either of its increase and decrease in the same
subject, or of its amount in different subjects, as compared one
with another. For since the Form of a thing is the very thing
itself, and the thing differs from the form no otherwise than
as the apparent differs from the real, or the external from the
internal, or the thing in reference to man from the thing in
reference to the universe; it necessarily follows that no nature
can be taken as the true form, unless it always decrease when
the nature in question decreases, and in like manner always
increase when the nature in question increases. This Table
therefore I call the *Table of Degrees* or the *Table of Comparison*.

*Table of Degrees or Comparison in Heat.*

I will therefore first speak of those substances which contain
no degree at all of heat perceptible to the touch, but seem to
have a certain potential heat only, or disposition and preparation for hotness. After that I shall proceed to substances
which are hot actually, and to the touch, and to their intens-
ities and degrees.

1. In solid and tangible bodies we find nothing which is in
its nature originally hot. For no stone, metal, sulphur, fossil,
wood, water, or carcass of animal is found to be hot. And the
hot water in baths seems to be heated by external causes;
whether it be by flame or subterraneous fire, such as is thrown
up from Aetna and many other mountains, or by the conflict
of bodies, as heat is caused in the dissolutions of iron and tin.
There is therefore no degree of heat palpable to the touch in
animate substances; but they differ in degree of cold, wood
not being equally cold with metal. But this belongs to the
Table of Degrees in Cold.

2. As far however as potential heat and aptitude for flame
is concerned, there are many inanimate substances found
strongly disposed thereto, as sulphur, naphtha, rock oil.

3. Substances once hot, as horse-dung from animal heat, and lime or perhaps ashes and soot from fire, retain some latent remains of their former heat. Hence certain distillations and resolutions of bodies are made by burying them in horse-dung, and heat is excited in lime by sprinkling it with water, as already mentioned.

4. In the vegetable creation we find no plant or part of plant (as gum or pitch) which is warm to the human touch. But yet, as stated above, green herbs gain warmth by being shut up; and to the internal touch, as the palate or stomach, and even to external parts, after a little time, as in plasters and ointments, some vegetables are perceptibly warm and others cold. . . .

### XIV.

How poor we are in history any one may see from the foregoing tables; where I not only insert sometimes mere traditions and reports (though never without a note of doubtful credit and authority) in place of history proved and instances certain, but am also frequently forced to use the words "Let trial be made," or "Let it be further inquired."

### XV.

The work and office of these three tables I call the Presentation of Instances to the Understanding. Which presentation having been made, Induction itself must be set at work; for the problem is, upon a review of the instances, all and each, to find such a nature as is always present or absent with the given nature, and always increases and decreases with it; and which is, as I have said, a particular case of a more general nature. Now if the mind attempt this affirmatively from the first, as when left to itself it is always wont to do, the result will be fancies and guesses and notions ill defined, and axioms that must be mended every day; unless like the schoolmen we have a mind to fight for what is false; though doubtless these will be better or worse according to the faculties and strength of the understanding which is at work. To God, truly, the Giver and Architect of Forms, and it may be to the angels and higher intelligences, it belongs to have an affirmative knowledge of forms immediately, and from the first contemplation. But this assuredly is more than man can do, to whom it is granted only to proceed at first by negatives, and at last to end in affirmatives, after exclusion has been exhausted.

XVI.

We must make therefore a complete solution and separation of nature, not indeed by fire, but by the mind, which is a kind of divine fire. The first work therefore of true induction (as far as regards the discovery of Forms) is the rejection or exclusion of the several natures which are not found in some instance where the given nature is present, or are found in some instance where the given nature is absent, or are found to increase in some instance when the given nature decreases, or to decrease when the given nature increases. Then indeed after the rejection and exclusion has been duly made, there will remain at the bottom, all light opinions vanishing into smoke, a Form affirmative, solid and true and well defined. This is quickly said; but the way to come at it is winding and intricate. I will endeavour however not to overlook any of the points which may help us towards it. . . .

XX.

And yet since truth will sooner come out from error than from confusion, I think it expedient that the understanding should have permission, after the three Tables of First Presentation (such as I have exhibited) have been made and weighed, to make an essay of the Interpretation of Nature in the affirmative way; on the strength both of the instances given in the tables, and of any others it may meet with elsewhere. Which kind of essay I call the *Indulgence of the Understanding*, or the *Commencement of Interpretation*, or the *First Vintage*.

*First Vintage concerning the Form of Heat.*

It is to be observed that the Form of a thing is to be found (as plainly appears from what has been said) in each and all the instances, in which the thing itself is to be found; otherwise it would not be the Form. It follows therefore that there can be no contradictory instance. At the same time the Form is found much more conspicuous and evident in some instances than in others; namely in those wherein the nature of the Form is less restrained and obstructed and kept within bounds by other natures. Instances of this kind I call Shining or Striking Instances. Let us now therefore proceed to the First Vintage concerning the Form of Heat.

From a survey of the instances, all and each, the nature of which Heat is a particular case appears to be Motion. This is displayed most conspicuously in flame, which is always in motion, and in boiling or simmering liquids, which also are in perpetual motion. It is also shown in the excitement or increase of heat caused by motion, as in bellows and blasts; on which see Tab. 3. Inst. 29.; and again in other kinds of motion, on which see Tab. 3. Inst. 28. and 31. Again it is shown in the extinction of fire and heat by any strong compression, which checks and stops the motion; on which see Tab. 3. Inst. 30. and 32. It is shown also by this, that all bodies are destroyed, or at any rate notably altered, by all strong and vehement fire and heat; whence it is quite clear that heat causes a tumult and confusion and violent motion in the internal parts of a body, which perceptibly tends to its dissolution.

## Note

1. Cf. Virgil, Georgics, Bk. 1. 92, 93. "Nor burns the sharp cold of the northern blast."

# Selection 35

## JOHN F. W. HERSCHEL

# Bacon's Method

# and the Role

# of Experiments

(141.) WHENEVER, THEREFORE, ANY phenomenon presents itself for explanation, we naturally seek, in the first instance, to refer it to some one or other of those real causes which experience has shown to exist, and to be efficacious in producing similar phenomena. In this attempt our probability of success will, of course, mainly depend, 1st, On the number and variety of causes experience has placed at our disposal; 2dly, On our habit of applying them to the explanation of natural phenomena; and, 3dly, On the number of analogous phenomena we can collect, which have either been explained, or which admit of explanation by some one or other of those causes, and the closeness of their analogy with that in question.

(142.) Here, then, we see the great importance of possessing a stock of analogous instances or phenomena which class themselves with that under consideration, the explanation of one among which may naturally be expected to lead to that of all the rest. If the analogy of two phenomena be very close and striking, while, at the same time, the cause of one is very obvious, it becomes scarcely possible to refuse to admit the action of an analogous cause in the other, though not so obvious in itself. For instance, when we see a stone whirled

Excerpt from J. F. W. Herschel, *A Preliminary Discourse on the Study of Natural Philosophy*. London: for Longman, Rees, Orme, Brown, & Green, 1830, pp. 148–154.

round in a sling, describing a circular orbit round the hand, keeping the string stretched, and flying away the moment it breaks, we never hesitate to regard it as retained in its orbit by the tension of the string, that is, by *a force* directed to the centre; for we feel that we do really exert such a force. We have here *the direct perception* of the cause. When, therefore, we see a great body like the moon circulating round the earth and not flying off, we cannot help believing it to be prevented from so doing, not indeed by a material tie, but by that which operates in the other case through the intermedium of the string,—a *force* directed constantly to the centre. It is thus that we are continually acquiring a knowledge of the existence of causes acting under circumstances of such concealment as effectually to prevent their direct discovery.

(143.) In general we must observe that motion, wherever produced or changed, invariably points out the existence of *force* as its cause; and thus the forces of nature become known and measured by the motions they produce. Thus, the *force* of magnetism becomes known by the deviation produced by iron in a compass needle, or by a needle leaping up to a magnet held over it, as certainly as by that adhesion to it, when in contact and at rest, which requires force to break the connection; and thus the currents produced in the surface of a quantity of quicksilver, electrified under a conducting fluid, have pointed out the existence and direction of forces of enormous intensity developed by the electric circuit, of which we should not otherwise have had the least suspicion.

(144.) But when the cause of a phenomenon neither presents itself obviously on the consideration of the phenomenon itself, nor is as it were forced on our attention by a case of strong analogy, such as above described, we have then no resource but in a deliberate assemblage of all the parallel instances we can muster; that is, to the formation of a class of facts, having the phenomenon in question for a head of classification; and to a search among the individuals of this class for some other common points of agreement, among which the cause will of necessity be found. But if more than one cause should appear, we must then endeavour to find, or, if we cannot find, to *produce, new facts*, in which each of these in succession shall be wanting, while yet they agree in the general point in question. Here we find the use of what Bacon terms "*crucial instances*," which are phenomena brought forward to decide between two causes, each having the same analogies in its favour. And here, too, we perceive the utility of *experiment* as distinguished from mere passive observation. We make an experiment of the crucial kind when we form combinations, and put in action causes from which some particular one shall be deliberately excluded, and some other purposely admitted; and by the agreement or disagreement of the resulting phenomena with those of the class under examination, we decide our judgment.

(145.) When we would lay down general rules for guiding and facilitating our search, among a great mass of assembled facts, for their common cause, we must have regard to the characters of that relation which we intend by cause and effect. Now, these are,—

1st, Invariable connection, and, in particular, invariable antecedence of the cause and consequence of the effect, unless prevented by some counteracting cause. But it must be observed, that, in a great number of natural phenomena, the effect is produced gradually, while the cause often goes on increasing in intensity; so that the antecedence of the one and consequence of the other becomes difficult to trace, though it really exists. On the other hand, the effect often follows the cause so instantaneously, that the interval cannot be perceived. In consequence of this, it is sometimes difficult to decide, of two phenomena constantly accompanying one another, which is cause or which effect.

2d, Invariable negation of the effect with absence of the cause, unless some other cause be capable of producing the same effect.

3d, Increase or diminution of the effect, with the increased or diminished intensity of the cause, in cases which admit of increase and diminution.

4th, Proportionality of the effect to its cause in all cases of *direct unimpeded* action.

5th, Reversal of the effect with that of the cause.

(146.) From these characters we are led to the following observations, which may be considered as so many propositions readily applicable to particular cases, or rules of philosophizing: we conclude, 1st, That if in our group of facts there be one in which any assigned peculiarity, or attendant circumstance, is wanting or opposite, such peculiarity cannot be the cause we seek.

(147.) 2d, That any circumstance in which all the facts without exception agree, *may* be the cause in question, or, if not, at least a collateral effect of the same cause: if there be but one such point of agreement, this possibility becomes a certainty; and, on the other hand, if there be more than one, they may be concurrent causes.

(148.) 3d, That we are not to deny the existence of a cause in favour of which we have a unanimous agreement of strong analogies, though it may not be apparent how such a cause can produce the effect, or even though it may be difficult to conceive its existence under the circumstances of the case; in such cases we should rather appeal to experience when possible, than decide *à priori* against the cause, and try whether it cannot be made apparent.

(149.) For instance: seeing the sun vividly luminous, every analogy leads us to conclude it intensely hot. How heat can produce light, we know not; and how such a heat can be maintained, we can form no conception. Yet we are not, therefore, entitled to deny the inference.

(150.) 4th, That contrary or opposing facts are equally instructive for the discovery of causes with favourable ones.

(151.) For instance: when air is confined with moistened iron filings in a close vessel over water, its bulk is diminished, by a certain portion of it being abstracted and combining with the iron, producing *rust*. And, if the remainder be examined, it is found that it will *not* support flame or animal life. This contrary fact shows that the cause of the support of flame and animal life is to be looked for in that part of the air which the iron abstracts, and which rusts it.

(152.) 5th, That causes will very frequently become obvious, by a mere arrangement of our facts in the order of intensity in which some peculiar quality subsists; though not of necessity, because counteracting or modifying causes may be at the same time in action.

(153.) For example: sound consists in impulses communicated to our ears by the air. If a series of impulses of equal force be communicated to it at equal intervals of time, at first in slow succession, and by degrees more and more rapidly, we hear at first a rattling noise, then a low murmur, and then a hum, which by degrees acquires the character of a musical note, rising higher and higher in acuteness, till its pitch becomes too high for the ear to follow. And from this correspondence between the pitch of the note and the rapidity of succession of the impulse, we conclude that our sensation of the different pitches of musical notes originates in the different rapidities with which their impulses are communicated to our ears.

(154.) 6th, That such counteracting or modifying causes may subsist unperceived, and annul the effects of the cause we seek, in instances which, but for their action, would have come into our class of favourable facts; and that, therefore, exceptions may often be made to disappear by removing or allowing for such counteracting causes. This remark becomes of the greatest importance, when (as is often the case) a single striking exception stands out, as it were, against an otherwise unanimous array of facts in favour of a certain cause.

# Selection 36

## NORWOOD RUSSELL HANSON

# The Origin

# of Hypothetico-Deductive

# Explanations

TYPICAL PHYSICAL LAWS are those of motion and gravitation, thermodynamics, electromagnetism, and of the conservation of charge in classical and quantum physics. These were not derived by Bacon's "Inductio per enumerationem simplicem, ubi non reperitur instantia contradictoria," but some philosophers have thought that they were. A second account treats these laws as high-level hypotheses in a hypothetico-deductive system (hereafter "H-D" system). It describes physical theory more adequately than did earlier accounts in terms of induction by enumeration, for it says what laws are, and what they can do, in the finished arguments of physicists. But it does not tell us how laws are come by in the first place; and the induction-by-enumeration story at least attempted this.

The two accounts are not alternatives: they are compatible. Acceptance of the second is no reason for rejecting the first. A law might have been arrived at by enumerating particulars; it could then be built into an H-D system as a higher order proposition. If there is anything wrong with the older view the H-D account does not reveal what the fault is.

Excerpts from N. R. Hanson, *Patterns of Discovery: An Inquiry into the Conceptual Foundations of Science.* New York: Cambridge University Press, 1958, pp. 70–73; 82–86. Reprinted by permission.

There *is* something wrong with the older view: it is false. Physicists rarely find laws by enumerating and summarizing observables. There is also something wrong with the H-D account, however. If it were construed as an account of physical practice it would be misleading. Physicists do not start from hypotheses; they start from data. By the time a law has been fixed into an H-D system, really original physical thinking is over. The pedestrian process of deducing observation statements from hypotheses comes only after the physicist sees that the hypothesis will at least explain the initial data requiring explanation. This H-D account is helpful only when discussing the argument of a finished research report, or for understanding how the experimentalist or the engineer develops the theoretical physicist's hypotheses; the analysis leaves undiscussed the reasoning which often points to the first tentative proposals of laws.

The inductive view is that the important inference is from the observation to the law, from the particular to the general. There is something true about this which the H-D account must ignore. Thus Newton wrote: "The main business of natural philosophy is to argue from phenomena." The simple inductive view, however, ignores what Newton never ignored: the inference is also from *explicanda* to an *explicans*. The reason for a bevelled mirror's showing a spectrum in the sunlight is not explained by saying that all bevelled mirrors do this. On the inductive account this latter generalization might count as a law: it would accord with Canon Raven's characterization of a natural law as a "summary of statistical averages." But only when it is *explained* why bevelled mirrors show spectra in the sunlight will we have a law of the type suggested, in this case Newton's laws of refraction. So the inductive view rightly suggests that laws are got by inference from data. It wrongly suggests that the law is but a summary of these data, instead of being what it must be, an explanation of the data.

H-D accounts all agree that physical laws explain data, but they obscure the initial connexion between data and laws; indeed, they suggest that the fundamental inference is from higher-order hypotheses to observation statements. This may be a way of setting out one's reasons for accepting an hypothesis after it is got, or for making a prediction, but it is not a way of setting out reasons for proposing or for trying an hypothesis in the first place. Yet the initial suggestion of an hypothesis is very often a reasonable affair. It is not so often affected by intuition, insight, hunches, or other imponderables as biographers or scientists suggest. Disciples of the H-D account often dismiss the dawning of an hypothesis as being of psychological interest only, or else claim it to be the province solely of genius and not of logic. They are wrong. If establishing an hypothesis through its predictions has a logic, so has the conceiving of an hypothesis. To form the idea of acceleration or of universal gravitation does require genius: nothing less than a Galileo or a Newton. But that cannot mean that the reflexions leading to these ideas are unreasonable or a-reasonable. Here resides the continuity in physical explanation from the earliest to the present times.

H-D accounts begin with the hypothesis as given, as Mrs. Beeton's recipes begin with the hare as given. A preliminary instruction in many cookery books, however, reads "First catch your hare." The H-D account tells us what happens after the physicist has caught his hypothesis; but it might be argued that the ingenuity, tenacity, imagination and conceptual boldness which has marked physics since Galileo shows itself more clearly in hypothesis-catching than in the deductive elaboration of caught hypotheses. Galileo struggled for thirty-four years before he was able to advance his constant acceleration hypothesis with confidence. Is this conceptually irrelevant? Will we learn much about Galileo's physical thinking if we just begin our analysis with the constant acceleration hypothesis as a basis for deduction? Was it only the predictions from this hypothesis which commended it to Galileo? The philosopher of science must answer "No."

Interpretation is not something a physicist works into a ready-made deductive system: it is operative in the very making of the system. He rarely searches for a deductive system *per se*, one in which his data would appear as consequences if only interpreted physically. He is in search, rather, of an explanation of these data; his goal is a conceptual pattern in terms of which his data will fit intelligibly alongside better-known data. Physics is not applied mathematics. It is a natural science in which mathematics can be applied.

In the thinking which leads to general hypotheses, there are

characteristics constant through the history of physics, from
Democritus and Heraclitus to Dirac and Heisenberg. Kepler
did not *begin* with the hypothesis that Mars' orbit was elliptical and then deduce statements confirmed by Brahe's observations. These latter observations were given, and they set the
problem—they were Johannes Kepler's starting point. He struggled back from these, first to one hypothesis, then to another,
then to another, and ultimately to the hypothesis of the elliptical orbit. Few detailed accounts have been given by philosophers of science of Kepler's achievements, although his discovery of Mars' orbit is physical thinking at its best. The
philosopher of physics should not neglect what Peirce calls
the finest retroduction ever made. . . .

Kepler blundered in his calculation of the planet's positions
at determined distances. Truth seemed very reluctant to de-
liver herself up to Kepler and he applied to her Virgil's

> Malo me Galataea petit, lasciva puella,
> Et fugit ad salices, et se cupit ante videri.

Again he failed; again Tycho's data resisted the pattern he
proposed. His distances, though observationally accurate,
were inconsistent with the elliptical form he ascribed to the
orbit. The observed distances would have required a new
oviform figure, again exceeding the ellipse in the first and
fourth quadrants and retiring within it in the second and third.
Kepler ascribed the error to his ancillary theory of librations,
wherein Mars was supposed to oscillate at right angles to its
orbit through its whole revolution. Though this gave accurate
distances, he abandoned it.

Now, with little conviction, Kepler returned to the ellipse.
This seemed the only way of preserving his other conceptually
valuable principles. He supposed that in doing this he was
appealing to something totally different from his theory of
librations. But slim hopes of success were held: "It is clear
therefore that the path has cheeks; it is not an ellipse. And
while an ellipse offers just equations, this cheeky figure offers
unjust ones." The reconsideration of the ellipse was thus
something into which Kepler retreated after finding no other
prospect for applying the principles he had established. The
ellipse as a physical hypothesis began to beckon. But now
Kepler became worried to distraction through trying to un-

derstand why Mars should abandon (in favour of an elliptical path) the librations, on whose assumption accurate calculations of distance were produced. He toiled on this problem like a man possessed, until finally his perplexities dissolved before an insight which transformed his data and all subsequent astronomy and physics. In his own words:

> Yet even this was not the crux of the matter. Indeed *the* great problem was this: that in considering and casting about to the very limits of my sanity, I could not discover why the planet, to which with such great probability and agreement of observed distances the libration *LE* in diameter *LK* could be attributed, why the planet should prefer an elliptical path indicated by the equations.
>
> Oh how ridiculous of me! As though the libration in diameter could not lead to the elliptical path. This idea carried no little persuasive force—the ellipse and the libration stand or fall together, as will be made clear in the next Chapter. There it will be demonstrated that there is no figure of a planet's orbit other than the perfect ellipse—the concurrence of reasons drawn from the principles of physics, with the experiences of observations and hypotheses being adduced in this chapter by anticipation.

This is a model of differences in conceptual organization. Before "O me ridiculum!" the libration theory and the elliptical theory were distinctly different for Kepler. The difference between "librations *v.* ellipse" and "librations = ellipse," is like the difference between the bird and the antelope, or the bear on the tree before and after identification.

The arithmetic blunder was quickly discovered, and the discovery led to the fullest confirmation of the hypothesis. As visual phenomena of ch. 1 were cast into patterns by the appreciation of some particular dot or line, so here the enormous heap of calculations, velocities, positions and distances which had set Kepler his problem now pulled together into a geometrically intelligible pattern. The elliptical areas were seen to be equivalent; similarly, the sums of the corresponding diametral distances were equal; the equations following from the ellipse were general expressions of Tycho's original data. All this made it clear that Mars revolves around the sun in an ellipse, describing around the sun areas proportional to its times.

Predictions to unobserved positions of Mars had not yet

been undertaken. Nor did Kepler feel it absolutely necessary to endure this before proposing the elliptical orbit hypothesis.

This was a physical discovery. Since the same physical conditions obtained throughout the solar system, the same equations ought to explain other planetary revolutions as well. These three great *explicantia* are the well-known result: (a) that planetary orbits are elliptical with the sun in their common focus (1609), (b) that they describe around the sun areas proportional to their times of passage (1609), and (c) that the squares of the times of their revolutions are proportional to the cubes of their greater axes, or their mean distances from the sun (1619). These are most important in the history of astronomy. They supplied the material for Newton's retroduction to the law of universal gravitation.

Of this monumental reasoning from *explicanda* to *explicans* could any account be more ludicrous than that of J. S. Mill, who argued that Kepler's law is just "a compendious expression for the one set of directly observed facts"? Mill had no real experience of theoretical astronomy. But he might have perused *De Motibus Stellae Martis*. Whewell is rightly uneasy about Mill's account. His alternative account, however, turns on Kepler's having got the hypothesis as a "colligating concept." This is little better than the modern hypothetico-deductive account which has it that Kepler succeeded "by thinking of general hypotheses from which particular consequences are deduced which can be tested by observation."

Kepler typifies all reasoning in physical science. Would it have required so much time, and genius, to "observe" the elliptical orbit in Tycho's data? *De Motibus Stellae Martis* is more than a compendious expression of Brahe's observations. Nor is it concerned with deducing geometrical consequences from the elliptical orbit hypothesis, "thought of" Kekulé-fashion. Kepler's task was: given Tycho's data, what is the simplest curve which includes them all? When he at last found the ellipse his work as a creative thinker was virtually finished. Any mathematician could then deduce further consequences not included in Tycho's lists. It required no genius to take Kepler's idea and try it for other planets.

Kepler never modified a projected explanation capriciously; he always had a sound reason for every modification he made. When he did make an adjustment which exactly satisfied the observations, it stood "upon a totally different logical footing from what it would if it had been struck out at random . . . and had been found to satisfy the observations. Kepler shows his keen logical sense in detailing the whole process by which he finally arrived at the true orbit. This is the greatest piece of Retroductive reasoning ever performed." What features of inference by retroduction does Kepler bring out? The reasoning from surprising data to an explanation binds together as "physics" centuries of inquiry, methods, techniques and problems.

Was Kepler's struggle up from Tycho's data to the proposal of the elliptical orbit hypothesis really inferential at all? He wrote *De Motibus Stellae Martis* in order to set out his reasons for suggesting the ellipse. These were not deductive reasons; he was working from *explicanda* to *explicans*. But neither were they inductive—not, at least, in any form advocated by the empiricists, statisticians and probability theorists who have written on induction.

Aristotle lists the types of inferences. These are deductive, inductive and one other called " ἀπαγωγή." This is translated as "reduction." Peirce translates it as "abduction" or "retroduction." What distinguishes this kind of argument for Aristotle is that

> the relation of the middle to the last term is uncertain, though equally or more probable than the conclusion; or again an argument in which the terms intermediate between the last term and the middle are few. For in any of these cases it turns out that we approach more nearly to knowledge . . . since we have taken a new term.

After describing deduction in a familiar way, Peirce speaks of induction as the experimental testing of a finished theory. Induction

> sets out with a theory and it measures the degree of concordance of that theory with fact. It never can originate any idea whatever. No more can deduction. All the ideas of science come to it by the way of Abduction. Abduction consists in studying facts and devising a theory to explain them. Its only justification is that if we are ever to understand things at all, it must be in that way. Abductive and inductive reasoning are utterly irreducible, either to the other or to Deduction, or Deduction to either of them. . . .

Deduction proves that something *must* be; Induction shows that something *actually is* operative; Abduction merely suggests that something *may be.*

. . . man has a certain Insight, not strong enough to be oftener right than wrong, but strong enough not to be overwhelmingly more often wrong than right. . . . An Insight, I call it, because it is to be referred to the same general class of operations to which Perceptive Judgments belong. . . . If you ask an investigator why he does not try this or that wild theory, he will say "It does not seem *reasonable.*"

Peirce regards an abductive inference (such as "The observed positions of Mars fall between a circle and an oval, so the orbit must be an ellipse") and a perceptual judgment (such as "It is laevorotatory") as being opposite sides of the same epistemological coin. *Seeing that* is relevant here. The dawning of an aspect and the dawning of an explanation both suggest what to look for next. In both, the elements of inquiry coagulate into an intelligible pattern. The affinities between seeing the hidden man in a cluster of dots and seeing the Martian ellipse in a cluster of data are profound. "What can our first acquaintance with an inference, when it is not yet adopted, be but a perception of the world of ideas?" But ". . . abduction, although it is very little hampered by logical rules, nevertheless is logical inference, asserting its conclusion only problematically, or conjecturally, it is true, but nevertheless having a perfectly definite logical form."

GERALD HOLTON

# On Themata

# in Scientific

# Thought

ALL MODERN ANALYSES of science agree that two types of propositions are scientifically *not* meaningless, namely, propositions concerning empirical matters of "fact," and propositions concerning the calculus of logic and mathematics that helps us to structure and analyze.

We can use these two generalized coordinates to define an x-y plane. It is the plane in which scientific discourse usually proceeds. A concept such as force may be considered as a point in the x-y plane. The projection on the x, or phenomenic, dimension corresponds to the empirical meaning of "force," i.e., its detection and measurement through, say, the distortion experienced by standard objects. The projection of the concept "force" on the y-dimension is its analytic meaning (vector property, e.g., parallelogram law of composition).

In the same manner, we analyze a statement, e.g., a hypothesis or the law of universal gravitational attraction, in terms of its phenomenic and heuristic-analytic components. Such an analysis is a "contingency analysis," because the value of a statement in the x-y plane is contingent on the possibility of (1) checking the phenomenic component (e.g., whether two masses do move closer in a Cavendish experiment) and (2)

Excerpt from G. Holton, *Thematic Origins of Scientific Thought: Kepler to Einstein*. Cambridge: Harvard University Press, 1973, pp. 190–92. Reprinted by permission.

checking the heuristic-analytic component (e.g., whether the analysis in terms of vectors in Euclidean space is more appropriate than, say, in terms of scalars). The x-y plane is thus the "contingent plane," where scientific concepts and propositions have both empirical and analytical relevance.

Now it has been the claim of modern positivism that statements are scientifically meaningful only insofar as they have components in the contingent plane. This attitude has also been the ruling one in the younger sciences such as psychology, and also in history, particularly the history of science. From Bacon, Kepler, and Newton on, all who have claimed not to feign hypotheses have been concerned with keeping the hypotheses they must use in the contingent plane. And this is one reason why science has grown so rapidly since the early part of the seventeenth century.

The fact, however, is that this aim is not and never can be fully achieved, and the analysis of historic cases in science should more widely begin to take into account that concepts and hypotheses used in science are meaningful not only in the contingent plane. The contingent components are merely two of three components, the results of a projection from an x-y-z space to the x-y plane. A concept such as force, for example, has also a *thematic* component, which is directly coupled neither to phenomena nor to tautological, analytic statements, but to the persisting theme of an active potency principle that stands behind the whole sequence of concepts from which our idea of force has developed: *energeia, anima, vis, Kraft*. Similarly, a hypothesis or a law such as the Law of Conservation of Momentum or of Energy can be more fully understood as a line element or configuration in three-dimensional "proposition space" with a projection on the third axis, that of themata corresponding to the delineation of the persistent motif or thema of "constancy" or "conservation."

Poincaré himself clearly expressed the role of themata in a passage in *La Science et l'hypothèse* (1902):

> Comme nous ne pouvons pas donner de l'énergie une définition générale, le principe de la conservation de l'énergie signifie simplement qu'il y a *quelque chose* qui demeure constant. Eh bien, quelles que soient les notions nouvelles que les expériences futures nous donneront sur le monde, nous sommes sûrs d'avance qu'il y aura quelque chose qui demeurera constant et

que nous pourrons appeler *énergie* [to which we now add: even when we used to call it only mass.] (Emphasis in original.)

A thematic position, or methodological thema, is a guiding theme in the pursuit of scientific work, such as the thema of expressing laws of constancy, of extremum, or of impotency, or quantification, or Rules of Reasoning. A thematic proposition, or thematic hypothesis, is a statement that is directly neither verifiable nor falsifiable, e.g., Einstein's principle of relativity in its modern sense and of the constancy of light velocity in free space—and Poincaré's basic belief in the ether.

The recognition of such thematic differences may help us understand the widespread feeling of paradox and outrage when a new thema is proposed in opposition to the prevalent ones—as was, of course, the case with relativity theory, so much so that Poincaré, to the end of his life in 1912, referred to Einstein's theory of relativity never once in print (and to Einstein, as far as I could discover, only once, on the subject of the photon, and in a derogatory way).

Finally, more can be said concerning the fact that it is the mark of a certain type of genius, in particular of the nonconservatory mind, to be "themata-prone"—at least for a time. On this point, nothing is more revealing than the transcript of the Solvay Congress of 1911, where Einstein and Poincaré met. The relativity theory was no longer an issue. The new topic was the quantum theory. It raised new, grave problems: for example, how could one understand probabilistic behavior on the atomic scale?

And right there and then, it was Einstein's turn to begin resisting a new theme, that of inherently probabilistic behavior. It now was he who began to warn that one should not feign a hypothesis located so far from the contingent plane. The next chapter in the history of science, from the point of view of thematic analysis, had opened.

HENRI POINCARÉ

# The Choice

# of Facts

THIS SHOWS US how we should choose: the most interesting facts are those which may serve many times; these are the facts which have a chance of coming up again. We have been so fortunate as to be born in a world where there are such. Suppose that instead of 60 chemical elements there were 60 milliards of them, that they were not some common, the others rare, but that they were uniformly distributed. Then, every time we picked up a new pebble there would be great probability of its being formed of some unknown substance; all that we knew of other pebbles would be worthless for it; before each new object we should be as the new-born babe; like it we could only obey our caprices or our needs. Biologists would be just as much at a loss if there were only individuals and no species and if heredity did not make sons like their fathers.

In such a world there would be no science; perhaps thought and even life would be impossible, since evolution could not there develop the preservational instincts. Happily it is not so; like all good fortune to which we are accustomed, this is not appreciated at its true worth.

Which then are the facts likely to reappear? They are first the simple facts. It is clear that in a complex fact a thousand circumstances are united by chance, and that only a chance still much less probable could reunite them anew. But are there any simple facts? And if there are, how recognize them?

Excerpts from H. Poincaré, *Science and Method* (originally published 1908), in *The Foundations of Science*. New York: The Science Press, 1929, pp. 363–67.

What assurance is there that a thing we think simple does not hide a dreadful complexity? All we can say is that we ought to prefer the facts which seem simple to those where our crude eye discerns unlike elements. And then one of two things: either this simplicity is real, or else the elements are so intimately mingled as not to be distinguishable. In the first case there is chance of our meeting anew this same simple fact, either in all its purity or entering itself as element in a complex manifold. In the second case this intimate mixture has likewise more chances of recurring than a heterogeneous assemblage; chance knows how to mix, it knows not how to disentangle, and to make with multiple elements a well-ordered edifice in which something is distinguishable, it must be made expressly. The facts which appear simple, even if they are not so, will therefore be more easily revived by chance. This it is which justifies the method instinctively adopted by the scientist, and what justifies it still better, perhaps, is that oft-recurring facts appear to us simple, precisely because we are used to them.

But where is the simple fact? Scientists have been seeking it in the two extremes, in the infinitely great and in the infinitely small. The astronomer has found it because the distances of the stars are immense, so great that each of them appears but as a point, so great that the qualitative differences are effaced, and because a point is simpler than a body which has form and qualities. The physicist on the other hand has sought the elementary phenomenon in fictively cutting up bodies into infinitesimal cubes, because the conditions of the problem, which undergo slow and continuous variation in passing from one point of the body to another, may be regarded as constant in the interior of each of these little cubes. In the same way the biologist has been instinctively led to regard the cell as more interesting than the whole animal, and the outcome has shown his wisdom, since cells belonging to organisms the most different are more alike, for the one who can recognize their resemblances, than are these organisms themselves. The sociologist is more embarrassed; the elements, which for him are men, are too unlike, too variable, too capricious, in a word, too complex; besides, history never begins over again. How then choose the interesting fact, which is that which begins again? Method is precisely the choice of facts;

it is needful then to be occupied first with creating a method,
and many have been imagined, since none imposes itself, so
that sociology is the science which has the most methods and
the fewest results.

Therefore it is by the regular facts that it is proper to begin;
but after the rule is well established, after it is beyond all
doubt, the facts in full conformity with it are erelong without
interest since they no longer teach us anything new. It is then
the exception which becomes important. We cease to seek
resemblances; we devote ourselves above all to the differences,
and among the differences are chosen first the most accen-
tuated, not only because they are the most striking, but because
they will be the most instructive. . . .

So when a rule is established we should first seek the cases
where this rule has the greatest chance of failing. Thence,
among other reasons, come the interest of astronomic facts,
and the interest of the geologic past; by going very far away
in space or very far away in time, we may find our usual rules
entirely overturned, and these grand overturnings aid us the
better to see or the better to understand the little changes
which may happen nearer to us, in the little corner of the
world where we are called to live and act. We shall better
know this corner for having traveled in distant countries with
which we have nothing to do.

But what we ought to aim at is less the ascertainment of
resemblances and differences than the recognition of likenesses hidden under apparent divergences. Particular rules
seem at first discordant, but looking more closely we see in
general that they resemble each other; different as to matter,
they are alike as to form, as to the order of their parts. When
we look at them with this bias, we shall see them enlarge and
tend to embrace everything. And this it is which makes the
value of certain facts which come to complete an assemblage
and to show that it is the faithful image of other known
assemblages.

I will not further insist, but these few words suffice to show
that the scientist does not choose at random the facts he observes. He does not, as Tolstoi says, count the lady-bugs, because, however interesting lady-bugs may be, their number is
subject to capricious variations. He seeks to condense much
experience and much thought into a slender volume; and that

is why a little book on physics contains so many past experiences and a thousand times as many possible experiences whose result is known beforehand.

But we have as yet looked at only one side of the question. The scientist does not study nature because it is useful; he studies it because he delights in it, and he delights in it because it is beautiful. If nature were not beautiful, it would not be worth knowing, and if nature were not worth knowing, life would not be worth living. Of course I do not here speak of that beauty which strikes the senses, the beauty of qualities and of appearances; not that I undervalue such beauty, far from it, but it has nothing to do with science; I mean that profounder beauty which comes from the harmonious order of the parts and which a pure intelligence can grasp. This it is which gives body, a structure so to speak, to the iridescent appearances which flatter our senses, and without this support the beauty of these fugitive dreams would be only imperfect, because it would be vague and always fleeting. On the contrary, intellectual beauty is sufficient unto itself, and it is for its sake, more perhaps than for the future good of humanity, that the scientist devotes himself to long and difficult labors.

It is, therefore, the quest of this especial beauty, the sense of the harmony of the cosmos, which makes us choose the facts most fitting to contribute to this harmony, just as the artist chooses from among the features of his model those which perfect the picture and give it character and life. And we need not fear that this instinctive and unavowed prepossession will turn the scientist aside from the search for the true. One may dream a harmonious world, but how far the real world will leave it behind! The greatest artists that ever lived, the Greeks, made their heavens; how shabby it is beside the true heavens, ours!

And it is because simplicity, because grandeur, is beautiful, that we preferably seek simple facts, sublime facts, that we delight now to follow the majestic course of the stars, now to examine with the microscope that prodigious littleness which is also a grandeur, now to seek in geologic time the traces of a past which attracts because it is far away.

We see too that the longing for the beautiful leads us to the same choice as the longing for the useful. And so it is that this economy of thought, this economy of effort, which is, according to Mach, the constant tendency of science, is at the same time a source of beauty and a practical advantage. The edifices that we admire are those where the architect has known how to proportion the means to the end, where the columns seem to carry gaily, without effort, the weight placed upon them, like the gracious caryatids of the Erechtheum.

# Part VIII

# Scientific Creativity (2): Experiment and Theory

# Introduction

THE POWER OF the scientific method is most fully realized when the investigator can (as Bacon recognized) observe the phenomena of interest precisely and completely, *and* (as Bacon may not have recognized) can develop theories which not only explain known data but push on into the unknown. Both activities—observing and theory making—are creative, and constitute important problems for the cognitive psychology of science. That theory construction is a creative enterprise is commonly acknowledged. But the special kind of observation that occurs in experimentation, and which greatly increases the power of the observational method (Mill 1843; Bernard 1865/1927), also puts heavy demands on the creativity and originality of the investigator. The best experiments—Pasteur's studies of bacterial infection, Pascal's barometric investigations, Pavlov's studies of conditioned reflexes, Rutherford's nuclear scattering experiments, Millikan's oil drop experiment—possess an elegant simplicity. Such simplicity stands as a tribute to the creativity of the experimenter.

Experiments are powerful because of their quality of "induced observation" (Bernard 1865/1927:19), and because control over extraneous factors is possible (Boring 1954). Thus, the experimental scientist is an active knower, not a passive one. That interaction between human activity and human thought is necessary in the production of knowledge is a relatively new idea in psychology, and only recently have its implications been recognized. It is a major theme in the works of thinkers as disparate as Piaget (e.g., 1970) and B. F. Skinner (1971). For both, no simple empiricism, and no simple rationalism, will account for human knowledge (see also Anzai & Simon 1979).

Knowledge is produced from the actions of the knowing organism. Blind actions are not enough to produce knowledge, however. Though chance may play a role (see Campbell 1974; Crovitz, selection 39; and Gruber, selection 40), it is *guided* chance. The observations must be planned. To produce knowledge, activity must have a goal, and must be implemented according to some plan (Miller, Galanter, and Pribram 1960). An empirical challenge to the psychology of scientific thinking is located here: Is it possible to characterize the cognitive mechanisms (beyond the traditional categories of hypotheses, and tests of hypotheses) which are used to guide the experimental and theoretical activity of the scientist? In other words, can we discover the psychological nature of any of the criteria implicated in the creation of hypotheses?

In this connection, the value and use of the principle of simplicity as a cognitive device merits the attention of cognitive psychologists. In theory construction and evaluation, simplicity has long been a guide, as it was to Copernicus (Kuhn 1957) and Einstein (Frank 1947). In the final selection of part 7, Poincaré voiced a preference for the pursuit of simple and elegant facts. Is the desire for simplicity purely an aesthetic preference, purely a matter of style? Or does it play a role in reducing the cognitive struggle with complex material? When the scientist says "There must be a simpler way," is there an implication that there must be an *easier* way as well? Considerations of this sort have not traditionally been addressed in studies of problem solving, except within the Gestalt tradition (Wertheimer 1945), but they will need to be a part of any complete cognitive psychology of science. What are the laws of "Good Gestalt" that govern the relative simplicity of a scientific formulation? The question is an empirical one. Is the ubiquity of analogy in scientific thought merely a strain-reducing device? Or is it more, something that reflects a fundamental presupposition about simplicity as a property of nature? This too is at least partly an empirical question.

In Part 7, we described certain ways in which the construction of taxonomies relied upon recognition of similarity and difference. Similar processes are to be found in the construction of experiments and theories, but in this domain, where the investigator plays a more active role, they are more likely to be associated with the use of metaphors, analogies, and models (in increasing order of formal exactness). The similarity and difference between a problem at hand and some more familiar domain serves to mediate the construction of those explicit links between elements that are needed to design good experiments or to build powerful theories.

The essential function of metaphors, analogies, and models is to bring to bear something relatively well understood on something less well understood, The use of a metaphor implies the fuzziest, or least formal, relation to the thing to be explained. The metaphor may in itself not be terribly well understood, and may fulfill explanatory functions for a wide variety of phenomena. Metaphoric thinking may thus be fairly close in its impact to that provided by fundamental presuppositions. The term "analogy" is usually reserved for more focused vehicles, again with the implication that the vehicle is borrowed from some area of human knowledge which is better known than that which the scientist is attempting to understand. Gruber (selection 40) uses the terms "image of wide scope" and "metaphor" in roughly equivalent ways, though certain of his metaphors, such as "the irregularly branching tree" and "the tangled bank" are metonyms (parts representing the whole) rather than metaphors. Metonymy may be a rather different psychological mechanism than either metaphor or analogy, since the scientist is coming from inside the system he or she is working to understand, rather than from outside. Finally, models are the most formal, most precisely specified of the three. Use of models becomes, at the extreme, inseparable from the construction of theory. Even so, models may contain features which are definitely not part of the phenomena to be explained. Thus, Einstein's epochal 1905 paper on quantum oscillations in black body radiation used the model of rigid spheres attached to the walls of the radiating cavity by little springs. The model led to correct predictions of Wien's Law, but no one, even for a moment, assumed that the springs were real.

All three devices are widely used in science (Hesse 1966; Leatherdale 1974). Maxwell's contributions in electromagnetic theory were, at least in part, founded on a mechanical model, a process admired by Kelvin and emulated by Helmholtz (Klein 1970:55ff). Dalton built little wooden models to aid in the conceptualization of the atom. Watson and Crick relied on cardboard molecular models in arriving at their hypothesis about DNA structure, and "verified" it using more precisely fabricated metal models (Watson 1968). Uhlenbeck and Goudsmit relied on a planetary analogy in formulating the idea of electron spin (Gutting 1978). Darwin (1859) used artificial selection as a model for natural selection. Recent work in cognitive psychology has used the structure of computer programs as a model of human thinking (e.g., Simon 1969), while work in artificial intelligence has used human thinking as a model for computer thinking (Raphael 1976).

There is no guarantee that the use of the devices we have been describing will lead to the truth. In that sense, metaphors, analogies, and models are heuristics, in exactly the same sense as Polya (1945) and Newell and Simon (1972) use that term. There is, in science, no calculus for discovery and no algorithm for creativity (contrary to a Renaissance belief—cf. Wilkins 1668). The implication is that the problem of understanding how we have come to know anything at all is a problem that amounts to characterizing the heuristics used in successful science, and characterizing the conditions under which they are used. That is a task for which contemporary cognitive science is well prepared.

The use of metaphors, analogies, and models is not unique to scientific thinking. Metaphor is a principle vehicle for understanding and for communication in art and literature. The ability to construct internal models can also be conceptualized as a fundamental feature of all human thinking. Craik describes thinking as involving construction of an internal model of a real process. The likeness of the internal model to external reality is a key point for him:

> If the organism carries a "small-scale model" of external reality and of its own possible actions within its head, it is able to try out various alternatives, conclude which is the best of them, react to future situations before they arise, utilise the knowledge of past events in dealing with the present and future, and in every way to react in a much fuller, safer, and more competent manner to the emergencies which face it. (1943:61)

The analogy between a brain and a calculating machine was explicitly used by Craik in developing his account.

The relevance of this model for science is shown by one of

Kuhn's later accounts (1974) of the role of paradigms in scientific thought.[1] In asking how scientists attain correspondence rules that relate scientific formalisms to phenomena, Kuhn pointed to the central role of **examples** of successful science. Thus, Huygens used Galileo's inclined plane experiments as a model for experiments on the pendulum. Similar modeling processes occur throughout science education. Kuhn suggests that "an acquired ability to see resemblances between apparently disparate problems plays in the sciences a significant part of the role usually attributed to correspondence rules. . . . Having seen the resemblance, one simply uses the attachments that have proved effective before" (Kuhn 1974:471). Recognition of such resemblances accounts for a significant part of the "mystery" of scientific creativity.

The distinction between scientific thinking and other manifestations of human thought deserves special comment. The literary use of analogy often adds layers of new meaning to a partly understood human experience. The description of experience with multiple metaphors and analogies may be highly desirable from a literary standpoint, since such descriptions successively enrich the experience. But for the scientist, the role of metaphors and analogies has an additional dimension. For the scientist, the metaphor, to be fully useful, must have (turning Ziman's phrase) contrivable consequences. These consequences must be assessable by public observations to be useful scientifically. Where the artist or poet uses metaphors to enrich understanding and to communicate, the scientist takes another step and examines empirical consequences in an attempt to assess something like the "truth value" of the metaphor.

If devices such as analogy and metaphor are essential in science, as seems likely, then the psychology of science must ask both why and how. It is apparent that construction of a model, analogy, or metaphor capitalizes upon a similarity relationship. Thus, as we argued in Part 7, an analysis of the cognitive process by which such likenesses are utilized will be of value. Philosophers of science (Polanyi, 1967; Hesse, 1966; Black, 1962, Bunge, 1962, Harré, 1970; Darden, 1978) have given attention to such issues, but there is a clear need for a systematic psychological analysis, as well. Boyd (1979), in a recent collection of important papers on metaphor (Ortony, 1979) has taken a first step by arguing that the use of metaphor in science is more than "merely" heuristic in that metaphors serve as a means by which real relationships are accommodated to the existing scientific vocabularly. Kuhn (1979), in a commentary on Boyd's paper, essentially agreed, and enlarged upon the similarity of models and metaphors in science. We have no quarrel with such an approach, except to point out that the "merely" heuristic is of central concern to the cognitive psychology of science.

The application of the processes of metaphor, analogy, and modeling to the construction of scientific theory and experiment is a major theme of the selections in this part. We begin with Crovitz's attempt (selection 39) to push to its extreme limits the belief that random combinations of ideas are critical to scientific creativity. Can processes based on the mere random rearrangement of ideas be used to account for problem solving in general, and science in particular? As an algorithm, Crovitz's technique is simple in the extreme, but it is important to note that it can only be effective when an appropriate prior choice of terms to use in the algorithm has been made. It is not based on pure chance.

In selection 40, Gruber argues that a kind of "chancy interaction" may be detected in the creative aspects of Darwin's thought. Serendipity arises, on this view, from a well-stocked mind in constant movement. Thus, where the usual associationist view of creativity conceives of the associated ideas as mere passive elements, Gruber suggests that they are, in fact, highly dynamic entities in their own right. The metaphor of a zoo, with exotic ideas noisily interacting, is more appropriate than the metaphor of silent chemical combinations among passive elements. Similar claims about the chanciness of ideational combinations have been made by William James (1880) and more recently by Donald Campbell (1974).

A brief, but greatly illuminating, consideration of metaphors, analogies, and models is made by Hesse in selection 41. Her argument should make clear that controversy surrounds the points we are making concerning the functional utility of models. As with philosophical issues surrounding hypothesis testing, it is a controversy which can be approached empirically (see also Weitzenfeld & Klein 1979). The second excerpt by Gruber (selection 42) provides an example of how simple, and yet how rich, metaphoric processes can be. In fact, they need not have a "one-time-only" character but can instead inform every aspect of a thinker's work, even across decades of inquiry.

Martin Deutsch (selection 43) provides an illuminating discussion of the use of model, analogy, and metaphor in physics. According to Deutsch, so much of what is studied in the modern laboratory is in principle unobservable, that the use of cognitive devices to render the phenomena mentally tractable is required. We can regard such devices as techniques by which the inner meaning of phenomena are "externalized" for human understanding.

In fact, in extreme cases, a mental construction is *all* that is necessary for a major advance in science. Selection 44, by Kuhn, makes this dramatically clear, but also relates such "thought experiments" to common (if no less extraordinary) processes in cognitive development. In science, thought experiments are useful because they are one means of allowing one to discover paradox, and to discover inconsistencies (see Born 1949). Inconsistencies, paradoxes, and anomalies have long been recognized as highly influential in scientific thought. Kuhn, of course, makes them the triggers of scientific revolutions. While we can be suspicious of claims of the discovery of empirical truths via thought experiments (as Tesla is said to have done—see Shepard 1978:142), there is no question that, as Kuhn shows, major advances have been made in science through such purely mental activities as that involved in Einstein's clock "experiments."

Thus having begun part 8 with Crovitz's almost (but not quite) "mindless" algorithm for scientific creativity, we close with Kuhn's consideration of the most mindful construction possible—the creative use of pure thought.

## Note

1. The paradigm concept has proved to have a multitude of meanings which we will not try to elaborate. See Kuhn 1962/1970; 1974; Lakatos and Musgrave 1970; and for a negative view, Toulmin 1972.

# Selection 39

HERBERT CROVITZ

# An Algorithm

# for Creativity

WE HAVE ALREADY quoted Gardner (1968) . . . as saying what has been said uncountable times before: that all there is to discovery and invention is putting a couple of old things in a new relation . . .

The elementary form of a statement of an action is "taking one thing in some relation to another thing." The same elementary sentence is the form of a sentence to describe the statement of a discovery, an invention, or, indeed, an idea. The "things" can be anything you like, as multiple as the things of the world can be, but the relations are small in number, in point of fact, if the elementary form of an action-sentence is "Take one thing dash another thing," there are only 42 words in the set of all Basic English words that can by the wildest stretch of imagination fit where the dash is. The words follow (from Crovitz 1967).

| about | at | for | of | round | to |
|---|---|---|---|---|---|
| across | because | from | off | still | under |
| after | before | if | on | so | up |
| against | between | in | opposite | then | when |
| among | but | near | or | though | where |
| and | by | not | out | through | while |
| as | down | now | over | till | with |

An old experiment in the psychology of problem solving is important now. It is the *Umweg* problem. Put some corn on the ground. Put a transparent barrier, like a sheet of glass, in front of the corn. Put a hungry chicken in front of the glass. The barrier is between the chicken and the corn. The chicken acts in a very stupid manner. It tries to get through the barrier that it cannot get through, to go to the corn which it cannot go to, but rather must go away from to go round the barrier to get to the corn. The chicken is wearing a relational-filter before its mind's eye; all it will act on is going straight to the corn.

Somebody described an analogous situation. There is a very long line of traffic in the turn-left lane of a highway, but none in the straight-through right lane. Drivers, however, wear their turn-left filters before their mind's eye, seldom doing the "creative, intelligent" thing of going straight across, then going round the block, thence wheeling down the road before any of those twenty cars ahead have got past the stoplight.

Success in problem solving comes from changing the relational-filter before the mind's eye, tacitly rippling through the set of actions that are possible, and then having the wit to recognize a solution when you see one. Or, if you are very stupid but very quick, like a computer, trying *all* the possible actions till the goal is attained and it is time to stop.

This computer maneuver has been simulated by busy inventive men whose biographers have found their methods out. Perhaps the most instructive case has to do with building a working model of the system you want to change, and then tinkering with it. Tinkering with a model of the thing is harmless; in an undisciplined run through the possible actions you may be lucky enough to come upon a solution. Thomas Edison was a great one for tinkering with models of things, and might be a pattern on which to build a discoverer-computer.

Edison himself was no slowpoke. The glory of his youth was that he had a fast, legible hand, and was capable of writing at a rapid rate. But, according to a son, later on he had little power of abstraction, being always in need of seeing a thing in its concrete form, and tinkering with it. One of the problems that is described well by Josephson (1959:123) is Edison's work on the quadraplex—his putative discovery of how to send four simultaneous messages along a telegraph wire, two in each direction at once. This was important at the time—it would quadruple the power of the telegraph without the need of having to build four times the number of wires. Edison built an "analog" of the electric wire, with pipes and valves and assorted gadgets for affecting the flow of the water in the pipe. Using the gadgets to force water back and forth, in the pattern of wires in the system that was planned, he tinkered and ended with separating the separable features of the flow of current, sending one message controlled by one, and another controlled by another (Josephson, p. 124).

Edison also had a workshop that does in space what notebooks do in time, allowing one project to infect a neighboring one, so that moves made here may also be tried out there. His lab was a big barn with worktables along the room holding separate projects in progress. He would inspect one, then another. Not a bad trick for an ancient to come upon for switching relational-filters. That and working a 20-hour day stood him in good stead indeed. Of course, by his lights he was a failure where inventions were concerned, for he never made any money from them except when he set up commercial ventures to control that practical part of things himself. Yet, when he visited Pasteur once, Edison did not make a very good impression, for Pasteur was horrified at glimpsing the distractions attached to Edison's commercial ventures.

Notice that the switching of relational-filters is hard for problem solvers to do naturally. The relevant history of intellectual precursors of this small work is reviewed engagingly by Mandler and Mandler (1964), who covered in some detail a history of changing emphasis that evolved from analysis of raw consciousness to inquiries into problem solving. They detected a growing tendency to probe the processes that intervene between problem and solution (p. 276). That review ends with Duncker.

Duncker studied this problem: to get rid of an inoperable stomach tumor without harming the healthy tissue of the body, by using rays that can be modulated in intensity; at a high enough intensity, they destroy. He gave a protocol, but not a good one. The experimenter kept butting in, giving hints, asking leading questions. Nonetheless, the remarks of the subject— his attempts to solve the problem aloud—are instructive when they are put into relational language. The sequence of comments, so translated, is this:

1. Take the rays *through* the esophagus.
2. Take the sensitivity *from* the tissues.
3. Take tissue *off* the tumor.
4. Take strong rays *after* weak rays.
5. Take a shield *on* stomach walls.
6. Take the tumor *across* the stomach.
7. Take a cannula *through* the stomach wall.
8. Take the power *from* the rays.
9. Take the tumor *to* the exterior.
10. Take strong rays *after* weak rays.
11. Take strong rays *after* weak rays.
12. Take the rays *from* the body, or take the power *from* the rays.
13. Take the tumor *at* the focus of rays.
14. Take one ray *across* another ray.

It is possible to devise an irreverent method for solving "practical" problems with the use of elementary sentences that list the *full* set of possible actions that that form of sentence can cover in the vocabulary of Basic English. That is the amusing task of the next chapter.

One thing that is wrong with the mind's eye is that it has relational-filters that get stuck. Let us speed up the exchange of filters.

With respect to the x-ray problem as given by Duncker, there are at least two domain-words, "ray" and "stomach-tumor." The Basic English translation of ray is "any of a number of straight lines going out in different directions from a common point." Stomach is "a bag-like expansion of the digestion-pipe." Tumor is "an abnormal mass from normal body materials." Thus the problem might become using lines to get rid of a mass in a bag. In this particular case, the translation into Basic English reduces the problem very quickly to a diagrammatic stark form of it. I leave the playing of such a diagram to the reader, merely noting that, given a mass in a bag and given lines emanating from a point, there are many possibilities other than framing the question of how to get rid of the mass, i.e., take the mass from the bag. There are words other than *from*.

We can learn a small but very important thing from some protocols we collected on the relational algorithm in Duncker's problems from college students, one of whom was poor Penelope. . . .

A group of 10 undergraduates was given the x-ray problem and the single sentence, "Take one ray across another ray." Everybody quickly saw the solution. But not Penelope. She rejected the sentence, never having her mind's eye put the tumor at the *intersection* of the two rays. Alas. It is not automatic; success is not assured; there is no royal road nor indeed any commoner's road that is sure to lead to knowledge.

Given the "laws of probability," chances are that a bunch of monkeys at a bunch of typewriters would peck out nonsense, but take the extreme case of their typing out the works in the British Museum. Miller (1951:809) treated this problem with great generality. Even given the sequential probabilities built into the English language, it would take quite a while for the monkeys to come upon anything that made much sense. Miller said that generating sense was *not* the interesting problem. Men or monkeys or machines could be given an algorithm to do it. The problem, said Miller, is that the monkeys would not know the wheat from the chaff.

What one needs are rules for matching output to a standard, the making of such a match being a signal that sense has been made and the typing may stop. However, we may note that the best of us do sometimes miss the significance of what we have before us—indeed, it has commonly been supposed that a rather mystic restructuring of significance underlies the sudden recognition that something is a solution to something that is a problem.

How is the monkey at the typewriter to know to stop? Say we chose to set up a computer to spin out randomness until it came to some, for us, recognizable order. How would we fix the machine to go and then stop? Stopping is the whole of the problem of creative problem solving if we can run off the set of possible alternatives. We need the wit to recognize a solution—or, in plain talk, we need to stop at the right alternative. How is that to be arranged?

Suppose we know the goal. In the x-ray problem it is most easily stated in two parts: first, the tumor is destroyed; second, the tissue that is not the tumor is unaffected. The second part is always satisfied while the first part is simultaneously satisfied when the goal is transformed into this: tumor is in body; body is around the tumor. The goal is satisfied when we can reach the sentence: the tumor is destroyed; the body around it is unaffected. We draw a figure. On the line *AB* the energy is constant and too low to destroy tissue. On the line *CD* the energy is also constant and too low to destroy tissue. So at *a,b,c,* and *d*—the body around the tumor—there is no effect. But at the crossing, the energy sums and whatever tissue is there is destroyed. Put the tumor there—the tumor is destroyed; the body around it is unaffected. The sentence representing the goal is matched! Stop!

We will plod along through Duncker's problems pairing elements from the domain of the problem through one of the 42 relation words that we took from Basic English. Often coming upon a solution possibility in this way is easy and swift. When one is involved in any problem the number of possible pairs of domain-words is large—it is precisely

(1) $x!$

where $x$ = the number of elements accessible in the domain of the problem. For a single run through the 42 relations in this case one must scan $42x!$ which in the case of a domain with as little as six elements is

(2) $42(6)(5)(4)(3)(2)(1) = 30{,}200$ sentences.

However, in the case in which an accessible element from the domain of the problem is duplicated so it can be related to *itself*, the total number of sentences that need be scanned is

(3) $42(1) = 42$ sentences.

Let us consider the separating of a tangled problem by a clean slice separating the domain to be considered into two identical halves. In the x-ray problem we can leave the esophagus, the throat, the tissues, the stomach wall, the x-ray, and the tumor out in the miasmal fog holding the bag with 30,200 possibilities therein. Instead we choose to take two—two of something: in this case, *fairly* obviously, it must be two rays.

The 42-algorithm is a labored thing except for speedy computers or when the slice is made. After all, in the case of a six-element domain, relating each of the six with another of itself will give as few as

(4) $42(x) = 42(6) = 252$ sentences,

although only the most muddle-headed or the most creative of us will have much choice choosing among relating two esophaguses versus two throats versus two tissues versus two stomach walls versus two x-rays versus two tumors. In the x-ray problem, where one of the relata, the x-ray, is always assumed to be one necessary element, if there are five other elements, the number is still

(5) $42(x) = 42(6) = 252$ sentences.

In terms of the take-two rhythm of this Gordian knot simplification: When each ancient had gone out from preparation into incubation, re some invention or discovery, could he have picked his solution out of a list of sentences that numbered 42 in all, when the single element of the problem domain to relate to itself had some special mark of relevance to pick it out? When Mendeleev set out *to arrange the elements* in a good sequence, it was "element dash element" that he needed. When Adams and Leverrier sought *to understand the peculiar orbit* of Uranus, it was "orbit dash orbit" that they needed. Should you seek to solve the problem of weakening the sonic boom, perhaps it is *"noise dash noise"* that you would need to solve that real problem—plus some verificational expertise: taking "a noise under a noise" (the new one under the source of the jet's boom) might deflect the boom or destroy the plane. Who will find out? . . .

In accordance with the view that the deep importance of simple answers might extend as far into fog as where we are, consider now a simple analogy for building a theory. The turnkey model supposes that there are a finite number of answers that are like keys; there is a large number of questions that are like locks. To "build" a theory is to find which locks are opened by which keys. It is very topsy-turvy, of course. The job of the theorist is to find some beautiful, or elegant, or merely simple, keys—and then scout around for beautiful questions that available keys unlock.

However, even a turnkey must keep his wits about him lest he fail to recognize when a key does fit a lock. One of the most inventive of men was Father Scheiner, a contemporary and, for good or ill, a reluctant enemy of Galileo. He measured the curvature of the anterior surface of the cornea of the eye. His method was really the same that Newton later used to invent the reflecting telescope. The fact of the matter is that warped mirrors give distorted images—nicely polished perfect spherical mirrors distort, or change, the size of the image only, proportional to the angle of curvature.

Father Scheiner had some marbles. The set of shiny round balls varied in regard to the diameter of the individual balls. Scheiner sat his people down facing a window with a cross on it—that cross, of course, being visible upon the cornea as a reflected image. Then Father Scheiner took his bag of marbles and placed one or another marble in the space between eye and nose, stopping when the image of the cross reflected from the eye was about equal in size to the image of the cross as reflected from the marble. It is easy to measure a marble; from it can easily be deduced the angle of curvature of the anterior surface of the cornea, *quid est demonstrandum*. When we look very closely at inventions, we have grim cause to recognize that, though the idea "take a marble on the nose" might "solve" the problem of measuring corneas, it is not at all insured that anyone has the wit unfailingly to recognize it as a solution—that is, to match it to a question. If anyone did, what else might he invent or discover in a bag of marbles?

From the point of view of the turnkey model of theory and discovery, the limit to the number of keys (answers) is unknown. Similarly, there is no charted course to arrive at locks (questions). Consider the analogous problem in physics for a moment. Galileo may have played with the law of falling bodies in thought, but in action he played with rolling marbles down inclined grooves. (Heaven help us to avoid hard and unclear things to do!) Given the rules of marbles rolling down grooves, Galileo generalized the rule. To generalize is to assert that *this* thing is just like all *those* things which are too far away to measure directly.

An aim of this book was to clarify the theoretic basis of the study of thought. While the intention to speak to the basis of the study of thought is obviously a very ambitious one, the work is limited to some very simple tactics. The nontrivial study of thought seems impossible to all very wise people, fruitless to all wise people, very difficult to all very intelligent people, and metaphysically hopeless to everyone else. What was needed is a chimneysweep to chop an open path for the escape of smoke and gases. Therefore, I asserted that there are three parts to thought. Each of these three parts was considered. For each a simplified theoretical basis was described; an algorithm was given for each, and protocols of the use of the algorithms were given.

The term "algorithm" means method, or procedure. Lately the word has been taken up by computer theorists and allied mathematicians. . . . The specific algorithms described, if they are used, are for the use of people; they would have to be drastically transformed for the use of machines, dolphins, honeybees, or geniuses.

Anyone who has wrestled with computer programming knows, to his sorrow, that programs and algorithms do not simplify inquiry. Rather they make the direction of the work *painfully* clear. A computer goes straight through a terribly tedious procedure to come at the end to a result; its glory is that it is *fast*. The writing of a program or an algorithm requires great alertness and patience—everything must be made painfully and haltingly explicit. Following an algorithm also requires patience. As Justus Buchler (1961:86) pointed out, groping is one price a finite creature has to pay in searching according to a method; I add that "repugnant docility" is another.

**Selection 40** _____

HOWARD E. GRUBER

# Chance,

# Choice

# and Creativity

TO UNDERSTAND THE relation between purpose, freedom, and creativity, we need to examine further the role of chance in creative work. For purposes of this discussion a thought process can be sub-divided into these two parts: the production of new ideas and the reorganization of ideas (both old and new) into new patterns. In saying this, of course, I am using *ideas* to refer to elements or sub-units of which larger patterns are composed.

We know very little as yet of the process by which a new idea is produced. Let us suppose that each new variant is not an isolated idea but a change in the properties of some larger mental structure of which it is a part. Let us suppose, furthermore, that those parts of a system of ideas that are free to vary at any given moment are variable only within certain limits. This would be analogous to what we might say of any other part of a living system—that it is variable in its functioning, but within limits that depend on its place in the system as a whole.

There are many things we do not know. How are individual ideas produced? Are new ideas common or rare? Are they

Excerpt from H. E. Gruber and P. H. Barrett, *Darwin on Man: A Psychological Study of Scientific Creativity.* New York: E. P. Dutton, 1974, pp. 248–254 (footnotes and page references to Darwin's journals deleted). Reprinted by permission of the publisher.

recurrent or unique? In spite of all this ignorance, there are a few definite things we can say. . . .

The belief that a useful theory of creativity absolutely requires us to understand the source of variant or new ideas is akin to Darwin's feeling, against which he struggled, that his theory of evolution must include an explanation of the cause of heritable variation. Of course, we would like to know the answers to these questions, but we can do useful work by taking the occurrence of variants as given, and then studying the process of search, selection, and reorganization in which a new argument is constructed.

If the occurrence of each new idea is a unique event, we can hardly apply the laws of chance to such events, since these laws deal only with statistical aggregates of many events that can all be treated as members of the same population. But even if new ideas are unique events, we still want some conception of the role of chance in the growth of thought. There is a way.

We need an approach that would do justice to our image of the creative person as well organized, purposeful, aware of the manifold possibilities that exist in the world, and therefore free to choose among them. We need an approach that respects the kinship of mental processes with other living systems: variation and novelty are not chaotic or unrelated to the organism's past, but express the degrees of freedom characteristic of a particular organization as it stands at one moment in its history. . . .

This is the crucial fact. When the paths of previously independent systems intersect, the result is something new, not predictable from our knowledge of the special laws governing each system in isolation.

We need only add that the very occurrence of any particular intersection is unpredictable from those special laws. It follows that the sum of these interactions is, like Darwin's irregularly branching tree of nature, a collection of occasions, each a lawful event, but their totality unpredictable from the knowledge of their parts. This is the way to conceive of the operation of chance in the panorama of unique events we know as wild nature.

Any time we want to search for some lawful relationships, we can submerge the differences among things unique, find some similarities and class like with like. Then the laws of such aggregates can be found. But if we want to conceive of living nature as a whole, we must also remember that it is made up of partially independent entities, each with a life of its own, interacting from time to time in irregular and unpredictable ways. . . .

Something similar can be said of the internal creative life of each person. We have already seen how Darwin's work was divided into a number of separate enterprises, each with a life of its own. This type of organization has several constructive functions. It permits the thinker to change his ideas in one domain without scrapping everything he believes. In this way he can go on working purposefully on a broad range of subjects without the disruptive effects that would ensue if every new idea and even every doubt immediately required a reorganization of the whole system of thought.

At the same time, the organization of thought in partially independent systems developing unevenly and each following somewhat different laws provides the setting for the chancy interactions that give each creative process its unique flavor. For example, as we have seen, in the summer of 1838 Darwin was studying biological variation, and he was searching for a mechanism of selection. Neither process alone would generate a workable theory of evolution. As it happened, on September 23 he wrote in his notebook: "Saw in Loddiges garden 1279 varieties of rosses!!! proof of capability of variation." On September 28 he recorded his reading of Malthus and his insight about superfecundity and natural selection: ". . . until the one sentence of Malthus no one clearly perceived the great check among men." Other sequences might have produced the same happy coupling of ideas; by "chance" this is the one that occurred.

There is a danger in this formulation. It seems to reintroduce the whole notion of chance recombination and sudden insight as the main feature of creative work. This is especially true because the particular example I have given uses only two elements, variation and selection, and therefore smacks of Koestler's combinatorial theory of "bi-sociation," in which it is always just *two* ideas that come together in a new juxtaposition. But some thinkers can keep more than two things in mind. In discussing the structure of Darwin's argument I have shown how many threads had to be woven together to make a coherent theory. Darwin's interest in variation and selection were not isolated events but growth products evolving from a changing argument governed by his ruling passion, the deliberate search for a theory of evolution.

The partially independent systems whose interactions produce the unique creative event are by no means limited to the internal thought processes of one person. He is engaged with others, and he is in touch with the world around him. Darwin went to the zoo to see what he could see, and he noticed something he could incorporate in his thinking about variation: "George the lion is extraordinarily cowardly.—the other one nothing will frighten—hence variation in character in different animals of same species." Evidently, if he were not thinking about variation he might have noticed something else. But it is also true that he would not have been thinking about variation in this way if the world were not full of such individual differences, or if he had not wandered enough to see many of them with his own eyes.

The evolution of enduring purposes partially de-couples the individual from his environment. But he is not entirely free, only free to choose among possible things: attainable goals and feasible paths to reach them. This was the meaning of the old idea that "freedom is the recognition of necessity." In a chancy world it needs some revision. If you have effectively organized your experience of the world and the record of your own actions in it, you can take better cognizance of the choices open to you. You are better able to choose one of the paths that increase the likelihood of reaching your goal. You can avoid wasting time looking for regularities that do not exist. The more you are aware of the actual probabilistic structure of your environment, the more able you are to search for a really favorable opportunity, rather than plunging through the first open door. For the creative person, *carpe diem* has a special meaning: since he is trying to do something that has never been done before, he must look for the rare opportunity and then seize it. To recognize what is rare, one must have the kind of knowledge of the world that is gained only by moving about in it.

But since we are talking about the freedom to do something, simply knowing the world is never enough. One must construct an image of the world and a plan of action. As we study

Darwin's thinking, we see him revising both his theories and his plans. His notebooks are mainly devoted to recording his ideas and useful information, but frequently enough he writes down clear-cut statements of intent: "If I be asked by what power the creator has added thought to so many animals of different types, I will confess my profound ignorance.—but seeing such passions acquired & hereditary & such definite thoughts, I will never allow that because there is a chasm between man . . . and animals that man has different origin."

On the winding path from intention to achievement all sorts of propitious occasions arise: those for valuable observations, timely recollections, fortunate personal contacts, disturbing confrontations, and seminal encounters with ideas. In a way, the individual manufactures them through his own activities, and they emerge from a very dense network of personal experience. Only hindsight tells us that some of these experiences are more important than others. Some are publicly observable and recordable, but for the most part they are extremely fleeting and private.

MARY B. HESSE

# The Function

# of Analogies

# in Science

IF A SCIENTIFIC theory is to give an "explanation" of experimental data, is it necessary for the theory to be understood
in terms of some model or some analogy with events or objects
already familiar? Does "explanation" imply an account of the
new and unfamiliar in terms of the familiar and intelligible,
or does it involve only a correlation of data according to some
other criteria, such as mathematical economy or elegance?

Questions of this sort have forced themselves upon scientists
and philosophers at various stages of the development of scientific theory, and particularly since the latter half of the nineteenth century, when physicists found themselves obliged to
abandon the search for mechanical models of the ether as
explanations of the phenomena of light and electromagnetism.
In 1914, in his book *La Théorie physique*, the French physicist
and philosopher Pierre Duhem contrasted two kinds of scientific mind, in which he also saw a contrast between the
Continental and English temperaments: on the one hand, the
abstract, logical, systematizing, geometric mind typical of Continental physicists, on the other, the visualizing, imaginative,
incoherent mind typical of the English—in Pascal's words, the

Excerpt from M. B. Hesse. *Models and Analogies in Science*. South Bend, Ind.: University of Notre Dame Press, 1966, pp. 1–5. Copyright © 1966, University of Notre
Dame Press.

"strong and narrow" against the "broad and weak." Correspondingly, Duhem distinguished two kinds of theory in physics: the abstract and systematic on the one hand, and on the other, theories using familiar mechanical models. He explains the distinction in terms of electrostatics:

> This whole theory of electrostatics constitutes a group of abstract ideas and general propositions, formulated in the clear and precise language of geometry and algebra, and connected with one another by the rules of strict logic. This whole fully satisfies the reason for a French physicist and his taste for clarity, simplicity and order. . . .
>
> Here is a book [by Oliver Lodge] intended to expound the modern theories of electricity and to expound a new theory. In it are nothing but strings which move around pulleys, which roll around drums, which go through pearl beads . . . toothed wheels which are geared to one another and engage hooks. We thought we were entering the tranquil and neatly ordered abode of reason, but we find ourselves in a factory. (Duhem 1954, ch. 4 & 5)

Duhem admits that such models drawn from familiar mechanical gadgets may be useful psychological aids in suggesting theories, although, he thinks, this happens less often than is generally supposed. But this admission implies nothing about the truth or significance of the models, for many things may be psychological aids to discovery, including astrological beliefs, dreams, or even tea leaves, without implying that they are of any permanent significance in relation to scientific theory. Duhem's main objection to mechanical models is that they are incoherent and superficial and tend to distract the mind from the search for logical order. He is not much concerned, as other writers have been, with the possibility that models may mislead by being taken too literally as explanations of the phenomena, and so he does not object to fundamental mechanical theories, such as that of Descartes, where the attempt is made to reduce all phenomena to a few mechanical principles in a systematic way. But for Duhem the essence of such a theory lies not so much in its analogies with familiar mechanical objects and processes, but rather in its economic and systematic character. The ideal physical theory would be a mathematical system with deductive structure similar to Euclid's, unencumbered by extraneous analogies or imaginative representations.

These and similar views were directly challenged by the English physicist N. R. Campbell, in his book *Physics, the Elements*, published in 1920. A footnote in which Campbell refers to national tendencies to prefer mechanical or mathematical theory suggests that he has Duhem among others in mind in mounting his attack, although he does not mention Duhem by name. Campbell's main target is the view that models are *mere* aids to theory-construction, which can be thrown away when the theory has been developed, and his attack is based on two main arguments. *First*, he considers that we require to be intellectually satisfied by a theory if it is to be an explanation of phenomena, and this satisfaction implies that the theory has an intelligible interpretation in terms of a model, as well as having mere mathematical intelligibility and perhaps the formal characteristics of simplicity and economy. The *second* and more telling argument presupposes the *dynamic* character of theories. A theory in its scientific context is not a static museum piece, but is always being extended and modified to account for new phenomena. Campbell shows in terms of the development of the kinetic theory of gases how the billiard-ball model of this theory played an essential part in its extension, and he argues perceptively that, without the analogy with a model, any such extensions will be merely arbitrary. Moreover, without a model, it will be impossible to use a theory for one of the essential purposes we demand of it, namely to make predictions in new domains of phenomena. So he concludes:

> . . . analogies are not "aids" to the establishment of theories; they are an utterly essential part of theories, without which theories would be completely valueless and unworthy of the name. It is often suggested that the analogy leads to the formulation of the theory, but that once the theory is formulated the analogy has served its purpose and may be removed or forgotten. Such a suggestion is absolutely false and perniciously misleading. (Campbell 1920:129)

Enough has been said to indicate the general tenor of the debate. The actual standpoints of Duhem and Campbell are affected by the state of their own contemporary physics, and there is no need to insist on the details of their arguments. Some of these certainly cannot survive actual evidence of the workability of new kinds of theory in modern physics, and in particular the restriction of the discussion to *mechanical* models (of which Duhem is more guilty than Campbell) requires to be modified. But many physicists would now hold in essentials with Duhem and would claim that Campbell's position has been decisively refuted by the absence of intelligible models in quantum physics: indeed, many would claim that something like Duhem's position must necessarily be the accepted philosophy underlying modern physical theory.

This essay has been written in the conviction that the debate has not been so decisively closed, and that an element of truth remains in Campbell's insistence that without models theories cannot fulfil all the functions traditionally required of them, and in particular that they cannot be genuinely predictive.

# Selection 42

## HOWARD E. GRUBER

# Images

# of Wide

# Scope

OVER THE YEARS, Darwin drew and redrew the tree diagram. I have paid attention to the scraps of paper in his manuscript on which these diagrams can be found, some dateable, others not. Some of them are hasty sketches, others painstakingly drawn and delicate traceries. On one such scrap there is the remark, "Tree not good simile—endless piece of seaweed dividing." He is probably not so much correcting himself as searching for the right variant of his image to express a particular idea that has caught his attention, just as in the *First Notebook*, after his first drawing he wrote, "The tree of life should perhaps be called the coral of life, base of branches dead, so that passages cannot be seen."

We have seen how Darwin's view of the functional significance of the life cycle is connected with his panoramic view of nature as a whole. It is not often enough brought out that there was a certain cosmological cast to Darwin's thinking. Influenced, perhaps intimidated, by the empiricism of his day, Darwin later suggested that he worked in a "Baconian" fashion, inductively from facts to theory. His notebooks do not

Excerpt from H. E. Gruber, Darwin's "Tree of Nature" and other images of wide scope. In J. Wechsler, ed., *On Aesthetics in Science*. Cambridge, Mass.: MIT Press, 1978, pp. 127–131. Reprinted by permission of the MIT Press. Figures reprinted by permission of the syndics of Cambridge University Library. Footnotes deleted.

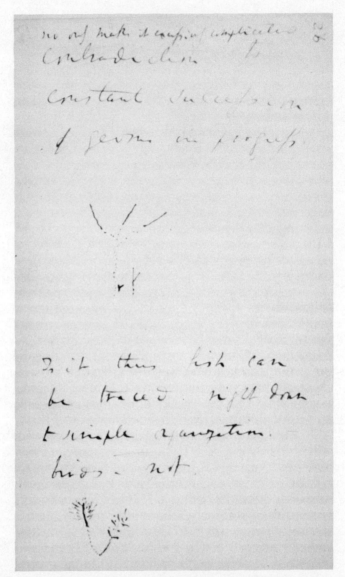

FIGURE 42.1. Darwin's first two tree diagrams, on page 26 of the *First Notebook*. Immediately preceding the upper tree the MS reads, "The tree of life should perhaps be called the coral of life, base of branches dead: so that passages cannot be seen.—[end of p. 25, beginning of p. 26] this again offers ((no only makes it excessively complicated)) contradiction to constant succession of germs in progress." Words in double parentheses were inserted above the line by Darwin.

Immediately preceding the lower tree the MS reads, "Is it thus fish can be traced right down to simple organization—birds—not." Courtesy of the Syndics of Cambridge University Library.

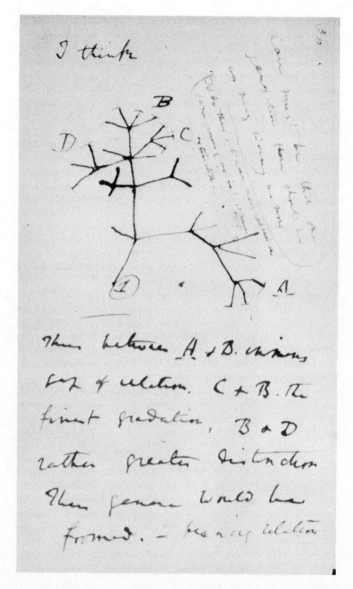

FIGURE 42.2. Darwin's third tree diagram, on page 36 of the *First Notebook*. The MS reads, "I think" followed by the diagram. Then, "Thus between A & B immense gap of relation, C & B, the finest gradation, B & D rather greater distinction. Thus genera would be formed,—bearing relation [end of p. 36, beginning of p. 37] to ancient types." The marginal insertion alongside the tree diagram reads, "Case must be that one generation then should have as many living as now. To do this & have many species in same genus (as is), *requires* extinction." Courtesy of the Syndics of Cambridge University Library.

bear this out. He sketched his ideas with a broad brush and often drew a long bow. Thus, "Astronomers might formerly have said that God ordered each planet to move in its particular destiny. In same manner God orders each animal created with certain form in certain country, but how much more simple and sublime power let attraction act according to certain law, such are inevitable consequences—let animal be created then by the fixed laws of generation, such will be their successors. Let the powers of transportal be such, and so will be the forms of one country to another.—Let geological changes go at such a rate, so will be the number and distribution of the species!!" In Darwin's image of the world, the life cycle and the evolving tree of life are nested in a larger view of the working of the cosmos.

It may be argued at this point that Darwin's diagrams are only conceptual tools for theoretical thought and have no aesthetic significance. Then why the evident pleasure in the actual drawings, the constant search for the right metaphor, the emotional excitement conveyed by his punctuation and frequent resort to a high-flown style? There is exactly that combination of feeling with concern for form and content that we have in mind when we speak of an aesthetic act. As well say that anamorphoses are not art, or that Dürer's use of instruments or Leonardo's studies of human anatomy have no aesthetic significance. Only if we presuppose a divorce between art and scientific thought would we be tempted to turn a blind eye to the aesthetic side of Darwin's imagery.

If the irregularly branching tree of life is Darwin's image of nature deployed in evolutionary time, the "tangled bank" of his eloquent closing paragraph of the Origin of Species represents his image of the same explosive vitality in all its complex interconnectedness at one moment in time. It was this passage that gave the title to Stanley Edgar Hyman's lovely book about the relation between intellectual work and literary imagination. Although foretastes of this passage occur in the notebooks and in the preliminary sketches of the Origin Darwin wrote in 1842 and 1844, the precise image of the tangled bank does not. Nevertheless, Darwin's fascination with the intricate web of contemporaneous relationships among organisms is evident as early as the notebooks of the Beagle voyage (1831–36) and appears in many forms long before the Origin.

In the carefully drawn-up table of contents of the *Origin* there occurs the striking phrase, "The relation of organism to organism the most important of all relations." This idea is spelled out in some detail in the text.

This is of course not an idea original with Darwin. It can be found in many places, notably Gilbert White's *Natural History and Antiquities of Selbourne*. It is nonetheless an idea essential to Darwin's thinking.

But interconnectedness does not by itself mean *imperfection*. Among Darwin's immediate precursors, in Lamarck's thought, in German *Naturphilosophie*, and in English Natural Theology, the idea of perfection was deeply embedded. In the first two it took the form of a scale of increasing perfection toward some limit or ideal type, that is, man, as in Lamarck's ladder of nature. The Natural Theologians could not accept this formulation because it meant that some of God's creation was less than perfect. In their view every organism was perfectly adapted to its place, seeming imperfections simply showed the limitations of our understanding of His work. Darwin's view of the natural order as inherently irregular, incomplete, and imperfect differed as radically from his predecessors' as did his view of the process by which this came about.

Thus two great and vital images, one of historical, the other of contemporary relations, form the substrate for the theory of evolution by natural selection. They are not, however, merely background, but are woven explictly into the theory as presented in the *Origin*.

The irregularly branching tree and the tangled bank represent the vital complexifying aspect of Darwin's thought. Other images must be sought that express the simplifying aspect. Of all those things that might occur in nature's incessant branching, some never do at all. The extinction of one evolutionary line makes impossible all the species that might have evolved from it. Of all those transient relations depicted in the tangled bank, some endure and others disappear. As we have already seen, the necessity for some principle of selection almost leaps out of these images.

MARTIN DEUTSCH

# Imagery

# and Inference

# in Physical Research

TIME AND AGAIN the question of evidence and inference in
experimental physics has been used as a starting point for the
discussion of scientific method. I shall discuss the question
in a rather personal way, for in my own work I have been
puzzled at times by the striking degree to which an experimenter's preconceived image of the process which he is investigating determines the outcome of his observations. The
image to which I refer is the symbolic anthropomorphic representation of the basically inconceivable atomic processes.

Modern microphysics deals with phenomena on a scale on
which there are no human sense organs to permit a direct
perception of the phenomena investigated. I do not mean simply that our organs are not sensitive enough to observe the
processes. This would be true in astronomy, where the observed celestial bodies may be too far away to be seen and
resolved by the human eye, and in geology, where the interior
of the earth is not accessible to direct observation. In quantum
physics, however, the inaccessibility is of a more fundamental
nature. While telescopes and microscopes and seismological
listening devices can refine our senses to the point where they

Excerpt from M. Deutsch, Evidence and inference in nuclear research. In D. Lerner,
ed., Evidence and inference. Glencoe, Ill.: The Free Press, 1959, pp. 96–106. Reprinted
by permission of Macmillan Publishing Company.

will give a sensory impression of the astronomical or microscopic objects which, although distorted, is still certainly relevant, there is no conceivable way of seeing, for example, an electron in its orbit in the hydrogen atom, or a neutron on its way out of a nucleus.

The human imagination, including the creative scientific imagination, can ultimately function only by evoking potential or imagined sense impressions. I cannot prove that the statement I have just made is absolutely true, but I have never met a physicist, at least not an experimental physicist, who does not think of the hydrogen atom by evoking a visual image of what he would see if the particular atomic model with which he is working existed literally on a scale accessible to sense impressions. At the same time the physicist realizes that in fact the so-called internal structure of the hydrogen atom is *in principle* inaccessible to direct sensory perception. This situation has far-reaching consequences for the method of experimental investigation. The use in the interpretation of physical phenomena of factors not directly perceivable is, of course, very old. But there is a basic difference between this procedure in classical phenomena and in microscopic quantum physics— a difference which I shall try to illustrate by two examples.

First let me take a classical example, the simple meteorological phenomenon of the rising and setting of the sun. There is probably immediate agreement among all rational observers of the sun's behavior, that we are dealing with the relative motion in space of a luminous body and the observer. (Only very rarely has the suggestion been made that we are really dealing with a stationary object and that only the light rays follow different paths in the course of the day.) The nature of this motion has been investigated by methods familiar to every schoolboy. The angle at which the sun is visible from various locations at various times has been recorded with great care and compared with the angle at which the moon and the stars are visible. Various kinematic descriptions have been achieved, such as the Ptolemaic and Copernican. Further refinement of such theories has led to a dynamic rather than kinematic description, in terms of the Newtonian system of gravitational mechanics appropriately modified by relativity. Whatever theory we may develop, there will never be any doubt that the relevant observations are those of the sun's position, of the length of shadows, etc. There is a direct and self-evident link between the sensory image and the scientific theory.

But now let us consider one theory of celestial mechanics which has at various times enjoyed great popularity: we assume that the sun is moved by a spirit which may be appeased or influenced to act in a favorable or unfavorable fashion. For example, the rising of the sun must be achieved by invoking the favor of this spirit. To test this theory our experimentation has to apply not only to the sun, the moon, the stars, and the shadows, but in principle also to the spirit which is not amenable to sensory perception. The whole context of relevant factors is now suddenly changed; if the sun is eclipsed it may become necessary to investigate whether the king did not violate a taboo. Furthermore, the question is really very difficult to settle by experimentation because the spirit might at one time be in a more tolerant mood than at another and we have, in principle, no way of making this mood directly accessible to our senses.

Bad though this situation is, it is in some respects more favorable than that obtaining in quantum physics. It is, after all, ultimately possible to decide between these theories with reasonable certainty by performing observations only on the celestial bodies themselves. Having accumulated a sufficient amount of data concerning the behavior of the sun over the centuries, we arrive at a complete description of past and future phenomena based entirely on direct sensory observations, and the animist theories collapse from the lack of evidence for their relevance rather than because they are explicitly disproved. The evidence is convincing for any person with an ordinary degree of imagination, i.e., an ordinary ability to evoke the sense impressions which he would receive from the sun under certain specified conditions.

In a typical modern physics experiment, on the other hand, the direct sensory impressions of the experimental situation are not only not very helpful to the solution of the problem investigated, but are not even likely to reveal the relevant factors. A typical high-energy physics laboratory abounds in impressive sights and sounds. There is likely to be the huge magnet of an accelerator, pounding periodically as it is energized, the whirr of the electric generators, perhaps clouds of condensation above containers of liquid air and the rushing cooling water. The control room and the experimental area are filled with colored lights, hundreds of knobs, moving meter needles, and flashing oscilloscope patterns to delight the heart of the space child.

I am sure that a scientist of 150 years ago, told to proceed with experimentation in this laboratory, would have no difficulty in making extensive observations on the change in the appearance of the lights as various knobs are turned, and on the variation in the pitch of the generator noise. But, of course, none of these changes directly accessible to sensory impression are really relevant to the experiment actually being carried out. Even in the very roughest, preliminary experiment the nuclear physicist does not look at the object of his investigation, or listen to it, or sniff it, or try to determine its temperature. He does not start with a simple impression of the phenomena, later refined by quantitative measurement. The only visible feature intimately connected with the actual experiment may be the position of a knob controlling the current through a magnet and a mechanical recording device indicating the rate of arrival of electrical signals from a small counter. The experimental procedure may consist simply in noting the reading on the recorder as a function of the magnet current. Both the change made—the position of the control knob—and the resulting effect—the reading on the recorder—seem almost negligible in the totality of material involved in the experiment.

This is also true if we search a little deeper and consider the total energy or total electric current involved in the actual observation, and compare it with the totals involved in the experiment. To make matters worse, there are probably several other knobs which could be turned with much more drastic effect on our recorder. In addition, there are literally dozens of similar recording devices and hundreds of other similar knobs connected with the experiment, some of which may show correlations much more marked than those under investigation. Yet, if you ask the experienced experimenter he may tell you that these other effects are either fully understood or certainly without importance. We see, then, a situation in which the experiment is carried out under conditions in which almost all sensory impressions concerning its operation are irrelevant to the question investigated and in which a large number of related phenomena remain uninvestigated and, therefore, unexplained.

How is it possible that important and reliable conclusions are drawn from this experimentation? The answer lies in the fact that the experimenter starts out with a well structured image of the actual connections between the events taking place. Far from approaching the problem with the completely open mind which is supposed to be characteristic of the scientist investigating in an unbiased fashion all possible connections, he starts with the conviction that all of the relevant occurrences except the one which he is actually investigating in the experiment are either fully understood or at least in principle explicable on the basis of the preconceived image. Without this image, the experiment certainly could not have been conceived in the first place.

In terms of our current image its description might be as follows: when the high-energy proton beam produced in the accelerator strikes the nuclei of the beryllium target, mesons are produced by the impact of these protons on the beryllium nuclei. They emerge from the target at various angles with various velocities. A narrow beam of these mesons is allowed to pass through a deflecting magnet and then through two particle detectors placed several meters apart. We determine the velocity of the mesons in terms of the measurable time interval between the electrical pulses indicating their passage through the first and second of the detecting counters mentioned. The angle through which a beam of charged particles is deflected by a magnet of given strength is determined by the particles' momentum (classically, the product of its mass and its velocity). Since the particles can enter the detecting counters only if they have been deflected through a specified angle, the counters record only particles of a fixed momentum. Having measured the momentum and the velocity, we can directly calculate the mass. By changing the magnet current we vary this momentum until it corresponds to the velocity selected by the counters.

If one does not already have some familiarity with the field, this very condensed description is probably not sufficient to form a clear image of the experiment. But I hope that it does give the feeling that with the aid of a few simple diagrams and perhaps a clarification of some of the concepts involved, there would be no difficulty in understanding the result of this kind of measurement. The only reason for the acceptance of this picture by non-physicists is, however, the good name of the physicist who presents the picture. I cannot actually show you the protons, the nuclei, the mesons or even the electrical impulses except in an indirect fashion which already involves a good part of the conclusions to be understood. . . .

There is no recipe, no rule of procedure which will permit the unimaginative experimenter to consider all of these possibilities by applying systematic logical analysis to all aspects of his apparatus. The only workable scheme is to operate within the framework of an anthropomorphic image. It is sometimes amusing to watch two highly competent physicists discuss the functioning of an apparatus if their images have for some reason taken somewhat different forms; despite the fact that they reach the same conclusions they may have some difficulty communicating with each other.

I should really say that there are two separate images involved in most experiments, one image representing the functioning of the apparatus, the other the basic physical phenomenon investigated. The procedure is probably more familiar in the instrumental aspect. We structure the behavior of the complicated physical phenomena according to a functional interpretation; that is, we consider each part of the apparatus as a unit having a definite function with respect to our purpose. The language which we apply to it is frequently ludicrously anthropomorphic. For example, we say without hesitation: "This amplifier does not *like* to operate at high counting rates," or "This circuit is designed to *reject* protons." The functional image which we construct is more sophisticated than this would imply, but it also ultimately reduces the functioning of the apparatus to terms analogous to phenomena directly amenable to sensory verification. This functional view of the physical phenomena does not cause any serious complications when applied to the apparatus. After all, the apparatus is designed to perform a certain function according to the desire of the experimenter, and while there may be aspects of its behavior which cannot be understood within this image, these are not necessarily relevant to the experiment.

The problem is very much more serious when we consider the image representing the fundamental physical phenomenon studied. The description which I gave at the outset of the experiment that has been discussed is certainly not the only possible one of the sequence of events. We could, for example, imagine the phenomenon as one of interaction between the beryllium target, struck by the protons in our accelerator, and certain atoms in our counters. The concept of an unobservable physical particle traveling in a well-defined path through the magnet would then not be part of the image. We clearly could not have designed our original experiment and measured the mass of the particle unless we had started out with an image of the phenomenon involving such a "particle." The very word used implies the image.

But a meson is clearly not an object with the general properties of a ball which we could see if we had a sufficiently good microscope, or feel impinging on our hand if our nerves were a little more sensitive. We are not forced by direct sensory perception to use this image. We have developed it because it allows us to reason from one experiment to the next by analogy; even in a mathematically sophisticated theory we deal with formal thought processes designed to connect sensory impressions. It, too, must proceed by analogy with the connections established between such perceivable events.

Now it is clear that if one is too strongly attached to one's preconceived model, one will of necessity miss all radical discoveries. It is amazing to what degree one may fail to register mentally an observation which does not fit the initial image. I could quote some sad examples from my own career. On the other hand, if one is too open-minded and pursues every hitherto unknown phenomenon, one is almost certain to lose oneself in trivia.

As I have tried to show in my illustration, in a typical experiment in modern physics the apparatus involves complex influences frequently of much greater order of magnitude than the phenomenon investigated. Each of these influences can occupy a lifetime of experimental investigation before it is fully understood, and is most likely to be of no particular scientific interest as it concerns only features of a particular piece of apparatus. The ability to distinguish between what we call colloquially a "dirt effect" and a basic phenomenon that is still an unknown is what constitutes the decisive intuition of the experimenter in this field. It is really the ability to recognize relevance in the evidence presented by the experiment.

# Selection 44

## THOMAS S. KUHN

# On Thought

# Experiments

THE HISTORICAL CONTEXT within which actual thought experiments assist in the reformulation or readjustment of existing concepts is inevitably extraordinarily complex. I therefore begin with a simpler, because nonhistorical, example, choosing for the purpose a conceptual transposition induced in the laboratory by the brilliant Swiss child psychologist Jean Piaget. Justification for this apparent departure from our topic will appear as we proceed. Piaget dealt with children, exposing them to an actual laboratory situation and then asking them questions about it. In slightly more mature subjects, however, the same effect might have been produced by questions alone in the absence of any physical exhibit. If those same questions had been self-generated, we would be confronted with the pure thought-experimental situation to be exhibited in the next section from the work of Galileo. Since, in addition, the particular transposition induced by Galileo's experiment is very nearly the same as the one produced by Piaget in the laboratory, we may learn a good deal by beginning with the more elementary case.

Piaget's laboratory situation presented children with two toy autos of different colors, one red and the other blue. During each experimental exposure both cars were moved uniformly in a straight line. On some occasions both would cover the same distance but in different intervals of time. In other ex-

Excerpt from T. S. Kuhn "A Function for Thought Experiments." In *L'aventure de l'esprit; Mélanges Alexandre Koyré.* Paris: Hermann, 1964, pp. 309–13, 316–20; 332–34.

posures the times required were the same, but one car would cover a greater distance. Finally, there were a few experiments during which neither the distances nor the times were quite the same. After each run Piaget asked his subjects which car had moved faster and how the child could tell.

In considering how the children responded to the questions, I restrict attention to an intermediate group, old enough to learn something from the experiments and young enough so that its responses were not yet those of an adult. On most occasions the children in this group would describe as "faster" the auto that reached the goal first or that had led during most of the motion. Furthermore, they would continue to apply the term in this way even when they recognized that the "slower" car had covered more ground than the "faster" during the same amount of time. Examine, for example, an exposure in which both cars departed from the same line but in which the red started later and then caught the blue at the goal. The following dialogue, with the child's contribution in italics, is then typical. "Did they leave at the same time?"—"*No, the blue left first.*"—"Did they arrive together?"—"*Yes.*"—"Was one of the two faster, or were they the same?"—"*The blue went more quickly.*" Those responses manifest what for simplicity I shall call the "goal-reaching" criterion for the application of "faster."

If goal reaching were the only criterion employed by Piaget's children, there would be nothing that the experiments alone could teach them. We would conclude that their concept of "faster" was different from an adult's but that, since they employed it consistently, only the intervention of parental or pedagogic authority would be likely to induce change. Other experiments, however, reveal the existence of a second criterion, and even the experiment just described can be made to do so. Almost immediately after the exposure recorded above, the apparatus was readjusted so that the red car started very late and had to move especially rapidly to catch the blue at the goal. In this case, the dialogue with the same child went as follows. "Did one go more quickly than the other?"—"*The red.*"—"How did you find that out?"—"*"I WATCHED IT."* Apparently, when motions are sufficiently rapid, they can be perceived directly and as such by children. (Compare the way adults "see" the motion of the second hand on a clock with the way they observe the minute hand's change of position.)

Sometimes children employ that direct perception of motion in identifying the faster car. For lack of a better word I shall call the corresponding criterion "perceptual blurriness."

It is the coexistence of these two criteria, goal-reaching and perceptual blurriness, that makes it possible for the children to learn in Piaget's laboratory. Even without the laboratory, nature would sooner or later teach the same lesson as it has to the older children in Piaget's group. Not very often (or the children could not have preserved the concept for so long) but occasionally nature will present a situation in which a body whose directly perceived speed is lower nevertheless reaches the goal first. In this case the two clues conflict; the child may be led to say that both bodies are "faster" or both "slower" or that the same body is both "faster" and "slower." That experience of paradox is the one generated by Piaget in the laboratory with occasionally striking results. Exposed to a single paradoxical experiment, children will first say one body was "faster" and then immediately apply the same label to the other. Their answers become critically dependent upon minor differences in the experimental arrangement and in the wording of the questions. Finally, as they become aware of the apparently arbitrary oscillation of their responses, those children who are either cleverest or best prepared will discover or invent the adult conception of "faster." With a bit more practice some of them will thereafter employ it consistently. Those are the children who have learned from their exposure to Piaget's laboratory.

But, to return to the set of questions which motivate this inquiry, what shall we say they have learned and from what have they learned it? For the moment I restrict myself to a minimal and quite traditional series of answers which will provide the point of departure for a later section. Because it included two independent criteria for applying the conceptual relation "faster," the mental apparatus which Piaget's children brought to his laboratory contained an implicit contradiction. In the laboratory the impact of a novel situation, including both exposures and interrogation, forced the children to an awareness of that contradiction. As a result, some of them changed their concept of "faster," perhaps by bifurcating it. The original concept was split into something like the adult's notion of "faster" and a separate concept of "reaching-goal-first." The children's conceptual apparatus was then probably richer and certainly more adequate. They had learned to avoid a significant conceptual error and thus to think more clearly.

Those answers, in turn, supply another, for they point to the single condition that Piaget's experimental situations must satisfy in order to achieve a pedagogic goal. Clearly those situations may not be arbitrary. A psychologist might, for quite different reasons, ask a child whether a tree or a cabbage were faster; furthermore, he would probably get an answer; but the child would not learn to think more clearly. If he is to do that, the situation presented to him must, at the very least, be relevant. It must, that is, exhibit the cues which he customarily employs when he makes judgments of relative speed. On the other hand, though the cues must be normal, the full situation need not be. Presented with an animated cartoon showing the paradoxical motions, the child would reach the same conclusions about his concepts, even though nature itself were governed by the law that faster bodies always reach the goal first. There is, then, no condition of physical verisimilitude. The experimenter may imagine any situation he pleases so long as it permits the application of normal cues. . . .

The relevant experiment is produced almost at the start of "The First Day" in Galileo's *Dialogue concerning the Two Chief World Systems*. Salviati, who speaks for Galileo, asks his two interlocutors to imagine two planes, CB vertical and CA inclined, erected the same vertical distance over a horizontal plane, AB. To aid the imagination Salviati includes a sketch like the one below. Along these two planes, two bodies are to be imagined sliding or rolling without friction from a common starting point at C. Finally, Salviati asks his interlocutors to concede that, when the sliding bodies reach A and B, respectively, they will have acquired the same impetus or speed, the speed necessary, that is, to carry them again to the vertical height from which they started. That request also is granted, and Salviati proceeds to ask the participants in the dialogue which of the two bodies moves faster. His object is to make them realize that, using the concept of speed then current, they can be forced to admit that motion along the perpendicular is simultaneously faster than, equal in speed to, and slower than the motion along the incline. His further object is, by the impact of this paradox, to make his interlo-

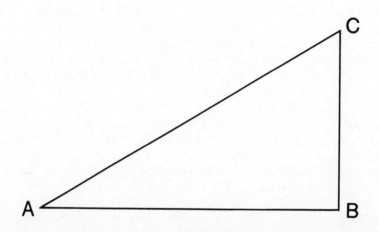

cutors and readers realize that speed ought not be attributed to the whole of a motion, but rather to its parts. In short, the thought experiments is, as Galileo himself points out, a propaedeutic to the full discussion of uniform and accelerated motion that occurs in "The Third Discourse" of his *Two New Sciences*. The argument itself I shall considerably condense and systematize since the detailed give and take of the dialogue need not concern us. When first asked which body is faster, the interlocutors give the response we are all drawn to though the physicists among us should know better. The motion along the perpendicular, they say, is obviously the faster.[1] Here, two of the three criteria we have already encountered combine. While both bodies are in motion, the one moving along the perpendicular is the "more blurred." In addition, the perpendicular motion is the one that reaches its goal first.

This obvious and immensely appealing answer immediately, however, raises difficulties which are first recognized by the cleverer of the interlocutors, Sagredo. He points out (or very nearly—I am making this part of the argument slightly more binding than it is in the original) that the answer is incompatible with the initial concession. Since both bodies start from rest and since both acquire the same terminal velocity, they must have the same mean speed. How then can one be faster than the other? At this point Salviati reenters the discussion, reminding his listeners that the faster of two motions is usually defined as the one that covers the same distance in a lesser time. Part of the difficulty, he suggests, arises from the attempt to compare two motions that cover different distances. Instead, he urges, the participants in the dialogue should compare the times required by the two bodies in moving over a common standard distance. As a standard he selects the length of the vertical plane CB.

This, however, only makes the problem worse. CA is longer than CB, and the answer to the question, which body moves faster, turns out to depend critically upon where, along the incline CA, the standard length CB is measured. If it is measured down from the top of the incline, then the body moving on the perpendicular will complete its motion in less time than the body on the incline requires to move through a distance equal to CB. Motion along the perpendicular is therefore faster. On the other hand, if the standard distance is measured up from the bottom of the incline, the body moving on the perpendicular will need more time to complete its motion that the body on the incline will need to move through the same standard distance. Motion along the perpendicular is therefore slower. Finally, Salviati argues, if the distance CB is laid out along some appropriate internal part of the incline, then the times required by the two bodies to traverse the two standard segments will be the same. Motion on the perpendicular has the same speed as that on the incline. At this point the dialogue has provided three answers, each incompatible with both the others, to a single question about a single situation.

The result, of course, is paradox, and that is the way, or one of them, in which Galileo prepared his contemporaries for a change in the concepts employed when discussing, analyzing, or experimenting upon motion. Though the new concepts were not fully developed for the public until the appearance of the *Two New Sciences*, the *Dialogue* already shows where the argument is headed. "Faster" and "speed" must not be used in the traditional way. One may say that at a particular instant one body has a faster instantaneous speed than another body has at that same time or at another specified instant. Or one may say that a particular body traverses a particular distance more quickly than another traverses the same or some other distance. But the two sorts of statements do not describe the same characteristics of the motion. "Faster" means something different when applied, on the one hand, to the comparison of instantaneous rates of motion at particular instants, and, on the other, to the comparison of the times required for the completion of the whole of two specified motions. A body may be "faster" in one sense and not in the other.

That conceptual reform is what Galileo's thought experiment helped to teach, and we can therefore ask our old questions about it. Clearly, the minimal answers are the same ones supplied when considering the outcome of Piaget's experiments. The concepts that Aristotle applied to the study of motion were, in some part, self-contradictory, and the contradiction was not entirely eliminated during the Middle Ages. Galileo's thought experiment brought the difficulty to the fore by confronting readers with the paradox implicit in their mode of thought. As a result, it helped them to modify their conceptual apparatus.

If that much is right, then we can also see the criterion of verisimilitude to which the thought experiment had necessarily to conform. It makes no difference to Galileo's argument whether or not bodies actually execute uniformly accelerated motion when moving down inclined and vertical planes. It does not even matter whether, when the heights of these planes are the same, the two bodies actually reach equal instantaneous velocities at the bottom. Galileo does not bother to argue either of the points. For his purpose in this part of the *Dialogue*, it is quite sufficient that we may suppose these things to be the case. On the other hand, it does not follow that Galileo's choice of the experimental situation could be arbitrary. He could not, for example, usefully have suggested that we consider a situation in which the body vanished at the start of its motion from C and then reappeared shortly afterwards at A without having traversed the intervening distance. That experiment would illustrate limitations in the applicability of "faster," but, at least until the recognition of quantum jumps, those limitations would not have been informative. From them, neither we nor Galileo's readers could learn anything about the concepts traditionally employed. Those concepts were never intended to apply in such a case. In short, if this sort of thought experiment is to be effective, it must allow those who perform or study it to employ concepts in the same ways they have been employed before. Only if that condition is met can the thought experiment confront its audience with unanticipated consequences of their normal conceptual operations. . . .

The outcome of thought experiments can be the same as that of scientific revolutions: they can enable the scientist to use as an integral part of his knowledge what that knowledge had previously made inaccessible to him. That is the sense in which they change his knowledge of the world. And it is because they can have that effect that they cluster so notably in the works of men like Aristotle, Galileo, Descartes, Einstein, and Bohr, the great weavers of new conceptual fabrics.

## Note

1. Anyone who doubts that this is a very tempting and natural answer should try Galileo's question, as I have, on graduate students of physics. Unless previously told what will be involved, many of them give the same answer as Salviati's interlocutors.

# Part IX

# Image

# and Reality

# in Science

# Introduction

PROBLEMS CAN BE worked upon in the "mind's eye" of the scientist. A common feature of the reports of many such solutions is that they attribute instrumentality to visual images. Some of the issues that flow from the attempt to characterize this property of scientific thinking will constitute the focus of part 9.

Ziman's (1978) treatment of the topic provides instructive examples. Comparing the relatively primitive, perceptual mode of pattern perception with the advanced and highly abstract quantitative procedures provided by high-speed computers, he points out that a scientist will often use a computer to do complex theoretical computations—then instruct it to print a picture which can be compared visually with a photograph (1978:55). Photographs themselves function as heuristic devices: consider Watson's unforgettable description (1968:107) of his excitement when he saw Rosalind Franklin's X-ray diffraction photographs of anhydrous DNA. Similarly, many discoveries in particle physics have relied on photographs of particle tracks in cloud chambers. Drawings of plants and molecules, periodic tables, normal distributions, and box-and-arrow flow-charts often serve to represent the current state of knowledge, and, more importantly, often function as tools for discovery. Darwin's drawing and redrawing of the irregularly branching tree (selection 42) was not idle doodling, but was the use of a figure, in a conscious effort at repetitive focusing of attention, to extract from that image (or perhaps to enrich that image with) all of its meaning. In just the same fashion, mental images may play key roles in creative thinking—whether or not they are ever put down on paper.

That concrete images are widespread in science is not difficult to show. Shepard (1978) has catalogued a large number of examples: Einstein, Maxwell, Faraday, Helmholtz, Galton, Hadamard, Watt, Tesla, Herschel, Snyder (a Manhattan Project scientist interviewed by Shepard), Watson, Kekulé, and Shepard himself. All of these individuals reported compelling instrumental images at critical points in the scientific enterprise. Such images may be present even in scientists who consider themselves nonvisual thinkers. Thus, we recently interviewed a molecular biologist who initially denied using visual images, but, later in the interview, described a vivid image used to conceptualize the process by which plasmid DNA was transferred from bacterial to host cells.

In recent years, a good deal of interest has been directed at the possibility that imagistic thinking may occur in a more-or-less distinct manner from nonimagistic thinking in general and verbal thinking in particular (e.g., Paivio 1971; Chase 1973; Neisser 1976; Shepard 1978). Laboratory studies have already demonstrated clear differences in the types of information processing that occur in visual and verbal modes (e.g., Bower 1972), lending credence to the distinction. Furthermore, the demonstration that visual processing is localized primarily in the right hemisphere of the brain, while verbal processing is localized primarily in the left hemisphere, has given the distinction a firm neurological basis (see Gazzaniga and LeDouz 1978, for an extensive review of the evidence). The relationship between type of thinking and hemispheric specialization has fueled speculation concerning the possible significance for human cognition (e.g., Ornstein 1977; Hofstatder 1979). The application to scientific thinking has suggested to some that different components of science reflect different sides of the brain: "The creative act has major right-hemisphere components. But arguments on the validity of the result are largely left-hemisphere functions" (Sagan 1977:183).

Before assenting too readily to such an extreme view on the separateness of imagery, however, it is important to review the arguments against it. Not everyone agrees that visual thinking is useful in science. Reliance on visual thinking may have heuristic value, but it may also function as a constraint. Heisenberg, for example, felt that concern for visualizability was inappropriate, and could hinder attainment of an appropri-

ately abstract representation (see Miller, selection 47). Bronowski (1973) provides an anecdote about John von Neumann
that illustrates the same point of view.

> He was the cleverest man I ever knew, without exception.
> And he was a genius, in the sense that a genius is a man who
> has *two* great ideas. . . . We once faced a problem together, and
> he said to me at once, "oh no, no, you are not seeing it. Your
> kind of visualizing mind is not right for seeing this. Think of
> it abstractly. What is happening on this photograph of an ex-
> plosion is that the first differential coefficient vanishes iden-
> tically, and that is why what becomes visible is the trace of the
> second differential coefficient." (1973:433–35)

In fact, von Neumann's claim that Bronowski's thinking was
limited by his imagery is consistent with certain laboratory
results. Brooks (1970) found that visualization of a problem
can either help or hurt performance, depending upon the task
context, just as verbalization can either help or hurt, depending upon context. Is the same true in science? Perhaps scientific issues dealing with questions of structure are more
tractable given a visual mode of thought, while those concerned with questions of process, in the absence of structural
knowledge, are more amenable to verbal, mathematical, or
other "imageless thought."

It is not universally agreed that images must have a specifically perceptual character. Pylyshyn (1973) argued that lists
of propositional statements can be constructed that are equivalent, in any given task, to the information purportedly provided by the image. For this reason, "image" may not be
needed as a distinct construct. List representations have, in
fact, been shown to be extremely powerful in the construction
of problem-solving routines (Reitman 1965; Newell & Simon
1972), lending further support to Pylyshyn's position.

In spite of these reservations, verbal modes of thinking, per
se, are not sufficient for all kinds of problem solving. Useful
though they are, verbal representations are at best incomplete,
"A thought may be compared to a cloud shedding a shower
of words" (Vygotsky 1962:150). Sometimes, as in the case of
Einstein (selections 23 and 46), the "shedding" is an act of
tedious labor, far removed from the original creative act. Unless we are to dismiss entirely the introspective accounts of
a large number of the most creative scientists, we are forced

to allow a constructive, functional role for nonverbal thought, at least some of which is experienced as explicitly visual. With respect to visual imagery, per se, in the absence of compelling data, we are equally forced to recognize the possibility that the differences between visual and nonvisual modes of thought, no matter how phenomenologically impressive, may *not* lead to significant differences in scientific accomplishment. The visual and nonvisual thinker may get to the same places intellectually, by a different route. Drawings and literal models, when used by the nonvisual investigator, may provide exactly the same data as the image does to the visualizer.

The most compelling evidence favoring visual images comes from just those studies in which subjects use visual information that would be difficult or cumbersome to represent propositionally—Shepard and Metzler's (1971) rotating geometric figures being the best example. For scientific thinking, such processes may permit simultaneous, nonlinear and nonlogical connections to be established between similar or related items, connections which would be difficult or impossible to establish verbally. The significance of Kekulé's dream, in this view, rests upon a relation of similarity between two imagined entities (rings and benzene molecules). There is no *a priori*, logical reason why these should be related. We speak of Kekulé's achievement as creative for just this reason.

The capacity for such links is essential in science, but so is linear, logical thought. In this respect, verbal processes complement visual processes. The grammatical rules of a language system allow the scientist to express propositionally the terms and predicates of the problem at hand. Further, if there exist rules of inference (what Braine [1978] has called inference schemata) that allow propositions to be simplified, altered, and combined, according to the rules of a logical syntax, then such processes may play a significant role in scientific thought, not logically as Carnap (1935) thought, but *psychologically*. On this view, even mathematics can be thought of as a grammatical system.

Thus, a solid-state physicist described to us a process whereby he visualized a three-dimensional molecule, and deliberately rotated it to see where a trace element could be added in the structure. Here, visualization was used integrally with a set of very complex logical constraints on structure.

The verbal and the nonverbal should thus be seen as complementary processes in scientific thinking. If the sequentially ordered, logical or "psycho-logical," processes of grammar are essential, so also are the nonlogical, simultaneous processes of imagery. In fact, images and words are not dichotomous entities. Visual signs can serve as the elements in a syntax as in sign languages (Tweney, Heiman, and Hoemann [1977]; Klima and Bellugi [1979]), just as syntax can be based upon nonlinguistic combinations (as in geometry). The more interesting dichotomy is that between representation and transformation, rather than between images and words. Thus, both imagery and verbalization relate terms, via similarity and difference relations, but they do so in different fashion. Both constitute *internalizations* of the sorts of processes discussed in Parts 7 and 8, as well as in this section.

Mention should be made of the argument that creative thinking is rooted in unconscious dynamic processes (e.g., Ruitenbeek 1965; Rothenberg & Hausman 1976; Rothenberg 1979). We agree that creative processes are not captured by logical models, but see no clear evidence as yet to suggest that psychodynamic models will explain the phenomena, yet be sufficiently explicit to make falsifiable predictions. Instead, models based on chancy interaction (e.g., Gruber 1974) or even on selective forgetting (Simon, selection 5), seem more promising. Some of the central evidence is beyond dispute. It is clear that creative solutions to problems are frequently reported as suddenly appearing from out of nowhere. But we do not believe that such accounts force any one explanation over any other; they simply force recognition of the limits of introspective data. Our approach to this issue is thus very much like the approach of Külpe and his followers in rejecting introspective methods (cf. Humphrey 1940, 1951). Once "imageless thoughts" were discovered in simple association tasks, it became clear that nothing more could be learned solely via introspective accounts. The cognitive psychology of science is facing the same issue.

Part 9 contains four selections. The first, by Planck (selection 45), serves as a reminder that the practicing scientist is a realist. While our theories about science may deny the possibility of obtaining absolute knowledge, while we may speak of "images" and "heuristics" and "metaphors," they are images of *something,* and heuristics and metaphors for *something.* We must keep in mind the fact that the scientist is, psychologically, concerned with "the truth." Planck's absolutism expresses a thema expressed by all scientists.

Selection 46 by Holton regards Einstein's thought from a perspective that matches some of the concerns outlined above. The reader should compare this selection with the earlier chapter on Einstein by Max Wertheimer (selection 23 in part 5). The complexity of the issues raised by Holton becomes even more apparent in selection 47 by Miller. His consideration of the early years of quantum theory reveals that visualizability itself became a disputed point among physicists. Miller's account should be compared with that of Deutsch (selection 43) in the preceding part.

The final selection (48) by Einstein, is the text of a famous letter written to Hadamard in response to a questionnaire. It is an introspective account which points to the multiplicity of possible avenues of scientific creativity, a multiplicity that covers all of the mechanisms we have discussed, and serves as a capstone for this book.

MAX PLANCK

# Absolutes

# in Science

MY ORIGINAL DECISION to devote myself to science was a direct result of the discovery which has never ceased to fill me with enthusiasm since my early youth—the comprehension of the far from obvious fact that the laws of human reasoning coincide with the laws governing the sequences of the impressions we receive from the world about us; that, therefore, pure reasoning can enable man to gain an insight into the mechanism of the latter. In this connection, it is of paramount importance that the outside world is something independent from man, something absolute, and the quest for the laws which apply to this absolute appeared to me as the most sublime scientific pursuit in life. . . .

In order to preclude a likely misunderstanding, I have to insert here a few explanatory remarks of general character. In the opening paragraph of this autobiographical sketch, I emphasized that I had always looked upon the search for the absolute as the noblest and most worth while task of science. The reader might consider this contradictory to my avowed interest in the Theory of Relativity. But it would be fundamentally erroneous to look at it that way. For everything that is relative presupposes the existence of something that is absolute, and is meaningful only when juxtaposed to something absolute. The oftenheard phrase, "Everything is relative," is

Excerpts from M. Planck. *A Scientific Autobiography and Other Papers*. New York: Philosophical Library, 1949, pp. 13; 46–47. Reprinted by permission of Ernest Benn Ltd.

both misleading and thoughtless. The Theory of Relativity, too, is based on something absolute, namely, the determination of the matrix of the space-time continuum; and it is an especially stimulating undertaking to discover the absolute which alone makes meaningful something given as relative.

Our every starting-point must necessarily be something relative. All our measurements are relative. The material that goes into our instruments varies according to its geographic source; their construction depends on the skill of the designer and toolmaker; their manipulation is contingent on the special purposes pursued by the experimenter. Our task is to find in all these factors and data, the absolute, the universally valid, the invariant, that is hidden in them.

This applies to the Theory of Relativity, too. I was attracted by the problem of deducing from its propositions that which served as their absolute immutable foundation. The way in which this was accomplished, was comparatively simple. In the first place, the Theory of Relativity confers an absolute meaning on a magnitude which in classical theory has only a relative significance: the velocity of light. The velocity of light is to the Theory of Relativity as the elementary quantum of action is to the Quantum Theory: it is its absolute core.

GERALD HOLTON

# "What,

# Precisely,

# Is 'Thinking'?"

WE KNOW FAR too little about Einstein as a child, but he is commonly reported to have been withdrawn, slow to respond, quietly sitting by himself at an early age, playing by putting together shapes cut out with a jigsaw, erecting complicated constructions by means of a chest of toy building parts. Before he was ten, he was making, with infinite patience, fantastic card houses that had as many as fourteen floors. He is said to have been unable or unwilling to talk until the age of three. In an (unpublished) biography of Einstein, his sister Maja wrote in 1924: "His general development during his childhood years proceeded slowly, and spoken language came with such difficulty that those around him were afraid he would never learn to talk." Many pediatricians and psychologists might consider such evidence to indicate an almost backward child.

But it is coming to be more widely agreed that an apparent defect in a particular person may merely indicate and imbalance of our normal expectations. While it is patently absured to think that a deficiency in one area "causes" or explains talent in another, a noted deficiency should at least alert us to look for a proficiency of a different kind in the exceptional person. The late use of language in childhood, the difficulty

Excerpt from G. Holton, On trying to understand scientific genius. *The American Scholar* (1971), 41:102–04. Reprinted by permission of the author.

in learning foreign languages—one remembers that Einstein
failed in foreign languages at the *Gymnasium* and again at the
entrance examination in Zurich (one of the reasons for his
having to go to the Kanton Schule in Aarau), that his vocabulary in English was fairly small—all this may indicate a polar-
ization or displacement in some of the skill from the verbal
to another area. That other, enhanced area is without doubt,
in Einstein's case, an extraordinary kind of visual imagery that
penetrates his very thought processes.

Although it seems to have been hardly noted so far, Einstein
himself plainly signals this point early in his *Autobiographical Notes.* He asks, rather abruptly:

> What, precisely, is "thinking"? When at the reception of
> sense-impressions, memory-pictures [*Erinnerungsbilder*] emerge,
> this is not yet "thinking." And when such pictures form series,
> each member of which calls forth another, this too is not yet
> "thinking." When, however, a certain picture [*Bild*] turns up
> in many such series, then—precisely through such return—it
> becomes an ordering element for such series, in that it connects
> series which in themselves are unconnected. Such an element
> becomes an instrument, a concept. . . . It is by no means nec-
> essary that a concept must be connected with a sensorily cog-
> nizable and reproducible sign (word). . . . All our thinking is
> of this nature of a free play [*eines freien Spiels*] with con-
> cepts. . . . For me it is not dubious that our thinking goes on
> for the most part without use of signs (words), and beyond that
> to a considerable degree unconsciously

It is not accidental at all that this surprising passage comes
just before Einstein tells of the two "wonders" experienced
in childhood. One of these was the experience with the compass that we have mentioned [not included here—Eds.]. The
other wonder, of a totally different nature, was the little book
dealing with Euclidean plane geometry which, he recalled,
was given to him at about the age of twelve. In this connection,
Einstein described an early, successful use of his particular
way of "thinking" in visual terms:

> I remember that an uncle told me the Pythagorean theorem
> before the holy geometry booklet had come into my hands.
> After much effort I succeeded in "proving" this theorem on the
> basis of the similarity of triangles; in doing so it seemed to me
> "evident" that the relations of the sides of the right-angled

triangle would have to be completely determined by one of the acute angles. Only something which did not in similar fashion seem to be "evident" appeared to me to be in need of any proof at all. Also, *the objects with which geometry deals seemed to be of no different type than the objects of sensory perception,* "*which can be seen and touched.*" (Italics supplied)

The objects of the imagination were to him evidently persuasively real, visual materials, which he voluntarily and playfully could reproduce and combine, analogous perhaps to the play with shapes in a jigsaw puzzle. The key words are *Bild* and *Spiel*; and once alerted to them, one finds them with surprising frequency in Einstein's writings. . . .

In Einstein's published work, his visual imagination sometimes breaks through vividly. One thinks here, for example, of passages where he describes thought experiments involving the picturesque tasks of coordinating watch readings, the arrival of light signals, the positions of locomotives and those of lightning bolts. Possibly Einstein's ability to deal with models and drawings in the patent office, and his own delight throughout his adult life with the workings of puzzle-toys, are additional clues of some significance. More important for our purposes, this ability of visualization is evident in the haunting thought experiment of the lightbeam, begun at Aarau, and even in the thought experiment proposed in his earlier essay of 1894–5, where he envisages probing the state of the ether in the vicinity of a current-carrying wire.

I have little doubt that the ability to make such clear visualizations of experimental situations was crucial in his task of penetrating to the relativity theory (for example, in the argument leading to the relativity of simultaneity). It is so to this day. As anyone knows who has tried to instruct students in relativity theory, the problem of getting a firm initial understanding of special relativity is not one of mathematics—at most, elementary calculus is required—but rather one of clear *Vorstellung*. What helps a beginner most is precisely the ability to imagine vividly some thought experiments involving the perceptions and reports of two observers who are moving relative to each other. The style of thinking necessary at the outset is quite different from working, for example, through the machinery of a formalism such as characterizes the work of a Sommerfeld.

But in the long run, the strength of Einstein's visual imagination was not limited only to such uses. Rather, the hypothesis here proposed is that the special ability to think with the aid of a play with visual forms has deeper consequences. It animates the consideration of symmetries and a corresponding distaste for extraneous complexities from the beginning to the end—from the result embedded in his 1905 paper that the Lorentz transformation equations yield "contractions" and "time dilations" that are, for the first time, symmetrical for all inertial systems, to his long wrestling with the task of constructing a relativistic theory of gravitation, basing its field structure on a symmetric tensor g and the symmetric infinitesimal displacement tensor $\Gamma$, and finally to his self-imposed labors of finding a theory of the total field through the generalization of giving up the symmetry properties of the g and $\Gamma$ fields. From the beginning to the end, Einstein's scientific thought was pervaded by questions of symmetry and the closely related concept of invariance.

ARTHUR I. MILLER

# Visualizability

# as a Criterion

# for Scientific

# Acceptability

THERE IS A domain of thinking where distinctions between conceptions in art and science become meaningless. For here is manifest the efficacy of visual thinking, and a criterion for selection between alternatives that resists reduction to logic and is best referred to as aesthetics. To demonstrate this domain we can examine a case study in the history of science—the genesis of quantum theory in the years 1913–27. The notions of aesthetics held by the dramatis personae of this period will not always be associated with the choice of a pleasing visualization, but sometimes with a choice of thema, such as continuity or discontinuity, with the choice of a particular mathematical framework, or with combinations of these criteria. And sometimes their aesthetic will change. . . .

A measure of how dramatic was the search for a consistent atomic theory, and of the ambiguities lurking at every turn, is the disagreement between the accounts of Bohr and Pauli. The path to the quantum theory of 1927 was not an orderly progression from visualizable models to a mathematical for-

Excerpts from A. I. Miller, Visualization lost and regained: The genesis of the quantum theory in the period 1913–1927. In J. Wechsler, ed., *On Aesthetics in Science*. Cambridge: MIT Press, 1978, pp. 73–96. Reprinted by permission of the publisher.

malism whose description of matter and phenomena in the atomic domain defied visualization in the ordinary sense of this word. Rather the situation was closer to the one recalled by Werner Heisenberg where the physicists experienced despair and helplessness because of their loss of visualization and of their distrust in customary intuition. It was a period when such time-honored concepts as space, time, causality, substance, and the continuity of motion were separated painfully from their classical basis. The rejection, one by one, of the tenets of classical physics was distressing because classical physics had become so commonsensical, so reflective of the world in which we live. Since it was a causal theory, a particle's motion was continuous and its position could be predicted with, in principle, perfect accuracy. There was no reason to doubt that our intuition could also be extended to phenomena involving the microscopic electron. So it was hoped with confidence that the discontinuities which emerged from Max Planck's theory of 1900 for the continuous spectrum of the radiation emitted from a hot substance could be removed somehow in a more fine-grained theory. But in 1913, the twenty-eight-year-old Niels Bohr proposed a new kind of theory for the free atom. It was *predicated* openly upon discontinuities and gross violations of classical physics. Yet despite its transgressions Bohr's theory retained the pleasing picture of Rutherford's atom as a miniature Copernican system. But Bohr did not consider the theory complete because it conflicted with classical physics. However, by 1918, the break between Bohr's theory and the classical physics became more striking when Bohr folded into his theory a kind of mathematical description for the unvisualizable quantum jump that was forbidden by the accepted view of the classical physics of particles: namely, a priori probabilities which renounce from the start any knowledge of the cause for phenomena. By 1923, the limitations of the picture of the atom as a miniature solar system had become forcefully apparent. . . .

So atomic physics began to slip into an abyss where the planetary electron went from a tiny sphere to an unvisualizable entity. What is so remarkable is that this surrender of visualization was precipitated only in part by empirical data, that is, problems with complex atoms, but more as a result of Bohr's aesthetic choice between the opposing themata of continuity versus discontinuity. Bohr's decision was to reject the light quantum in order to retain the traditional dichotomy between the continuous radiation field and discrete matter. This limited duality was his aesthetic in 1923–25 and it was not linked to visualization. Heisenberg, at age twenty-three a contemporary of Pauli, spent late 1924 to early 1925 at Bohr's Institute for Theoretical Physics at Copenhagen; he was already making his mark in physics. Whereas for Bohr the loss of visualization was painful, Heisenberg found it to be congenial to his nonvisual mode of thinking. . . .

Even more remarkable than the loss of visualization was how it was regained in the period 1926–27 when empirical data played almost no role. Rather the path to regaining visualization is characterized by the high drama of the intense personal struggles among the dramatis personae over their choices of the themata in conflict—continuity versus discontinuity, the usefulness of mathematical models versus mechanistic-materialistic models, and whether to maintain causality. . . .

A new figure enters in 1926. Erwin Schrödinger at forty-three was a Professor at Zurich and an outsider both geographically and in thinking from the small band of physicists concentrated mainly in Copenhagen and Göttingen who were almost entirely responsible for the development of the quantum mechanics. Schrödinger's sentiments were closer to those of the continuum-oriented physicists at Berlin, in particular Einstein and Planck. Schrödinger felt so "repelled" both by the lack of visualization and the mathematics used in the quantum mechanics that he formulated the wave mechanics. His aesthetic was linked to a continuum-based physics which visualized that atomic entities were composed of packets of waves. Nevertheless, he claimed that this was better than no picture at all. . . .

Heisenberg, from the beginning, had found the wave mechanics "disgusting" and was further enraged at Born's desertion of the cause of quantum mechanics. This situation and conversations with Bohr in late 1926 reinforced both Heisenberg's aesthetic and his nonvisual modes of thinking, leading him to boldly *demarcate between* intuition and visualization. The results in early 1927 were the uncertainty relations and the rejection of the only causal law that this viewpoint could consider—classical causality.

To Heisenberg's consternation and eventual distress Bohr pressed him relentlessly not to publish his results. For Bohr

had realized that only the complete wave-particle duality for
matter and light could lead to a consistent interpretation of
the quantum mechanics containing visualization. However,
the causal laws could no longer be associated with space-time
pictures, but with the conservation laws of energy and momentum. This viewpoint is embodied in Bohr's complementarity principle of 1927. The analysis leading to the complementarity principle plumbed the very depths of knowledge
down to the formation of ideas themselves. Bohr emphasized
that the necessary prerequisite to this analysis was the discovery that visual thinking preceded verbal thinking, and
linked to visual thinking could only be the aesthetic of the
symmetry offered by the complete wave-particle duality.

The roots of complementarity are as varied and deep as its
ramifications into other disciplines and so I believe it is ap-
ropos here to mention Bohr's lifelong interest in art, especially
in cubism. The Danish artist Mogens Anderson, a friend of
Bohr, recollected what most impressed Bohr in cubism: ". . .
face and limbs depicted simultaneously from several angles. . . .
That an object could be several things, could change, could
be seen as a face, a limb and a fruitbowl." We shall see that
this motif has striking parallels to that offered by complementarity where the atomic entity has two sides—wave and particle—and depending on how you look at it (that is, what experimental arrangement is used), that is what it is.

So in late 1927, Bohr raised physics from the darkness of
the abyss and the drama was completed. We are reminded
here of one of Bohr's favorite sayings from Schiller:

> Only fullness leads to clarity.
> And truth lies in the abyss.

## Visualization

## Lost

## (1923–25)

By the first decade of the twentieth century it was taken for
granted that physical theory should be a reflection of the continuity that we observe in the world of our perceptions. . . .

The newly discovered electron was visualized as having a definite shape, and when accelerated it becomes the source of electromagnetic waves (light) spreading out like the circular ripples caused by a stone thrown into water. This dichotomy between discrete matter and the continuous or wave picture of light was acceptable in classical physics. Even though the electron introduced a discontinuity into the substratum of nature, the laws governing physical theory were required to be continuous. . . .

Although the renunciation of a picture of the bound electron was a necessary prerequisite to the discovery of the new quantum mechanics, nevertheless the lack of *Veranschaulichung* or *Anschaulichkeit* or an intuitive [*anschauliche*] interpretation was of great concern to Bohr, Born, and Heisenberg, and this concern emerges from their scientific papers of the period 1925–27. For example, in the important paper of Born, Jordan, and Heisenberg of late 1925, Heisenberg writes in the "Introduction" that the present theory labors "under the disadvantage of not being directly amenable to a geometrically intuitive interpretation [*anschauliche interpretiert*] since the motion of electrons cannot be described in terms of the familiar concepts of space and time." Heisenberg continues, "In the further development of the theory, an important task will lie in the closer investigation of the nature of this correspondence between classical and quantum mechanics and in the manner in which symbolic quantum geometry goes over into intuitive classical geometry [*anschaulich klassiche Geometrie*]." In the section, "The Zeeman effect," probably also written by Heisenberg, the notion of planetary stationary states arises when he writes of the inability of the new quantum mechanics to resolve the problem of the anomalous Zeeman effect as perhaps due to the result of an "intimate connection between the innermost and outermost orbits . . . "; however, Heisenberg hoped that the recent hypothesis of an electron spin by Uhlenbeck and Goudsmit might provide an alternate route. Yet this fourth degree of freedom for the electron could not be visualized because a point on a spinning electron of finite extent could move faster than light.

Born also felt uncomfortable without a means to visualize the concepts in the mathematical formalism of the new quantum mechanics. In late 1925, Born writes that "we have the right to use the terms 'orbit' or even 'ellipse,' 'hyperbola,' etc.

in the new theory. . . . Our imagination is restricted to a limiting case of possible physical processes."

Andrade, in his book published in late 1926, tempered his praise of the new quantum mechanics by pointing out that "its weakness is that we have no geometrical or mechanical picture" of the physical processes to which it is applied of the type to which "we are accustomed." . . .

Acknowledging Heisenberg's new quantum mechanics, Bohr writes that it may at first "seem deplorable" that in atomic physics "we have apparently met with such a limitation in our usual means of visualization"; nevertheless, "mathematics in this field too, presents us with tools to prepare the way for further progress." Thus, where visualization has been lost, mathematics must be the guide toward "further progress." In summary, by the latter part of 1925, empirical data and theoretical results caused Bohr to demarcate between the conservation laws and pictures in space and time.

## Visualization

## Regained

## (1926–27)

Erwin Schrödinger, in the third of the four "communications" of 1926, left no room for doubt that a sense of aesthetics inspired him to formulate the wave mechanics:

> My theory was inspired by L. de Broglie, *Ann. de Physique* (10) 3, p. 22, 1925 (Thèses, Paris, 1924) and by short but incomplete remarks by A. Einstein, *Berl. Ber.* (1925) pp. 9 ff. No genetic relationship whatever with Heisenberg is known to me. I knew of his theory, of course, but felt discouraged, not to say repelled, by the methods of transcendental algebra, which appeared very difficult to me and by the lack of visualisability [*Anschaulichkeit*].

In a more objective tone one of his principal criticisms against the quantum mechanics is that it appeared to him "extraordinarily difficult" to approach such processes as collision phenomena from the viewpoint of a "theory of knowledge" in which we "suppress intuition [*Anschauung*] and operate only with abstract concepts such as transition probabilities, energy levels, and the like." For although, he continues, there may exist "things" which cannot be comprehended by our "forms of thought," and hence do not have a space and time description, "from the philosophic point of view" Schrödinger was sure that "the structure of the atom" does not belong to this set of things.

Another of Schrödinger's reasons for preferring a wave-theoretical approach is his preference for a continuum-based theory in which he claimed there are no quantum jumps, over the "true discontinuum theory" of Heisenberg. In addition, Schrödinger pushed his proof of the mathematical equivalence of the wave and quantum mechanics to the conclusion natural to his viewpoint—when discussing atomic theories he "could properly also use the singular."

But what sort of picture did Schrödinger offer? He maintains that no picture at all is preferable to the miniature Copernican atom, and in this sense the purely positivistic standpoint of the quantum mechanics is preferable because of "its complete lack of visualization"; however, this conflicts with Schrödinger's philosophic viewpoint. Schrödinger bases his visual representation of bound and free electrons on the comparison with classical electrodynamics of the solution to the fundamental wave equation of the wave mechanics, that is, the wave function. The electron in a hydrogen atom is represented as a distribution of electricity around the nucleus. . . .

To summarize the state of quantum physics in the first half of 1926: While no adequate atomic theory existed as of July 1925, by mid-1926, there were two seemingly dissimilar theories. Quantum mechanics was purported to be a "true discontinuum theory." Although a corpuscular-based theory, it renounced any visualization of the bound corpuscle itself (at this time quantum mechanics could discuss only bound state problems). However, its mathematical apparatus was unfamiliar to physicists and also difficult to apply. On the other hand, there was Schrödinger's wave mechanics which was a continuum theory focusing entirely upon matter as waves, offering a visual representation of atomic phenomena and accounting for discrete spectral lines. Its more familiar mathematical apparatus set the stage for a calculational breakthrough. The wave mechanics delighted the more continuum-based portion of the physics community, particularly Einstein and Planck. Although the final experimental verification of the complete wave-particle duality of matter would not appear until 1927, many already subscribed to it. . . .

The tension between the quantum and wave mechanics increased with the appearance of Born's quantum theory of scattering in the latter part of 1926. Born's analysis of data from the scattering of electrons from hydrogen atoms convinced him of the need for a quantum-theoretical description of scattering consistent with the conservation of energy and momentum; yet, Born writes, neither scattering problems nor transitions in atoms can be "understood by the quantum mechanics in its present form"—here, under quantum mechanics, Born includes both Heisenberg's and Schrödinger's mechanics. His reasons are: Heisenberg's quantum mechanics denies an "exact representation of the processes in space and time"; Schrödinger's wave mechanics denies visualization of phenomena with more than one particle. In Born's view treating problems concerning scattering and transitions requires the "construction of new concepts," and his vehicle would be Schrödinger's version because it allows the use of the "conventional ideas of space and time in which events take place in a completely normal manner," that is, the possibility of visualization. . . .

But what kind of "intuitive content" can a physicist offer who denies visualization of physical processes? In the very first sentence Heisenberg begins his bold reply to this question by stating two criteria for a theory "to be understood intuitively": in all simple cases the theory's experimental consequences can be thought of in a qualitative manner; and its application should lead to no internal contradictions. Although the mathematical scheme of quantum mechanics requires no revisions, Heisenberg continues, "Heretofore, the intuitive [anschauliche] interpretation of the quantum mechanics is full of internal contradictions which become apparent in the struggle of the opinions concerning discontinuum- and continuum-theory, waves and corpuscles." Thus, the quantum mechanics satisfies only one of Heisenberg's criteria "to be understood intuitively." Heisenberg writes that just as in the general relativity theory, where the extension of our usual conceptions of space and time to very large volumes follows from the mathematics of the theory, a revision of our usual kinematical and mechanical concepts "appears to follow directly from the fundamental equations of the quantum mechanics." Heisenberg deduced from these equations the result that, unlike in classical physics, in the atomic domain the uncertainties in measuring the position and momentum of an atomic particle cannot be simultaneously reduced (even in principle) to zero. Rather, the product of the uncertainties is a small but nonzero number—Planck's constant. For example, the more precisely the particle's position is measured, the less precisely can its momentum be ascertained. Heisenberg attributed the cause of this uncertainty relation to be the "typical discontinuities" which contradict our customary intuition. He concludes that the mathematical formalism of quantum mechanics determines the restrictions in the atomic domain of such classical concepts as position and momentum. Clearly, in order to maintain his aesthetic of a limited wave-particle duality which is not linked to visualization, Heisenberg has boldly demarcated the notion of "to be understood intuitively" from the visualization of atomic processes. He has chosen to resolve the "struggle of the opinions" between thema-antithema couples with a corpuscular-based theory that lacks visualization of the corpuscle and severely restricts visualization of physical processes.

This viewpoint led him to assert that in the case of "the strong formulation of the causal law: 'if we know the present exactly, then we can calculate the future,' it is not the consequent that is false but the presupposition." The "strong formulation of the causal law" is the one from classical physics, and it is dependent upon visualization and the absence of discontinuities. However, the uncertainty relation between position and momentum places limits upon the precision with which the initial conditions of an atomic particle can be specified; its path cannot be traced to any arbitrary degree of accuracy as in classical physics. Heisenberg's rejection of the causal law from classical physics is therefore not unexpected.

With these results, Heisenberg concludes with a remarkable passage, "one will not have to any longer regard the quantum mechanics as unintuitive [unanschaulich] and abstract."

Heisenberg's strong predisposition to his successful quantum mechanics predicated upon a lack of visualizability is a reflection of his preference for nonvisual thinking, and is undoubtedly the root of his redefinition of intuition. His lack of trust in visual thinking to understand quantum mechanics could very well have been further reinforced by discussions with Bohr during the period of their intense struggle to understand the riddles of the quantum theory. For imagine using visual representations, as Heisenberg writes in the uncertainty principle paper, to think qualitatively of the experimental results in all simple cases—for example, the determination of the slit through which an electron passes in the diffraction of a low intensity beam of electrons by a double slit grating, or what it means for a light quantum to be polarized. Heisenberg recalled that "we couldn't doubt that this [i.e., quantum mechanics] was the correct scheme but even then we didn't know how to talk about it"; these discussions left them in "a state of almost complete despair." The loss of visualization must have been especially difficult for Bohr whose essays are filled with visual words—for example, picture [*Bild*], visual ideas [*Vorstellungen vor Augen*], mechanical models. Arnheim, in an essay on the psychology of art, writes of the "apprehension" that develops in a scientist during a transition from a corpuscular theory with "determined contour line" to more complex models. The states of mind of both Bohr and Heisenberg fit this description, for they were set adrift, Bohr lacking visualization and both Bohr and Heisenberg distrusting their intuitions.

Heisenberg recalled that their approaches to gedanken experiments differed. Bohr, by late 1926, had accepted the duality in the quantum theory and its reflection in nature as the complete wave-particle duality, even though the wave aspect of matter had not yet been definitely established experimentally. For Bohr the wave-particle duality was the "central point in the whole story," because it permitted him to use visual thinking once again, that is, to play with pictures of waves and particles.

Heisenberg relied solely upon the mathematical scheme of quantum mechanics until December of 1926 when he became aware of the Dirac and Jordan transformation theories, proving to his satisfaction the equivalence between the wave and quantum mechanics. Then Heisenberg could understand Bohr's interchangeable use of wave and quantum mechanics because he could mathematize Bohr's visual arguments. Nevertheless,

Heisenberg steadfastly refused to believe that there was a complete dualism in quantum theory, rather the transformation theory showed how "very flexible" was the quantum mechanics.

In the light of their different viewpoints, Bohr's dissatisfaction with Heisenberg's uncertainty principle paper should come as no surprise. Bohr insisted that one cannot allow the mathematical formalism to restrict words like position and momentum because, despite the uncertainty relation, you have to use them "just because you haven't got anything else." An unpleasant atmosphere developed in which Heisenberg refused to change the content of the paper but did acquiesce to add a "Nachtrag bei Korrectur" (Postscript with Corrections). There Heisenberg writes that Bohr's recent investigations led him to conclude that uncertainty in observation is not rooted "exclusively upon the presence of discontinuities," but in the wave-particle duality of matter.

On 16 September 1927 at the International Congress of Physics at Como, Italy, Bohr presented a viewpoint which he hoped would "be helpful" to "harmonize the apparently conflicting views taken by different scientists." He realized that the "classical mode of description must be generalized" because our customary intuition cannot be extended into the atomic domain. There Planck's constant links the measuring apparatus to the physical system under investigation in a way that is "completely foreign to the classical theories," and this is the root of the unavoidable statistics of the quantum theory. While in the classical theories this interaction can be neglected, in the atomic domain it cannot. Then, in the atomic domain, the notion of an undisturbed system developing in space and time is an abstraction and "there can be no question of causality in the ordinary sense of the word," that is, strong causality. Bohr's viewpoint on this situation is the complementarity principle:

> The very nature of the quantum theory thus forces us to regard the space-time coordination and the claim of causality, the union of which characterizes the classical physical theories, as complementary but exclusive features of the description, symbolizing the idealization of observation and definition respectively.

Bohr's response is to separate the causal law from a space-time description; the union of the two was the classical or strong

causality. The causal law and space-time pictures are complementary because they are both necessary for a complete
description of phenomena.

Bohr next emphasizes that his deliberations upon the relationship between visual thinking and the wave-particle duality of light were central to the genesis of the complementarity
viewpoint. . . .

According to the viewpoint of complementarity, the pictures
of light and matter as waves and particles are not contradictory, as had been thought previously, but are "complementary
pictures" because they are both necessary for a complete description of atomic phenomena; they are mutually exclusive
because in a given experimental arrangement atomic entities
can exhibit only one of their two sides. The scheme of complementarity permits a self-consistent description of how the
quantum theory relates to the simple experiments that had
driven Bohr and Heisenberg to despair. For in the complementarity paper Bohr emphasizes that "every word in the language refers to our ordinary perception," and according to our
ordinary perception there are only two kinds of phenomena—
corpuscular and undulatory—just as our intuition tells us that
"things" are either discontinuous or continuous. The failure
of our ordinary intuition in the atomic domain, Bohr writes,
is rooted in "the general difficulty in the formation of human
ideas, inherent in the distinction between subject and object."

This completes Bohr's analysis of 1927, which contains the
viewpoint soon to be referred to as the *Kopenhagener Geist
der Quantentheorie*. It is an extraordinary analysis because
Bohr's method to arrive at a contradiction-free interpretation
of the quantum theory led him to ever-deeper levels of analysis: from a purely scientific analysis, to an epistemological
analysis, to an analysis of perceptions, and then to the origins
of scientific concepts. A necessary prerequisite to this analysis
was the acceptance of the complete wave-particle duality in
nature. This permitted Bohr to use the symmetry of the pictures of waves and particles which are familiar from our customary intuition. He could then discuss all simple experiments. Visual thinking preceded verbal thinking, and linked
with visual thinking was Bohr's new aesthetic of the symmetry
of pictures afforded by accepting the complete wave-particle
duality. Bohr's viewpoint stands in contrast to the views of

Heisenberg and Born. Although Heisenberg's viewpoint maintained the validity of the conservation of energy and momentum, it was corpuscular-based and severely restricted visualization as well as the use of words such as position and momentum. Heisenberg considered the classical law of causality which is linked to the pictures to be invalid in the atomic domain. Born's viewpoint of 1926 contained energy and momentum conservation, visualization, waves and particles, but the waves were not observable quantities. . . .

In conclusion, the importance for creative thinking of the domain where art and science merge has been emphasized by the great philosopher-scientists of the twentieth century—Bohr, Einstein, and Poincaré. For in their research the boundaries between disciplines are often dissolved and they proceed neither deductively through logic nor inductively through the exclusive use of empirical data, but by visual thinking and aesthetics.

# Selection 48

ALBERT EINSTEIN

# A Testimonial

My Dear Colleague:

In the following, I am trying to answer in brief your questions as well as I am able. I am not satisfied myself with those answers and I am willing to answer more questions if you believe this could be of any advantage for the very interesting and difficult work you have undertaken.

(A) The words or the language, as they are written or spoken, do not seem to play any role in my mechanism of thought. The psychical entities which seem to serve as elements in thought are certain signs and more or less clear images which can be "voluntarily" reproduced and combined.

There is, of course, a certain connection between those elements and relevant logical concepts. It is also clear that the desire to arrive finally at logically connected concepts is the emotional basis of this rather vague play with the above mentioned elements. But taken from a psychological viewpoint, this combinatory play seems to be the essential feature in productive thought—before there is any connection with logical construction in words or other kinds of signs which can be communicated to others.

(B) The above mentioned elements are, in my case, of visual and some of muscular type. Conventional words or other signs have to be sought for laboriously only in a secondary stage, when the mentioned associative play is sufficiently established and can be reproduced at will.

(C) According to what has been said, the play with the mentioned elements is aimed to be analogous to certain logical connections one is searching for.

(D) Visual and motor. In a stage when words intervene at all, they are, in my case, purely auditive, but they interfere only in a secondary stage as already mentioned.

(E) It seems to me that what you call full consciousness is a limit case which can never be fully accomplished. This seems to me connected with the fact called the narrowness of consciousness (*Enge des Bewusstseins*).

Remark: Professor Max Wertheimer has tried to investigate the distinction between mere associating or combining of reproducible elements and between understanding (*organisches Begreifen*); I cannot judge how far his psychological analysis catches the essential point.

With kind regards . . .

Albert Einstein

# Epilogue

THE SELECTIONS IN this book document the claim that scientific thinking can be illuminated by empirical psychological research. The most definitive results are those concerning the use of hypotheses (see parts 2, 3, and 4). It has been manifest for centuries that hypotheses are used extensively in the process of scientific inquiry. It is now empirically clear that both confirmatory and disconfirmatory strategies are instrumental in testing hypotheses, even though questions remain concerning the best mix of these strategies. These findings present a serious challenge to normative philosophical models of science, as we shall argue below.

To assert that mathematics and statistics represent powerful heuristic devices is to reaffirm what every scientist believes. But, as parts 5 and 6 make clear, there are many specifically psychological questions concerning their use which have hardly been explored, questions about their functions and about their consequences. The value, and the danger, of such heuristics seems to lie in their semiautonomous character. Mathematical representations, like statistical models, or attractive analogies, can provide a sense of certainty, but they can also lead the investigator farther and farther from empirical realities. This possibility is especially dangerous where error-ridden data make the precise test of hypotheses difficult, and where the intuitions of even experienced scientists have been shown to be often awry (Tversky and Kahneman, selection 31). Scientific knowledge of the consequences of using mathematical or statistical models is very scant, though there is no lack of speculation.

The concept of cognitive heuristics such as metaphors and analogies may have value even in that wooliest of domains, the consideration of scientific creativity (parts 7, 8, 9). Case studies and historical evidence suggest that heuristics play an important role in creativity. A good deal has been written about the psychological value of, for example, classification in science. Not much research exists which approaches these issues in a rigorous manner, in spite of the rich array of available research techniques.

**Theories**

**of Scientific**

**Thinking**

The cognitive psychology of science is too new to have generated much in the way of formal theory based upon existing empirical knowledge. Nonetheless, a number of more general theoretical perspectives can be described, which have guided inquiry into the nature of thought, and which can be extended to the particular case of scientific thought. A review of some of those perspectives will highlight the possibilities for more specific, testable theories in the future. Among the major approaches are: associationism, Gestalt theory, information processing theory, genetic epistemology, and psychodynamic theory. Some major accounts (e.g., Koestler 1964) cut across several categories. The reader should note that our characterization of each theory is brief and partial.

TRADITIONAL THEORIES
Classical association theories (e.g., that of James Mill, 1829) were meant to be applicable to all human inquiry, including science, but more recent psychological models have not been framed so broadly. Anderson and Bower (1973), for example, have proposed an associative model of human long-term memory, but the scope of the theory is restricted to semantic knowledge of an everyday sort. The power of such models is still a major source of controversy in cognitive psychology; their elegant simplicity is sometimes said to be achieved at the expense of fidelity to the richness of mental phenomena (see

Bartlett 1932:304–08). In any case, they can be extended to science only with extensive modifications—as we will see when Campbell's (1974) theory is examined.

Gestalt theory is represented by Wertheimer's 1945 chapter on Einstein (selection 23). The nature of science was an important problem for Gestalt psychology in the 1930s, manifested in, for example, Köhler's book *The Place of Value in a World of Facts* (1938), Lewin's well-known 1931 paper on the difference between Aristotelian and Galileian science, and Duncker's (1935/1945) monograph on problem solving.

Several themes relevant to the psychology of science run through Gestalt accounts of problem solving: the importance of restructuring problem elements, the existence of "functional fixedness," (Luchins 1942) and the importance of understanding the problem space (which may include value-laden elements). These concepts have become integral parts of every current theory of problem solving, and will likely become part of the cognitive psychology of science as well.

It is proper to ask, however, if a satisfactory account of science could be constructed on the basis of Gestalt concepts. In particular, does Gestalt theory permit predictions of impending restructurings, or is it useful only after the fact, as a description of what happened? The answer to this question may depend in part on whether or not there can be "laws of good Gestalt" applicable to scientific statements. One domain where such laws might emerge is in considerations of the relative simplicity of explanations.

Similarity is a primitive concept in the Gestalt tradition. Good figure, a basic principle of perceptual organization, can be thought of in terms of the relative simplicity of alternative perceptual organizations of the same field. Both terms, similarity and simplicity, have been used freely in accounts of scientific thinking, the latter also under other names—parsimony, Occam's Razor, Lloyd Morgan's canon. Perhaps the link between scientific thinking and perceptual organization is more than merely semantic. The personal documents of scientists lead one to the inescapable conclusion that imagery, often *visual* imagery, is instrumental in scientific discovery. If so, then the laws of perceptual organization may lead to insight into the creative function of imagery. The very phrase "perceptual organization" implies a unity of the perceptual field. Bronowski (1965) suggested a complex analogy between the Gestalt approach to visual perception and its possible extension to science. He wrote "The discoveries of science . . . are extrapolations—more, are explosions of hidden likeness" (p. 19), and referred to the "unity" found in hidden likenesses (p. 13).

What is the relation between simplicity and similarity? When Niels Bohr proposed that the electron orbits of the hydrogen atom were quantized, his somewhat preposterous analogy to a planetary system was tolerated because of the overwhelming similarity between the predicted spectral lines and the empirical spectral lines. When it became clear that the same numerical constant (Planck's Constant, $h$) was implicated in the spectral lines of hydrogen *and* in a variety of other quantized phenomena, then an "explosion of hidden likeness" appeared which suggested the underlying simplicity of quantum mechanics. Is similarity, therefore, one basis of simplicity? Perhaps so, but it is important to remember that simplicity is often the result of a judgment that is aesthetic in character. We do not know *why* nature ought to be simple, though we strongly prefer it that way.

The Gestalt psychologists recognized that mind imposed order on nature. That fact is nowhere clearer than in the early experiments on the phi phenomenon (Wertheimer 1912) which led to the dictum, "The whole is different from the sum of its parts." But they also saw that nature imposed order on mind. This principle also suffuses their work, perhaps most explicitly in the elucidation of the concept of isomorphism. The mutual constraints of self and other and their mutual imposition of order upon one another suggest that we may observe much about the laws of thought from our constructions of reality, and much about the laws of scientific thought from our scientific constructions of reality.

Relatively little attention has been paid to scientific thinking by information-processing theorists, in spite of a recent thrust toward the study of "natural cognition" (e.g., Allport 1975; Neisser 1976), and in spite of some distrust of traditional laboratory-based paradigms for the study of information processing (see Tulving 1979; Estes 1979). As with Gestalt theory, many of the processes important in information processing theories will likely be part of a comprehensive theory of science. Thus, the characterization of strategies for problem solution (e.g., Simon & Reed 1976) and the importance of problem representation (e.g., W. Reitman 1965; J. Reitman 1976; Simon & Hayes 1976) are directly applicable in the more complex domain of scientific problem solving.

The major difficulty faced by the approach is the probable inapplicability of its most powerful method: chronometric decomposition of mental tasks (Sternberg, 1969; Blumenthal 1977; Posner 1978). Chronometric analysis is not likely to be workable in the very complex situations characteristic of science. This means, we believe, that such techniques must be applied to analogues of science, rather than to science itself, and an attempt must then be made to determine empirically whether or not the results generalize to the more complex situation. Thus, R. Sternberg's componential analysis of analogical tasks will face a severe test before it can be said to "be general to other kinds of inductive reasoning tasks" (1977:375). Process tracing, on the other hand, (Newell & Simon 1972) may have some utility, but the complexity of the scientific process makes it unlikely that detailed computer simulations will be possible. As the tasks studied become more open, the needed programs become larger. Again, recourse to analogues is essential.

In recent years, a great deal has been learned about human memory by psychologists working within an information-processing framework. Insofar as remembered knowledge is critically important in scientific cognition, then such work is relevant to the understanding of science. However, it should be noted that most of the work has centered on the deliberate acquisition of experimenter-provided material, rather than on the potentially more relevant processes involved in the utilization of already acquired material (exceptions to this generalization can be found in, e.g., Norman and Rumelhart 1975).

Genetic epistemology has been characterized earlier (selection 3). Piaget's approach to the nature of science bears certain resemblances to the neo-Kantian approach of Ernst Cassirer (1923), and of Cassirer's follower Heinz Werner (1940). In both, there is an emphasis on the genesis of scientific knowledge which goes beyond Piaget's work in placing heavier emphasis on the role of symbolic representational systems. Werner, like Piaget, emphasized the continuity of scientific and nonscientific thought, though neither man devoted specific research attention to the case of science as such.

A more complete understanding of science may be facilitated by a developmental framework along either Piagetian or Wernerian lines. For example, Shweder (1977) has shown that the illusory correlation phenomenon (Chapman & Chapman 1969) can be related to certain magical thinking practices—i.e., to primitive scientific thinking. Placing such problems in a comparative framework, as Werner would have advocated, should make it possible to see further similarities. Further, a developmental approach may reveal similarities between children's thinking at various ages and various types of scientific thinking. Piaget is not the only one to have made such claims (similar notions can be found even in Wundt 1912/1916), but he is the only one to have attempted systematic research along such lines. Furthermore, we consider it likely that some of the most fruitful educational applications of the psychology of science will emerge from developmental studies (see e.g., Siegler 1978).

Freud argued that human creativity was based upon the sublimation and substitute gratification of instinctual needs (Freud 1917/1965). In the intervening decades, there have been many attempts to elaborate specific accounts of human creativity based upon psychoanalytic concepts. Most such attempts have relied upon case studies of artistic creativity, though Rothenberg (1979) has recently applied his concept of "Janusian thinking" (a secondary process for handling two ideas at once) to the case of Albert Einstein, and Westman (1975) had considered Copernicus and Rheticus. Furthermore, few psychodynamic accounts have considered cognitive, as opposed to motivational, aspects of science.

The problem with psychoanalytic accounts of scientific creativity is the same as that faced by psychoanalysis generally, namely, the inability to generate testable hypotheses (but see Grünbaum 1979). Nevertheless, an analytic approach has, on occasion, proved to be a useful tool in the analysis of artistic creativity. We expect that similar advantages might accrue if it were applied to particular scientists. More to the point, much of the "specialness" of scientific creativity resides in those sudden flashes of insight that reveal, for example, mathematical truth to a Poincaré. It is likely that interview and free-association techniques will be useful adjuncts in the exploration of such instances. Much of what scientists report about their own thinking might well appear in a new light if subjected to such examination.

RECENT ATTEMPTS AT A THEORY OF SCIENTIFIC THINKING
Attention must be given to three recent attempts to construct explicit theories of scientific thought. Each captures important aspects of science, and it is hoped that each will generate further research.

Campbell's (1974) evolutionary epistemology posits an associative linkage of scientific ideas and a selection mechanism that rejects "unfit" ideas, allowing "fit" ideas to survive. The selection process can occur either intraindividually, when the scientist seeks out the limitations of his or her own ideas, or interindividually, in a competitive social milieu. Campbell's theory implies that chance plays a large role in scientific progress, a view shared by Feyerabend (1975). The potential power of Campbell's deceptively simple theory is suggested in his 1977 William James Lectures at Harvard University, in which he argued that evolutionary epistemology could provide a larger framework in which the operation of normative social processes could be studied alongside positivistic aspects of science, like the importance of experimental data. While Campbell's theory has not so far inspired much empirical research, it holds promise as a possible integrative framework for the psychology of science.

A very different framework for the study of science has been presented by Mitroff (selection 20). Building on Churchman's (1971) description of science as an "inquiry system," Mitroff characterized several distinct approaches to scientific inquiry. He distinguished between Apollonian science, which emphasizes precise and orderly research, and Dionysian science, which emphasizes intuition, and tries for unexpected approaches. This distinction appeared to be related to personality differences in the scientists studied by Mitroff. Further empirical research into the basis of such differences may permit us to understand more fully the relation between scientific activity and the emotional and motivational bases of the scientist (see Wilkes 1979, for further work along similar lines).

Koestler (1964) presented a theory of creative activity which deals with the sources of artistic and scientific creativity, and relates both to a theory of humor. For Koestler, the unexpected combinations of ideas that produce laughter, as in jokes or puns, are similar to the unexpected associations that result in creative new solutions to problems. In particular, for Koestler as for Gruber (1974), the most fruitful ideas occur when associative chains from different content domains intersect in a common term; the term is then said to be "bisociated" with two normally independent frames of reference. Koestler's theory has inspired little research, but deserves, we feel, to be taken more seriously by psychologists. Its potential value was brought home to us in a recent interview with a highly creative physicist. When under pressure to solve a particular problem in a creative way, he habitually isolates himself in his study, and surrounds himself with a large number of books and papers, *all* of which are irrelevant to the problem at hand. He then reads a page here, a section there, moving from work to work, until ideas relevant to the problem at hand begin to occur to him. It is as neat an illustration of the power of Koestler's formulation as one could hope for!

We must confess what is probably obvious to the reader: Our predilection is not at present toward the development of a broad theory of scientific thinking. The current state of the psychology of science is such that far more empirical data is needed before the construction of any such theory is likely to be fruitful. Of course, all empirical investigation is theory laden, and *must* be theory laden. Thus, what we are arguing is a question of relative emphasis. In the long run, knowledge will be produced by the social aggregation of results derived from a variety of emphases.

Nonetheless, in closing this section, we propose (not entirely facetiously) three "Laws of Scientific Thought":

I. *Every hypothesis continues in a state of belief unless acted upon by a force.* This follows from the empirical work on confirmation and disconfirmation.

II. *The degree of belief in a particular hypothesis is directly proportional to the similarity between the hypothesis' prediction and the empirical facts.* In most cases, paradoxically, degree of belief fails to be inversely proportional to any lack of similarity.

III. *For every hypothesis there is at least one alternate hypothesis.* In most cases, this will be provided by the social context.

**Methods**

**for Studying**

**Scientific**

**Thinking**

In advocating future research by psychologists on the nature of science, it should be understood that the greatest progress will result if a great variety of approaches are used. We will attempt to identify and briefly characterize a few of the more promising ones.

HISTORICAL RESEARCH

The cognitive psychologist interested in scientific thinking should be conversant with methods of historical analysis and with the history of science. History of science provides, of course, a rich lode of hypotheses for the psychologist. It also provides one criterion for the goodness of theory, and for the generalizability of laws. While science often progresses only after the decision is made to stop "saving the appearances," (see Butterfield 1957) certainly one criterion for scientific sensibility is the extent to which its claims are in accord with examined experience. If such accord is not present, then it is incumbent to show why not. How can the psychologist approach such questions? The issue is not new. Psychology has long known that special methods are needed for relating single cases to general laws (cf. James 1890; Spranger 1928; Allport 1937). In the cognitive psychology of science, this effort must take on special characteristics, because the "single case" can be so complex, and already possessed of an overt, articulated theory about its own nature and methods.

CASE STUDY METHODS
The individual cognitive case study of a historical figure is exemplified by Gruber's work (1974) on Charles Darwin, which we have excerpted in selections 21, 40, and 42. Gruber examined Darwin's notebooks intensively, maintaining the perspective of a cognitive psychologist. One can study a living figure just as intensely (Gruber and Voneche 1977; Goodfield, 1981). Similarly, a recent study by Latour and Woolgar (1979) is essentially a case study, though of a laboratory group rather than an individual.

There is a long tradition of case study research in clinical psychology and psychiatry. Psychologists should thus be sensitized to the limitations of the approach, especially where generalizability is concerned. The problem is serious in the cognitive psychology of science. In particular, there is an enormous self-selection bias. Only the most successful scientists, and the most successful works, are likely to be studied. Thus, the very phenomena selected by case study researchers as instrumental in success (e.g., Einstein's vivid visual imagery) may appear equally often in careers filled with frustration and failure. The retrospective, individual case study method is not sensitive to such a possibility, nor is the *in vivo* individual case study.

A more satisfactory approach is based upon the multiple case study. For example, Shemberg and Leventhal (1972) studied the effectiveness of a particular brand of reality-oriented psychotherapy with a number of outpatient schizophrenics, and applied a similar therapeutic technique to a group of people with a common neurotic disorder. Thus they took steps toward the standardization of conditions and opened the possibility of generalizing across people. Some of the worst aspects of the case study approach—the possibility that change is due to idiosyncratic features of the patient, and the retrospective selection of the successful case—were thereby ameliorated (if not eliminated). Similar strategies must be applied in the analysis of science.


EXPERIMENTAL METHODS
An essential feature of the experimental method, one that provides both its power and its weakness, is its artificiality. The

experimentalist can create new conditions at will, and thus can put hard questions to nature, even where the nature in question is the mind of the scientist. But the question of generalizability is critical; generalizability over tasks as well as over people. Typically only a single task is studied, and the relevance of that task to "real science" is often tenuous. Very interesting behavior may manifest itself in, for example, the 2 4 6 task (Wason 1968a; Tweney et al., 1980). But is the behavior relevant to that of scientists behaving as scientists?

MULTIPLE METHODS

Given the desirability of an experimental approach for many questions about the cognitive activities that mark science, we believe that a more nearly adequate method will be one that is centered on experimental approaches, but that takes advantage of the power of other methodologies.

The experimental method is defined by manipulation and control of variables and by randomization of assignment of subjects to treatments. The power of the method can be more fully exploited if cognitive psychologists combine these aspects with the intensive study of individuals. This fruitful union certainly characterizes the work of Piaget, and is embodied in his "clinical concentric" method. It is central in Newell and Simon's work, in which process tracing is applied to different individuals attempting to solve the same problems. A multiple approach also characterizes some of our own work, for example, in the analysis of the protocols of individual subjects in selection 18.

Given the exploitation of standardized conditions and the intensive study and description of the individual, some classical methodological problems in psychology become moot. The hoary, though unresolved, problem of whether idiographic or nomothetic methods are preferred virtually disappears. The related question of whether individual differences are the object of study or should be relegated to "error" also disappears—it depends on how the data are analyzed. Thus, we believe that the concatenation of multiple approaches within empirical studies holds out considerable promise, allowing simultaneously for the standardization of conditions, the richness of the multiple case study, and the insight into cognitive processes afforded by the tools developed by contemporary cognitive psychology.


INTERVIEWS, SURVEYS, AND DIARIES

Hadamard (1945) conducted extensive surveys of mathematicians and exploited such data well (see selection 48). Not many surveys have dealt with cognitive aspects of scientific activity (but see Mahoney 1976; Bean 1979), though well-designed surveys and interviews can produce valuable insights. Both interviews and surveys suffer, of course, from a virtually total reliance on the recollections of the scientist. There are thus problems of forgetting, of selective recall, of self-deception—all problems endemic to the use of introspective reports (Brentano 1874/1973; Külpe 1893/1895; Titchener 1912; Boring 1953). If eyewitness testimony cannot be trusted (Loftus 1975); if introspective accounts of one's own behavior are suspect (Nisbett & Wilson 1977); if personal accounts of scientific discovery are often contradictory and inconsistent (Watson 1968; Chargaff 1978); then we can be justifiably suspicious about any particular statement gathered in an interview. Nonetheless, such data cannot simply be dismissed. If common themes emerge from interviews, and, most important, if the resulting account of scientific inference is in accord with that derived from other sources (as it often will not be, of course), then we may regard such accounts as useful. At worst, self-report data will constitute part of the appearances which must be "explained away." At best, they will illuminate, validate, and support other data. If so, then a kind of logical triangulation is possible: interview data support and lead to experiments, which support and lead to studies of diaries, and so on. The potentially most useful record is the scientific diary; long recognized as a rich source for those interested in the psychology of the scientist. Even further, careful observation of the day-by-day activity of scientists will be essential for a full understanding (cf. Goodfield, 1981). In the general introduction to this book, we have pointed to some of the possibilities for the especially rich notebooks of Michael Faraday as a source of data. Such applications hold great future promise.

THE METHODS OF COGNITIVE PSYCHOLOGY
The psychologist interested in scientific cognition is most
likely to make progress if multiple methods are used, if those
methods inform one another, and if the entire enterprise is
carried out with the perspective of contemporary cognitive
psychology. To date, there are no examples of work published
in this spirit, though an illuminating example produced in an
exactly contrary spirit exists. Latour and Woolgar (1979) examined the day-to-day activities of an immunological laboratory from an anthropological perspective. While certain insights were gained, the study is unsatisfactory from a cognitive
psychological perspective because it does not come to grips
with the intellectual bases for the activity of the workers in
the lab. By not dealing with the cognitive underpinnings of
the activity of their scientist-subjects, they missed the most
fundamental aspect of science—that it is, *in fact*, problem-solving activity. The cognitive component cannot be ignored unless one is willing to deny the possibility of scientific knowledge and to write off science as a mere cultural artifact. We
do not accept this stance, and therefore feel that unless one
determines why a particular scientist adopts a hypothesis, or
rejects an example as irrelevant, one can not ultimately understand the behavior. The cognition of the individual scientist is fundamental; the social and cultural aspects of science
are derivative.

Not all of cognitive psychology is automatically relevant to
the psychology of scientific thinking. One could argue that
scientific thinking is after all just human thinking, applied to
a particular class of problems. But scientific thinking may well
differ from ordinary thinking in important ways. Scientists
self-consciously bring a store of knowledge to bear on the task
at hand, as well as a highly developed set of intellectual tools.
They may use extensive note-taking, carefully organized records of data, files, and libraries, as "external memories." They
use blackboards, mathematics, even formal logic. Latour and
Woolgar (1979) noted that nearly all of the behavioral activity
in a major laboratory consisted of manipulation of symbols,
and only a tiny fraction involved direct contact with the phenomena under investigation. Cognitive psychologists have
typically studied "prescientific man." The typical subject in

a psychological laboratory has access only to presented stimuli and almost never to memory aids or other heuristics. The intent has been to study basic cognitive processes, unencumbered by cultural artifacts or aids (Neisser 1967; Posner 1978). A cognitive psychology of science will have to focus instead on *aided* cognition, on the psychology of scientists when all the tools of the trade are at their disposal. We see the results of such research as potentially contributory to the study of all human cognition.

## Toward

## Empirical Norms

## for Science

At various places in this book we have hinted at the possibility of an empirically based normative model of science. Of all the consequences which might flow from a psychology of science, this is perhaps the most exciting and far-reaching.

What are the minimum requirements of an empirically based normative model? How might it differ from nonempirical normative models? A normative model (empirically based or otherwise) can be defined as a set of procedures which, when followed, is intended to maximize the likelihood of attaining a desired end-state. Any such model requires:

1) A finite set of alternative options or responses which might produce the end-state,
2) The ability to specify the probability that a given option will produce the end-state under given conditions, and
3) A well defined end-state.

For one class of normative models the specification of (1), (2), and (3) can be arrived at a priori. For example, once one knows the rules of blackjack and the composition of a deck of playing cards, it is possible to specify procedures of play which maximize the likelihood of winning any given hand. These procedures are logical consequences of the rules of blackjack. They can be shown to be valid using formal proofs in probability theory.

Other types of normative rules do not admit of a priori specification. For example, it is widely held in baseball that right-
handed pitchers have an advantage when facing right-handed
batters. It follows that one maximizes one's chance of winning
by starting a right-handed pitcher against a predominantly
right-handed team. This normative rule cannot be derived a
priori from the rules of baseball. Instead, it evolved after empirical comparisons between the batting averages of left-handers against right-handed pitchers, left-handers against left-
handed pitchers, etc. One must play, or observe, baseball
games in order to formulate the rule. It could be argued that
in principle the rule should be derivable a priori from knowledge of anatomy and ballistics. However, it is unlikely that
such a rigorous derivation could at present be given, and in
any case the rule was not in fact formulated in such a way.

Until recently most normative models of science have been
based upon a priori considerations. There are two broad
classes of such models—one basically inductivist, the other deductivist. Inductivist models conceive of science as a process
wherein general laws and theories grow out of an accumulation of empirical evidence. The more ambitious versions (such
as Bacon's) assert that if evidence is collected and conceptualized in the right way it *guarantees* the truth of the general
laws. Less ambitious versions (e.g., Carnap's) assert that evidence only makes the general laws *probable*. Deductivist
models like Popper's deny that evidence can ever show a general law or theory to be either true or probably true. Evidence
can, however, be used to show that a theory is false. Progress
in science occurs because of the elimination of incorrect theories rather than the gradual accumulation of correct ones.
Both models are based upon assumptions said to have a higher
order of validity than science itself. The claim that science is
a "rational" process by which to acquire knowledge is thus
dependent upon the extent to which scientific activity is in
accord with the assumptions and derived rules.

Recently, many philosophers of science (e.g., Feyerabend
1975; Laudan 1977) have expressed pessimism about the ability of either inductive or deductive models to provide a satisfactory account of scientific activity; to provide a set of in-
ternally consistent rules which will guarantee (or make

probable) true laws and theories or the elimination of false laws and theories.

There is another problem, however, which is of more direct relevance to the concerns raised in this book. Historical studies of science have made it abundantly clear that *no current normative philosophical model is in even approximate accordance with actual scientific practice.* This fact poses a seemingly insoluble problem for the normative philosopher. If the philosophical norms are valid, how is it that much excellent science, at least as excellence is judged by scientists and historians, proceeds in violation of these norms?

The problem arises if it is assumed that scientific activity can be evaluated only by a priori criteria. If the process of evaluation is broadened to include empirical criteria, the problem dissolves. Philosophical norms need not be antecedent to and independent of actual scientific practice. Each may be used to evaluate and modify the other. Thus, even philosophers strongly inclined toward a priori normative analysis have begun to admit that philosophical models that label most scientific activity "irrational" are untenable (e.g., Giere 1973).

Laudan (1977) claims that there is "a subclass of cases of theory-acceptance and theory-rejection about which most scientifically educated persons have strong (and similar) normative intuitions, for example, . . . it was rational to accept Newtonian mechanics and to reject Aristotelian mechanics, by, say, 1800 [and] . . . it was irrational after 1830 to accept the biblical chronology as a literal account of earth history" (p. 160). He goes on to argue that such examples "can function as decisive *touchstones* for appraising and evaluating different normative models . . ." (p. 160), and that ". . . *the degree of adequacy of any theory of scientific appraisal is proportional to how many of* [these examples] *it can do justice to.*" (p. 161, italics in the original).

Laudan is arguing that normative philosophical models must be validated empirically, using historical evidence. A priori analyses have failed because philosophers have insisted that scientific knowledge meet an impossible standard—that it be true, or probably true. Unfortunately, ". . . these efforts have floundered because no one has been able to demonstrate that a system like science, with the methods it has at its disposal, can be guaranteed to reach the 'Truth,' either in the short or in the long runs" (p. 125). Nor has anyone ". . . been able to say what it would mean to be 'closer to the truth' let alone to offer criteria for determining how we could assess such proximity" (p. 126).

Philosophers have therefore been unable to define an appropriate end-state for science. Laudan's solution is to replace "Truth" with consensual validity about scientific progress. The difficulty is that there is no longer an objective standard for evaluation. Judgments about what constitutes "good science" are ultimately dependent upon the intuitions of scientifically educated persons. Why the intuitions of this group should be preferable to the intuitions of other groups, theologians or poets, say, is not clear. Laudan responds to this problem by pointing out that there is only a very small set of historical episodes about which there is general consensus. Thus, it is not claimed that any and all scientific activity is normatively valid. This at least avoids the empty circularity wherein "good science" is defined as "whatever good scientists do."

It does not, however, as Barnes (1979) has noted, address the central problem—the subjectivity inherent in using historical cases to validate scientific activity. Because of the extraordinary richness of the data base, historical evidence used for such a purpose lends itself too readily to confirmation bias (Fischoff 1975). It is not very surprising that scientifically educated people believe that Newtonian mechanics was a better system than Aristolelian physics. What else are they likely to believe given the very scientific education that qualifies them to have their intuitions taken seriously? Scientific history, no less than political history, is written by winners.

Normative criteria are needed which are less difficult to explicate than "Truth" but more objective than consensual validity. We believe that it is possible that such criteria could be developed from research on scientific problem solving of the sort discussed in Part 4. The aspect of much of this research (Wason 1968a; Mynatt, Doherty, and Tweney 1978) that is central here is that *the experimenters know the truth*. That is, the experimenters set up a problem in which the correct solution is completely specified. Thus, it is possible, at least in principle, to evaluate different approaches in terms of objective indices of problem solving efficiency. For example, one should be able to determine for a given class of problems whether people who ignore falsification do better or worse than those who act in a more "Popperian" manner; whether multiple hypotheses facilitate problem solving; whether highly speculative hypotheses are useful.

The normative criteria likely to emerge from such a research program are not likely to be as simple as a priori philosophical models. Many factors, including task and subject characteristics, interact to determine optimal behavior in a given situation. Under some conditions it may be rational to ignore falsification, or to use multiple hypotheses, or to use speculative hypotheses. Under other conditions it may be just as rational to do something different. Nevertheless, recognition of such complexity is preferable to an insistence that *all* scientific inquiry be conducted in one prescribed way.

One serious objection to such a program concerns the generalizability or external validity of such research. What assurance is there that strategies that prove effective in laboratory settings (no matter how complex or sophisticated) will also be effective in the conduct of real science? The objection, while serious, is not fatal. If science is amenable, in principle, to empirical investigation one need not be pessimistic about the application of laboratory methods to its understanding. Whether laboratory findings can generalize to real science is itself an empirical, and hence testable, issue. This is an appropriate place for historical evidence, interview data, autobiographical accounts, and diaries. If laboratory research were to indicate that falsification is nearly always an effective strategy, one would, on the basis of historical data, be skeptical about the generalizability of this finding. If, on the other hand, laboratory research were to suggest that falsification were an effective strategy only under some types of conditions, and an examination of historical data in light of the laboratory data indicated that scientists who employed falsification under those conditions did in fact make more progress than scientists who did not employ falsification, then one could be more confident about the import of *both* the laboratory and the historical data.

Let us be absolutely clear about what is proposed. It is not that philosophical considerations should be ignored in formulating normative models of science. It is rather that empirical evidence, including the appropriate use of historical evidence, is crucial to certain fundamental problems of philosophical import. What is being proposed is that useful normative models of science will result from the interpenetration and mutual fertilization of the philosophy of science and the newly emerging sciences of science. The science of the cognitive activity of individual scientists will be one of the keystone sciences of this new joint venture.

## Conclusion

There is no question, for us, that a cognitive psychology of science is possible and necessary. The question, more properly, should concern the exact nature of such a discipline, and the most fruitful research to pursue in the future. If the readings in this book, and our attempts to place those readings in context, have convinced the reader of this, and serve to inspire further inquiry, then we will have achieved our major purpose.

# References

Agassi, J. *Faraday as a Natural Philosopher*. Chicago: University of Chicago Press, 1971.

Allport, D. A. Critical notice: The state of cognitive psychology. *Quarterly Journal of Experimental Psychology* (1975), 27:141–52.

Allport, G. W. *Personality: A Psychological Interpretation*. New York: Holt, 1937.

Anderson, J. R. and G. H. Bower. *Human Associative Memory*. Washington, D.C.: Winston, 1973.

Anderson, N. H. and D. O. Cuneo. The height and width rule in children's judgments of quantity. *Journal of Experimental Psychology: General* (1978), 107:335–78.

Andreski, S. *Social Science as Sorcery*. London: Deutsch, 1972.

Anzai, Y. and H. A. Simon. The theory of learning by doing. *Psychological Review* (1979), 86:124–40.

Aristotle. *On the soul*. In R. McKeon, ed., *The Basic Works of Aristotle*. New York: Random House, 1941. J. A. Smith, trans. Originally written c.350 B.C.

Austin, J. H. *Chase, Chance, and Creativity*. New York: Columbia University Press, 1978.

Bacon, F. *Novum Organum*. In J. Spedding, R. L. Ellis, and D. D. Heath, eds., *The Works of Francis Bacon*. Popular edition. New York: Hurd & Houghton, 1878. Originally published 1620.

Bakan, D. *On Method: Toward a Reconstruction of Psychological Investigation*. San Francisco: Jossey-Bass, 1967.

Barnes, B. Vicissitudes of belief: Essay review of Laudan's *Progress and Its Problems*. *Social Studies of Science* (1979), 9:247–63

Bartlett, F. C. *Remembering: A Study in Experimental and Social Psychology*. Cambridge: Cambridge University Press, 1932.

——*Thinking: An Experimental and Social Study*. London: George Allen & Unwin, 1958.

Bean, L. H. The logical reasoning abilities of scientists; An empirical investigation. Master's thesis, Ohio State University, 1979.

Bernard, C. *An Introduction to the Study of Experimental Medicine*. New York: Macmillan, 1927. H. C. Greene, trans. Originally published 1865.

Beth, E. W. and J. Piaget. *Mathematical Epistemology and Psychology*. Dordrecht: Reidel, 1966.

Bhaskar, R. and H. A. Simon. Problem solving in semantically rich domains: An example from engineering thermodynamics. *Cognitive Science* 1977, 1:193–215.

Black, M. *Models and Metaphors*. Ithaca, N.Y., Cornell University Press, 1962.

Blumenthal, A. L. *The Process of Cognition*. Englewood Cliffs, N.J.: Prentice-Hall, 1977.

Bongard, M. *Pattern Recognition*. New York: Spartan Books, 1970. J. K. Hawkins, ed., T. Cheron, trans. Originally published 1967.

Boring, E. G. *A History of Experimental Psychology*. 2d ed. New York: Appleton-Century-Crofts, 1950.

—— A history of introspection. *Psychological Bulletin* (1953), 50:169–86.

—— The nature and history of experimental control. *American Journal of Psychology* (1954), 67:573–89.

Born, M. *Natural Philosophy of Cause and Chance.*, Oxford: Oxford University Press, 1949.

Bourne, L. E. *Human Conceptual Behavior*. Boston: Allyn and Bacon, 1966.

Boyd, R. Metaphor and theory change: What is "Metaphor" a metaphor for? In A. Ortony, ed., *Metaphor and Thought*. Cambridge: Cambridge University Press, 1979.

Bower, G. H. Mental imagery and associative learning. In L. W. Gregg, ed., *Cognition in Learning and Memory*. New York: Wiley, 1972.

Bower, G. and T. Trabasso. Concept identification. In R. C. Atkinson, ed., *Studies in Mathematical Psychology*. Stanford, Calif.: Stanford University Press, 1964.

Braine, M. D. S. On the relation between the natural logic of reasoning and standard logic. *Psychological Review* (1978), 85:1–21.

Bransford, J. D. and N. S. McCarrell. A sketch of a cognitive approach to comprehension: Some thoughts about understanding what it means to comprehend. In W. B. Weimer and D. S. Palermo, eds., *Cognition and the Symbolic Process*. Hillsdale, N.J.: Lawrence Erlbaum Associates, 1974.

Braun, J. R., ed. *Clinical Psychology in Transition*. Rev. ed. Cleveland: World, 1966.

Brehmer, B. Preliminaries to a psychology of inference. *Scandinavian Journal of Psychology* (1979), 20:193–210.

Brentano, F. *Psychology from an Empirical Standpoint*. New York: Humanities Press, 1973. A. C. Rancurello, D. B. Terrell, and L. McAlister, trans. Originally published 1874.

Bringmann, W. G. and R. D. Tweney, eds. *Wundt Studies: A Centennial Collection*. Toronto: C. J. Hogrefe, 1980.

Bronowski, J. *Science and Human Values*. New York: Harper Torchbooks, 1965.

—— *The Ascent of Man*. Boston: Little, Brown, 1973.

Brooks, L. R. An extension of the conflict between visualization and reading. *Quarterly Journal of Experimental Psychology* (1970), 22:91–96.

Bruner, J. S., J. J. Goodnow, and G. A. Austin. *A Study of Thinking*. New York: Wiley 1956.

Buchanan, B. G. and J. Lederberg. *The Heuristic Dendral Program for Explaining Empirical Data*. Stanford Artificial Intelligence Project Memo AIM-141. Stanford University, Computer Science Department, 1971.

Buchanan, B. G., G. L. Sutherland, and E. A. Feigenbaum. Heuristic Dendral: A program for generating explanatory hypotheses in organic chemistry. In B. Meltzer and D. Michie, eds., *Machine Intelligence 4*, Edinburgh: Edinburgh University Press, 1969.

Buchler, J. *The Concept of Method*. New York: Columbia University Press, 1961.

Bunge, M. *Intuition and Science*. Englewood Cliffs, N.J.: Prentice-Hall, 1962.

Bunge, M., ed. *The Critical Approach to Science and Philosophy: Essays in Honor of Karl R. Popper*. New York: Free Press of Glencoe 1964

Butterfield, H. *The Origins of Modern Science*. Rev. ed. New York: The Free Press, 1957.

Butts, R. E. and J. W. Davis, eds. *The Methodological Heritage of Newton*. Toronto: University of Toronto Press, 1970.

Campbell, D. T. Evolutionary epistemology. In P. A. Schilpp, ed., *The Philosophy of Karl Popper*. La Salle, Ill: Open Court, 1974.

—— Descriptive epistemology: Psychological, sociological, and evolutionary. Preliminary draft of the William James Lectures, Harvard University, 1977.

Campbell, N. R. *Physics, The Elements*. Cambridge: Cambridge University Press, 1920.

Carnap, R. *Philosophy and Logical Syntax*. London: Kegan Paul, 1935.

—— *Logical Foundations of Probability*. 2d ed. Chicago: University of Chicago Press, 1962.

Cartwright, D. Determinants of scientific progress: The case of research on the risky shift. *American Psychologist* (1973), 28:222–31.

Cassirer, E. *Substance and Function*. Chicago: Open Court, 1923. W. C. Swabey and M. C. Swabey, trans.

Chamberlin, T. C. The methods of the earth-sciences. *Popular Science Monthly* (1904), 66:66–75.

Chapanis, N. P. and A. Chapanis. Cognitive dissonance: Five years later. *Psychological Bulletin* (1964), 61:1–22.

Chapman, L. J. and J. P. Chapman. Illusory correlation as an obstacle to the use of valid psychodiagnostic signs. *Journal of Abnormal and Social Psychology* (1969), 74:271–80.

Chargaff, E. *Heraclitean Fire: Sketches from a Life before Nature.* New York: Rockefeller University Press, 1978.

Chase, W. G. *Visual Information Processing.* New York: Academic Press, 1973.

Chase, W. G. and H. A. Simon. The mind's eye in chess. In W. G. Chase, ed., *Visual Information Processing.* New York: Academic Press, 1973.

Churchman, C. W. *The Design of Inquiring Systems.* New York: Basic Books, 1971.

Cohen, I. B. *Franklin and Newton.* Philadelphia: American Philosophical Society, 1956.

Cohen, J. The statistical power of abnormal-social psychological research. *Journal of Abnormal and Social Psychology* (1962), 65:145–53.

—— *Statistical Power Analysis in the Behavioral Sciences.* New York: Academic Press, 1969

Cole, J. R., and S. Cole. The Ortega hypothesis. *Science* (1972), 178:368–75.

Comte, A. *The Positive Philosophy of Auguste Comte.* London: Bell, 1896. H. Martineau, trans. Originally published 1830–1842.

Craik, K. J. W. *The Nature of Explanation.* Cambridge: Cambridge University Press, 1943.

Cronbach, L. J. The two disciplines of scientific psychology. *American Psychologist* (1957), 12:671–84.

Crovitz, H. F. The form of logical solutions. *The American Journal of Psychology* (1967), 80:461–62.

—— *Galton's Walk: Methods for the Analysis of Thinking, Intelligence, and Creativity.* New York: Harper & Row, 1969.

Crovitz, H. F. and R. W. Horn. A method for studying the time course of illuminations. Presented at the 85th Annual Convention of the American Psychological Association, San Francisco, August 1977.

Danziger, K. Wundt's psychological experiment in the light of his philosophy of science *Psychological Research* (1980), 42:109–122.

Darden, L. Theory construction in genetics. Paper presented at the Leonard Conference on Scientific Discovery, Reno, October 1978.

Darnton, R. *Mesmerism and the End of the Enlightenment in France.* Cambridge: Harvard University Press, 1968.

Darwin, C. *The Origin of Species.* New York: The Modern Library, no date. Originally published 1859.

—— *The Autobiography of Charles Darwin.* N. Barlow, ed. London: Collins, 1958.

Descartes, R. *The Philosophical Works of Descartes Rendered into English by E. S. Haldane and G. R. T. Ross.* Vol. 1. Cambridge: Cambridge University Press, 1969. This edition first published in 1911. Originally published 1628, 1637.

Deutsch, M. Evidence and inference in nuclear research. In D. Lerner, ed., *Evidence and Inference,* pp. 96–106. Glencoe, Ill.: The Free Press, 1959.

Digby, K. *Of the Sympathetick Powder. A Discourse in Solemn Assembly at Montpellier.* London: for John Williams, 1669. Originally published 1657.

Doherty, M. E., C. R. Mynatt, R. D. Tweney, and M. D. Schiavo. Pseudo-diagnosticity. *Acta Psychologica* (1979), 43:111–21.

Duhem, P. *The Aim and Structure of Physical Theory.* Princeton N.J.: Princeton University Press, 1954. First published 1914.

Duncan, C. P. Induction of a principle. In C. P. Duncan, ed., *Thinking: Current Experimental Studies.* Philadelphia: Lippincott, 1967.

Duncker, K. *On Problem Solving. Psychological Monographs,* 1945, *58,* whole no. 270. L. S. Lees, trans. Originally published 1935.

Dyson, F. W., A. S Eddington, and C. Davidson. A determination of the deflection of light by the sun's gravitational field, from observations made at the total eclipse of May 29, 1919. *Philosophical Transactions of the Royal Society of London, Series A* (1919), 220:291–333.

Eccles, J. C. Under the spell of the synapse. In F. G. Worden, J. P. Swazey, and G. Adelman, eds., *The Neurosciences: Paths of Discovery,* pp. 162–63. Cambridge, Mass. MIT Press, 1975.

Edge, D. O. and M. J. Mulkay. *Astronomy Transformed.* New York: Wiley, 1976.

Eiduson, B. T. *Scientists: Their Psychological World.* New York: Basic Books, 1962.

Einhorn, H. J. and R. M. Hogarth. Confidence in judgement: Persistence of the illusion of validity. *Psychological Review* (1978), 85:395–416.

Einstein, A. Über einen die erzengung und verwandlung des lichtes detreffenden heuristischen gesichtspunkt. *Annalen der Physik* (1905), 17:132–48.

—— Autobiographical notes. In P. A. Schilpp, ed., *Albert Einstein: Philosopher-Scientist.* New York: Tudor, 1949.

—— *Out of My Later Years.* Secaucus, N.J. Citadel Press, 1977. First published 1956.

Eisberg, R. M. *Fundamentals of Modern Physics.* New York: Wiley, 1961.

Elkana, Y., J. Lederberg, R. K. Merton, A. Thackray, and H. Zuckerman. *Toward a Metric of Science.* New York: Wiley, 1978.

Elstein, A. S., L. S. Shulman, and S. A. Sprafka. *Medical Problem Solving: An Analysis of Clinical Reasoning.* Cambridge: Harvard University Press, 1978.

Ericsson, K. A. and H. A. Simon. Verbal reports as data. *Psychological Review* (1980), 87:215–51.

Estes, W. K. Probability learning. In A. W. Melton, ed., *Categories of Human Learning.* New York: Academic Press, 1964.

—— On the descriptive and explanatory functions of theories of memory. In L. G. Nilsson, ed., *Perspectives on memory Research: Essays in Honor of Uppsala University's 500th Anniversary.* Hillsdale, N.J.: Lawrence Erlbaum Associates, 1979.

Evans, J. St. B. T. and J. S. Lynch. Matching bias in the selection task. *British Journal of Psychology* (1973), 64:391–97.

Faraday, M. *Faraday's Diary. Being the Various Philosophical Notes of Experimental Investigation Made by Michael Faraday during the Years 1820–1862 and bequeathed by him to the Royal Institution of Great Britain. . . .* T. Martin, ed., with a foreword by Sir H. Bragg. 7 vols. + index volume. London: G. Bell and Sons, 1932–1936.

Fearing, F. *Reflex Action: A Study in the History of Physiological Psychology.* Baltimore: Williams & Wilkins, 1930.

Fechner, G. *Elemente der Psychophysik.* 2 vols. Leipzig: Breitkopf & Hartel, 1860.

Feldman, J. Some decidability results on grammatical inference and complexity. *Information and control* (1972), 20:244–62.

Feldman, J., F. M. Tonge, Jr., and H. Kanter. Empirical explorations of a hypothesis-testing model of binary choice behavior. In A. C. Hoggatt and F. E. Balderston, eds., *Symposium on Simulation Models.* Cincinnati, Ohio: South-Western, 1963.

Feyerabend, P. K. Against method: Outline of an anarchistic theory of knowledge. In M. Radner and S. Winokur, eds., *Analyses of Theories and Methods of Physics and Psychology. Minnesota Studies in the Philosophy of Science,* vol. 4. Minneapolis: University of Minnesota Press, 1970.

—— *Against Method.* London: NLB, 1975.

Fisch, R. Psychology of science. In I. Spiegel-Rosing and D. deSolla Price, eds., *Science, Technology, and Society: A Cross-disciplinary Perspective.* London: Sage, 1977.

Fischoff, B. Hindsight ≠ foresight: The effect of outcome knowledge on judgment under uncertainty. *Journal of Experimental Psychology: Human Perception and Performance* (1975), 3:288–99

Fisher, R. A. Statistical Methods and Scientific Influence. Edinburgh: Oliver & Boyd, 1956.

—— *The Design of Experiments.* Edinburgh: Oliver & Boyd, 1966. First published 1935.

—— *Statistical Methods for Research Workers.* Darien, Conn.: Hafner, 1970. First published 1925.

Fiske, D. W. The limits of the conventional science of personality. *Journal of Personality* (1974), 42:1–11.

Fourier, J. *Theory of Heat. Great Books of the Western World,* vol. 45. Chicago: Encyclopaedia Britannica, 1952. by A. Freeman, trans. Originally published 1822.

Fox, J. Making decisions under the influence of memory. *Psychological Review* (1980), 87:190–211.

Frank, P. *Einstein: His Life and Times.* New York: Knopf, 1947.

Freedman, D., R. Pisani, and R. Purves. *Statistics.* New York: Norton, 1978.

Freud, S. *Introductory Lectures on Psychoanalysis*. New York: Norton, 1965. J. Strachey, ed. and trans. Originally published 1917.

Galileo, G. *Dialogue on the Great World Systems*. Chicago: University of Chicago Press, 1953. T. Salusbury, trans. rev. & annotated by G. de Santillana. Originally published in 1632.

Galton, F. *Inquiries into Human Faculty and its Development*. London: Dent, 1907. First published 1883.

—— *Natural Inheritance*. London: Macmillan, 1889.

Gardner, M. *Logic Machines, Diagrams and Boolean Algebra*. New York: Dover, 1968.

Garner, W. R. *The Processing of Information and Structure*. Potomac, Lawrence Erlbaum Associates, 1974.

Garner, W. R., H. W. Hake, and C. W. Eriksen. Operationism and the concept of perception. *Psychological Review* (1956), 63:149–59.

Gazzaniga, M. S. and J. E. LeDoux. *The Integrated Mind*. New York: Plenum Press, 1978.

Geller, E. S. and G. F. Pitz. Confidence and decision speed in the revision of opinion. *Organizational Behavior and Human Performance* (1968), 3:190–201.

Gergen, K. J. Social psychology as history. *Journal of Personality and Social Psychology* (1973), 26:309–20.

Gerwin, D. G. Information processing, data inferences, and scientific generalization. *Behavioral Sciences* (1974), 19:314–25.

Gerwin, D. G., and P. Newsted. A comparison of some inductive inference models. *Behavioral Sciences* (1977), 22:1–11.

Getzels, J. W., and M. Csikszentmihalyi. *The Creative Vision: A Longitudinal Study of Problem Finding in Art*. New York: Wiley, 1976.

Ghiselin, B. *The Creative Process*. Berkeley: University of California Press, 1952.

Giere, R. History and philosophy of science. Intimate relationship or marriage of convenience? *British Journal of the Philosophy of Science* (1973), 24:282–97.

Gilbert, G. K. The origin of hypotheses. *Science* (1896), n.s. 3, pp. 1–3.

Goodfield, J. *An Imagined World*. New York: Harper, 1981.

Greenwald, A. G. Consequences of prejudice against the null hypothesis. *Psychological Bulletin* (1975), 82:1–20.

Gregg, L. W. Internal representation of sequential concepts. In B. Kleinmutz, ed., *Concepts and the Structure of Memory*. New York: Wiley, 1967.

Gregg, L. W., & Simon, H. A. Process models and stochastic theories of simple concept formation. *Journal of Mathematical Psychology* (1967), 4:246–76.

Gregson, R. A. M. *Psychometrics of Similarity*. New York: Academic Press, 1975.

Gruber, H. E. and P. H. Barrett. *Darwin on Man*. New York: Dutton, 1974.

Gruber, H. E. Darwin's "Tree of nature" and other images of wide scope. In J. Wechsler, ed., *On Aesthetics in Science*. Cambridge, Mass.: MIT Press, 1978.

Gruber, H. E. and J. J. Voneche, eds. *The Essential Piaget: An Interpretive Reference and Guide*. New York: Basic Books, 1977.

Grunbaum, A. The special theory of relativity as a case study of the importance of philosophy of science for the history of science. In B. Baumrin, ed., *Philosophy of Science*, vol. 1. New York: 1963.

—— Is Freudian psychoanalytic theory pseudo-scientific by Karl Popper's criterion of demarcation? *American Philosophical Quarterly* (1979), 16:131–41.

Gutting, G. The logic of discovery. Paper presented at the Leonard Conference on Scientific Discovery. Reno, Nevada, 1978.

Hadamard, J. *The Psychology of Invention in the Mathematical Field*. Princeton, N.J.: Princeton University Press, 1945.

Hanson, N. R. *Patterns of Discovery*. London: Cambridge University Press, 1958.

—— Hypotheses fingo. In R. E. Butts and J. W. Davis, eds., *The Methodological Heritage of Newton*. Toronto: University of Toronto Press, 1970.

Harlow, H. F. Mice, men monkeys, and motives. *Psychological Review*, 1953, 60, 23–32.

Harre, R. Philosophy of science, history of. In P. Edwards, ed., *The Encyclopedia of Philosophy*, vol. 6, S.V. New York: Macmillan, 1967.

—— *The Principles of Scientific Thinking*. Chicago: University of Chicago Press, 1970.

Hartley, D. *Observations on Man: His Frame, his Duty, and his Expectations*. London: for Thomas Tegg, 1843, First published 1749.

Hawkins, D. Review of Kuhn's structure of scientific revolutions. *American Journal of Physics* (1963), 31:554–55.

Henle, M. On the relation between logic and thinking. *Psychological Review* (1962), 69:366–78.

Herschel, J. F. W. *A Preliminary Discourse on the Study of Natural Philosophy*. London: for Longman, Rees, Orme, Brown, & Green, 1830.

Hesse, M. B. *Models and Analogies in Science*. South Bend, Indiana: University of Notre Dame Press, 1966.

Hofstadter, D. R. *Godel, Escher, Bach: An Eternal Golden Braid*. New York: Basic Books, 1979.

Hogan, R., C. B. DeSoto, and C. Solano. Traits, texts, and personality research. *American Psychologist* (1977), 32:255–64.

Holton, G. On trying to understand scientific genius. *American Scholar* (1971), 41:102–4.

—— Mach, Einstein and the Search for Reality. In G. Holton, ed., *The Twentieth-Century Sciences: Studies in the Biography of Ideas*. New York: Norton, 1972.

—— Thematic Origins of Scientific Thought: Kepler to Einstein. Cambridge: Harvard University Press, 1973.

—— The Scientific Imagination: Case Studies. London: Cambridge University Press, 1978.

Huesmann, L. R. and C. Cheng. A theory for the induction of mathematical functions. Psychological Review (1973) 60:126–38.

Hume, D. An Inquiry Concerning Human Understanding. Indianapolis: Bobbs-Merrill, 1955. First published 1748.

Humphrey. G. The problem of the direction of thought. British Journal of Psychology (1940), 30:183–96.

—— Thinking: An Introduction to its Experimental Psychology. London: Methuen, 1951.

Hunt, E. B., J. Marin, and P. J. Stone. Experiments in Induction. New York: Academic Press, 1966.

Hunter, R., and I. Macalpine. Three Hundred Years of Psychiatry, 1535–1860: A History Presented in Selected English Texts. London: Oxford University Press, 1963.

Inhelder, B., and J. Piaget. The Growth of Logical Thinking from Childhood to Adolescence. New York: Basic Books, 1958.

James, W. Great Men, Great Thoughts, and the Environment. Atlantic Monthly (1880), 46:441–59.

—— Principles of Psychology. 2 vols. New York: Holt, 1890.

Jeffries, R., P. G. Polson, L. Razran, and M. E. Atwood. A process model for missionaries-cannibals and other river-crossing problems. Cognitive Psychology (1977), 9:412–40.

Johnson, W. E. Logic, Part I. New York: Dover, 1964. Originally published 1921.

Johnson-Laird, P. N., P. Legrenzi, and S. Legrenzi, Reasoning and a sense of reality. British Journal of Psychology (1972), 63:395–400.

Johnson-Laird, P. N. and M. Steedman. The psychology of syllogisms. Cognitive Psychology (1978), 10:64–99.

Johnson-Laird, P. N. and P. C. Wason, eds., Thinking: Readings in Cognitive Science. Cambridge: Cambridge University Press, 1977.

Josephson, M. Edison. New York: McGraw-Hill, 1959.

Kant, I. Anthropology from a Pragmatic Point of View. Carbondale, Ill.: Southern Illinois University Press, 1978. V. L. Dowdell, trans. First published 1798.

Karmiloff-Smith, A. and B. Inhelder. If you want to get ahead get a theory. Cognition (1975), 3:195–212.

Katz, D. Animals and Men: Studies in Comparitive Psychology. London: Longmans, Green, 1937.

Kepler, J. The Harmonies of the World, Book V. Great Books of the Western World, vol. 16. Chicago: Encyclopaedia Britannica, 1952. C. G. Wallis, trans. Originally published in 1619.

Kevles, D. J. *The Physicists: The History of a Scientific Community in Modern America*. New York: Knopf, 1978.

Klein, M. J. *Paul Ehrenfest*. Vol. 1: *The Making of a Theoretical Physicist*. New York: Elsevier, 1970.

Klima, E. and U. Bellugi, *The Signs of Language*. Cambridge: Harvard University Press, 1979.

Koestler, A. *The Act of Creation*. New York: Macmillan, 1964.

Köhler, W. *The Place of Value in a World of Facts*. New York: Liveright, 1938.

Kotovsky, K. and H. A. Simon, Empirical tests of a theory of human acquisition of concepts for sequential patterns. *Cognitive Psychology* (1973), 4:399–424.

Kuhn, T. S. *The Copernican Revolution: Planetary Astronomy in the Development of Western Thought*. Cambridge: Harvard University Press, 1957.

—— The role of measurement in the development of physical science. *Isis* (1958), 49:161–93.

—— The function of measurement in modern physical science. In H. Woolf, ed., *Quantification: A History of the Meaning of Measurement in the Natural and Social Sciences*. Indianapolis: Bobbs-Merrill, 1961.

—— *The Structure of Scientific Revolutions*. Chicago: University of Chicago Press, 1962.

—— A function for thought experiments. In *L'aventure de l'esprit; Mélanges Alexandre Koyré*. Paris: Hermann, 1964. (Reprinted in T. S. Kuhn, Ed., *The essential tension*. Chicago: University of Chicago Press, 1977.)

—— Second thoughts on paradigms. In F. Suppe, ed., *The Structure of Scientific Theories*. 1st ed. Urbana: University of Illinois Press, 1974.

—— The history of science. In T. S. Kuhn, ed., *The Essential Tension: Selected Studies in Scientific Tradition and Change*. Chicago: University of Chicago Press, 1977.

—— Metaphor in science. In A. Ortony, ed., *Metaphor and Thought*. Cambridge: Cambridge University Press, 1979. pp. 409–419.

Külpe, O. *Outlines of Psychology: Based upon the Results of Experimental Investigations*. London: Swan Sonnenschein & Co., 1895. E. B. Titchener, trans. Originally published in 1893.

Lakatos, I. Proofs and refutations. *British Journal for the Philosophy of Science* (1963), 14:1–25; 120–39; 221–45; 296–342.

—— Falsification and the methodology of scientific research programme. In I. lakatos and A. Musgrave, eds., *Criticism and the Growth of Knowledge*. Cambridge: Cambridge University Press, 1970.

—— *The Methodology of Scientific Research Programmes*. Cambridge: Cambridge University Press, 1978.

Lakatos, I. and A. Musgrave, eds., *Criticism and the Growth of Knowledge*. London: Cambridge University Press, 1970.

Landauer, T. K. Memory without organization: Properties of a model with random storage and undirected retrieval. *Cognitive Psychology* (1975), 7:495–531.

Latour, B. and S. Woolgar. *Laboratory Life: The Social Construction of Scientific Facts*. Beverly Hills, Calif.: Sage, 1979.

Laudan, L. Theories of scientific method from Plato to Mach: A bibliographical review. *History of Science* (1969), 7:1–63.

—— *Progress and its Problems: Towards a Theory of Scientific Growth*. Berkely: University of California Press, 1977.

Leatherdale, W. H. *The Role of Analogy, Model and Metaphor in Science*. New York: Elsevier, 1974.

Leeuwenberg, E. L. L. Quantiative specification of information in sequential patterns, *Psychological Review* (1969), 76, 216–20.

Leventhal, D. B. and K. M. Shemberg. Psychotherapy: Theory, experience and personalized actuarial tables. *British Journal of Medical Psychology* (1977), 50:361–65.

Levine, M. *A cognitive Theory of Learning*. Hillsdale, N.J.: Lawrence Erlbaum Associates, 1975.

Lewin, K. The conflict between Aristotelian and Galileian modes of thought in contemporary psychology. *Journal of General Psychology* (1931), 5:141–77.

Locke, J. *An Essay concerning Human Understanding*. P. H. Nidditch, eds. Oxford: Oxford University Press, 1975. First published 1690.

Loftus, E. F. Leading questions and the eyewitness report. *Cognitive Psychology* (1975), 7:560–72.

Losee, J. *A Historical Introduction to the Philosophy of Science*. London: Oxford University Press, 1972.

Luchins, A. S. *Mechanization in problem-solving*. *Psychological Monographs* (1942), 54:whole no. 248

Lykken, D. T. Statistical significance in psychological research. *Psychological Bulletin* (1968), 70:151–59.

McGuire, W. J. The yin and yang of progress in social psychology: Seven koans. *Journal of Personality and Social Psychology* (1973), 26:446–56.

Mach, E. *Knowledge and Error: Sketches on the Psychology of Inquiry*. Boston: D. Reidel, 1975. T. J. McCormack and P. Foulkes, trans. Originally published in 1905.

Mahoney, M. J. *Cognition and Behavior Modification*. Cambridge, Mass. Ballinger, 1974.

——*Scientist as Subject*. Cambridge, Mass. Ballinger, 1976.

Mahoney, M. J. and B. G. DeMonbreun. Psychology of the scientist: An analysis of problem-solving bias. *Cognitive Therapy and Research* (1978), 1(3):229–38.

Mandler, J. M., and G. Mandler. *Thinking: From Association to Gestalt*. New York: Wiley, 1964.

Manktelow, K. I. and J. St B. T. Evans. Facilitation of reasoning by realism: Effect or non-effect? *British Journal of Psychology* (1979), 70:477–88.

Mannheim, K. *Ideology and Utopia: An Introduction to the Sociology of Knowledge.* New York: Harcourt, Brace, 1936. Wirth & E. Snils, trans. Originally published 1929–1931.

Maslow, A. H. *The Psychology of Science: A Reconnaissance.* New York: Harper & Row, 1966.

Medawar, P. B. Is the scientific paper a fraud? In D. Edge, ed., *Experiment: A Series of Scientific Case Histories First Broadcast in the BBC Third Programme.* London: British Broadcasting Corporation, 1964.

Meehl, P. E. Theory-testing in psychology and physics: A methodological paradox. *Philosophy of Science* (1967), 34:103–115.

—— The cognitive activity of the clinician. In P. E. Meehl, *Psychodiagnosis: Selected papers.* Minneapolis: University of Minnesota Press, 1973 (a). Originally published 1960.

—— Some ruminations on the validation of clinical procedures. In P. E. Meehl, *Psychodiagnosis: Selected papers.* Minneapolis: University of Minnesota Press, 1973 (b). Originally published 1959.

—— Theoretical risks and tabular asterisks: Sir Karl, Sir Ronald and the slow progress of soft psychology. *Journal of Consulting and Clinical Psychology* (1978), 46:806–34.

Merton, R. K. *Science, Technology and Society in Seventeenth-Century England.* New York: Harper & Row, 1970. First published 1938.

—— Priorities in scientific discovery: A chapter in the sociology of science. *American Sociological Review* (1957), 22:635–59.

—— "Recognition" and "excellence." Instructive ambiguities. In A. Yarmolinsky, ed., *Recognition of Excellence: Working Papers.* New York: The Free Press, 1960.

—— The ambivalence of scientists. *Bulletin of the Johns Hopkins Hospital* (1963), 112:77–97.

—— *On the Shoulders of Giants: A Shandean Postcript.* New York: Harcourt, Brace & World, 1965.

—— *The Sociology of Science: Theoretical and Empirical Investigation.* N. W. Storer, ed. Chicago: University of Chicago Press, 1973.

Merz, J. T. *A History of European Thought in the Nineteenth Century.* 4 vols. 3d ed. Edinburgh: William Blackwood and Sons, 1907. Originally published 1896.

Mill, James. *Analysis of the Phenomena of the Human Mind.* 2 vols. J. S. Mill, ed. London: Longmans, Green, Reader, and Dyer, 1869. Originally published in 1829.

Mill, J. S. *A System of Logic, Ratiocinative and Inductive.* London: 1843.

Miller, A. I. Albert Einstein and Max Wertheimer: A Gestalt psychologist's view of the genesis of special relativity theory. *History of Science* (1973), 13:75–103.

—— Visualization lost and regained: The genesis of the quantum theory in the period 1913–1927. In J. Wechsler, ed., *On Aesthetics in Science* Cambridge: MIT Press, 1978.

Miller, G. A. Speech and language. In S. S. Stevens, ed., *Handbook of Experimental Psychology*. New York: Wiley, 1951.

Miller, G. A., E. Galanter and K. H. Pribram. *Plans and the Structure of Behavior*. New York: Holt, Rinehart and Winston, 1960.

Miller, G. A. Project grammarama. *The Psychology of Communication*. New York: Basic Books, 1967.

Minsky, M. and S. Papert. *Perceptrons*. Cambridge: MIT Press, 1969.

Mischel, T. Kant and the possibility of a science of psychology. *The Monist* (1967), 51:

Mischel, W. On the future of personality measurement. *American Psychologist* (1977), 32:246–54.

Mitroff, I. *The Subjective Side of Science*. Amsterdam: Elsevier, 1974. (a)

—— Norms and counter-norms in a select group of the Apollo moon scientists: A case study of the ambivalence of scientists. *American Sociological Review* (1974), 39:579–95. (b)

—— A Brunswik lens model of dialectical inquiry systems. *Theory and Decision* (1974), 5:45–67 (c).

Mitroff, I. I. and R. H. Kilmann. *Methodological Approaches to Social Science*. San Francisco: Jossey-Bass, 1978.

Mowrey, J. D., M. E. Doherty, and S. M. Keeley. The influence of negation and task complexity on illusory correlation. *Journal of Abnormal Psychology* (1979), 88:334–37.

Murphy, G. *An Historical Introduction to Modern Psychology*. New York: Harcourt, Brace, 1930.

Mynatt, B. T. and K. H. Smith. Constructive processes in linear order problems revealed by sentence study times. *Journal of Experimental Psychology: Human Learning and Memory* (1977), 3:357–74.

Mynatt, C. R., M. E. Doherty, and R. D. Tweney. Confirmation bias in a simulated research environment: An experimental study of scientific inference. *Quarterly Journal of Experimental Psychology* (1977), 29:85–95.

—— Consequences of confirmation and disconfirmation in a simulated research environment. *Quarterly Journal of Experimental Psychology* (1978), 30:395–406.

Nabi, I. *On the Tendencies of Motion*. Unpublished manuscript, no date, no place.

Nambu, Y. The confinement of quarks. *Scientific American* (1976), 235:48–60.

Neisser, U. *Cognitive Psychology*. New York: Appleton-Century-Crofts, 1967.

—— *Cognition and Reality: Principles and Implications of Cognitive Psychology*. San Francisco: Freeman, 1976.

Newell, A. You can't play 20 questions with nature and win: Projective comments on the papers of this symposium. In W. Chase, ed., *Visual Information Processing*. New York: Academic Press, 1973.

Newell, A. and H. A. Simon. GPS, A program that simulates human thought. In H. Billings, ed., *Lernende Automaten*. Munich: Oldenbourg, 1961.

—— *Human Problem Solving.* Englewood Cliffs, N. J.: Prentice-Hall, 1972.

Newton, I. *Mathematical Principles of Natural Philosophy.* Berkeley: University of California Press, 1947. A. Motte, trans., revised by F. Cajori. First published 1687.

—— *Opticks.* London: Bell & Hyman, 1931.

—— *Papers and Letters on Natural Philosophy, and Related Documents.* I. B. Cohen, ed. Cambridge: Harvard University Press, 1958. First published 1671–1728.

Nisbett, R. E., and T. D. Wilson. Telling more than we can know: Verbal reports on mental processes. *Psychological Review* (1977), 84:231–59.

Nisbett, R. E. and Ross, L. *Human Inference: Strategies and Shortcomings of Social Judgement* Englewood Cliffs, N. J.: Prentice-Hall, 1980.

Norman, D. A. *Memory and Attention.* New York: Wiley, 1976.

Norman, D. A., and D. E. Rumelhart. eds. *Explorations in Cognition.* San Francisco: Freeman, 1975.

Olby, R. *The Path to the Double Helix.* Seattle: University of Washington Press, 1974.

Ornstein, R. E. *The Psychology of Consciousness.* 2d ed. New York: Harcourt Brace Jovanovich, 1977. First published 1972.

Ortony, A., ed. *Metaphor and Thought.* Cambridge: Cambridge University Press, 1979.

Overall, J. E. Classical statistical hypothesis testing within the context of Bayesian theory. *Psychological Bulletin* (1969), 71:285–92.

Paivio, A. *Imagery and Verbal Processes.* New York: Holt, 1971.

Pearson, K. *The Life, Letters and Labours of Francis Galton.* 3 vols. Cambridge: Cambridge University Press, 1914–1930.

Penrose, J. An investigation into some aspects of problem-solving behaviour. Doctoral dissertation, University of London, 1962.

Piaget, J. Les deux directions de la pensée scientifique. *Archives des Sciences Physiques et Naturelles* (1929), 11:145–62.

—— *Insights and Illusions of Philosophy.* New York: World, 1971. First published 1965.

—— *Genetic Epistemology.* New York: Columbia University Press, 1970. E. Duckworth, trans.

Pitz, G. An inertia effect (resistance to change) in the revision of opinion. *Canadian Journal of Psychology* (1969), 23:24–33.

—— Bayes' theorem: Can a theory of judgment and inference do without it? In F. Restle, R. M. Shiffrin, N. J. Castellan, J. R. Lindman, and D. B. Pisoni, eds., *Cognitive Theory,* vol. 1. Hillsdale, N. J.: Laurence Erlbaum Associates, 1975.

Pitz, G. F., L. Downing, and H. Reinhold. Sequential effects in the revision of subjective probabilities. *Canadian Journal of Psychology* (1967), 21:381–93.

Pivar, M. and M. Finkelstein. Automation, using LISP, of inductive inference on sequences. In E. C. Berkeley and D. G. Bobrow, eds., *The Programming Language LISP*. Cambridge, Mass. Information International, 1964.

Planck, M. *A Scientific Autobiography and Other Papers*. New York: Philosophical Library, 1950.

Platt, J. R. Strong inference. *Science* (1964), 146:347–53.

Plotkin, G. D. A further note on inductive generalization. In B. Meltzer and D. Michie, eds., *Machine Intelligence 6*. New York: American Elsevier, 1971.

Poincaré, H. *The Foundations of Science*. New York: Science Press, 1913.

—— *La science et l'hypothese*. Paris: Ernest Flammerion, 1902.

Polanyi, M. *The Tacit Dimension*. New York: Doubleday, 1966.

Polya, G. *How to Solve it*. Princeton, N. J.: Princeton University Press, 1945.

—— *Mathematics and Plausible Reasoning*. 2 vols. Princeton, N. J.: Princeton University Press, 1954.

Popper, K. R. *The Logic of Scientific Discovery*. London: Hutchinson, 1959. Originally published 1934.

—— Philosophy of science: A personal report. In C. A. Mace, ed., *British Philosophy in Mid-century*, pp. 155–91. London: George Allen & Unwin, 1957.

—— *Conjectures and Refutations*. London: Routledge & Kegan Paul, 1978. Originally published 1962.

—— *Objective Knowledge*. Oxford: Oxford University Press, 1972.

Posner, M. I. *Chronometric Explorations of Mind*. Hillsdale, N. J.: Lawrence Erlbaum Associates, 1978.

Pylyshn, Z. W. What the mind's eye tells the mind's brain: A critique of mental imagery. *Psychological Bulletin* (1973), 80:1–24.

Pyne, S. J. Methodologies for geology: G. K. Gilbert & T. C. Chamberlin. *Isis* (1978), 69:413–24.

Raphael, B. *The Thinking Computer: Mind inside Matter*. San Franciso: Freeman, 1976.

Reichenbach, H. *The Rise of Scientific Philosophy*. Berkeley: University of California Press, 1951.

Reitman, J. S. Skilled perception in GO: Deducing memory structures from inter-response times. *Cognitive Psychology* (1976), 8:336–56.

Reitman, W. R. *Cognition and Thought: An Information Processing Approach*. New York: Wiley, 1965.

Restle, F. A. The selection of strategies in cue learning. *Psychological Review* (1962), 69:329–43.

—— Theory of serial pattern learning: Structural trees. *Psychological Review* (1970), 77:481–95.

Roe, A. *The Making of a Scientist*. New York: Dodd, Mead, 1953.

Rokeach, M. *The Open and Closed Mind*. New York: Basic Books, 1960.

Rosch, E., C. B. Mervis, W. D. Gray, D. M. Johnson, and P. Boyes-Braem. Basic objects in natural categories. *Cognitive Psychology* (1976), 8:382–439.

Rose, A. and Tweney, R. The Child as Intuitive Scientist: Confirmation and Disconfirmation in Balance Scale Task. Bowling Green (Ohio) State University, unpublished ms. 1980.

Rosenthal, R. *Experimenter Effects in Behavioral Research*. New York: Appleton-Century-Crofts, 1966.

Ross, L. D. The intuitive psychologist and his shortcomings: Distortions in the attribution process. In L. Berkowitz, ed., *Advances in Experimental Social Psychology*, vol. 10. New York: Academic Press, 1977.

Ross, L., M. R. Lepper, F. Strack, and J. Steinmetz. Social explanation and social expectation: Effects of real and hypothetical explanations on subjective likelihood. *Journal of Personality and Social Psychology* (1977), 35 (11):817–29.

Rostand, J. *Error and Deception in Science: Essays on Biological Aspects of Life*. A. J. Pomerans, trans. London: Hutchinson, 1960.

Rothenberg, A. Einstein's creative thinking and the general theory of relativity: A documented report. *American Journal of Psychiatry* (1979), 136:38–43.

Rothenberg, A. and C. R. Hausman, eds. *The Creativity Question*. Durham, N. C.: Duke University Press, 1976.

Rucci, A. and R. D. Tweney. Analysis of variance and the "Second Discipline" of scientific psychology: An historical account. *Psychological Bulletin* (1980), 87:166–84.

Ruitenbeek, H. M., ed. *The Creative Imagination: Psychoanalysis and the Genius of Inspiration*. Chicago: Quadrangle Books, 1965.

Sagan, C. *The Dragons of Eden: Speculations on the Evolution of Human Intelligence*. New York: Random House, 1977.

Salmon. W. C. *The Foundations of Scientific Inference*. Pittsburgh: University of Pittsburgh Press, 1967.

deSantillana, G. *The Crime of Galileo*. Chicago: University of Chicago Press, 1955.

Sarton, G. *Sarton on the History of Science, Essays*. D. Stimson, ed. Cambridge: Harvard University Press, 1962.

Schilpp, P. A., ed. *The Philosophy of Karl Popper*. LaSalle, Ill.: Open Court, 1974.

Schlenker, B. R. Social psychology and science. *Journal of Personality and Social Psychology* (1974), 29:1–15.

Schofield, R. E. *Mechanism and Materialism: British Natural Philosophy in an Age of Reason*. Princeton, N. J.: Princeton University Press, 1969.

Schuck, J. R. Factors affecting reports of fragmenting visual images. *Perception and Psychophysics* (1973), 13:382–90.

Sekuler, R. Seeing and the nick in time. In M. H. Siegel and H. P. Zeigler, eds., *Psychological Research: The Inside Story*, pp. 177–97. New York: Harper & Row, 1976.

Shemberg, K. M. and D. B. Leventhal. Outpatient treatment of schizophrenic students in a university clinic. *Bulletin of the Menninger Clinic* (1972), 36:617–40.

Shepard, R. N. Externalization of mental images and the act of creation. In B. S. Randhawa and W. E. Coffman, eds., *Visual Learning, Thinking, and Communication.* New York: Academic Press, 1978.

Shepard, R. N. and P. Arabic. Additive clustering: Representation of similarities as combinations of discrete overlapping properties. *Psychological Review* (1979), 86:87–123.

Shepard, R. N. and G. W. Cermak. Perceptual-cognitive explorations of a toroidal set of free-form stimuli. *Cognitive Psychology* (1973), 4:351–77.

Shepard, R. N. and J. Metzler. Mental rotation of three-dimensional objects. *Science* (1971), 171:701–3.

Shepard, R. N., A. K. Romney, and S. B. Nerlove, eds. *Multidimensional Scaling: Theory and Applications in the Behavioral Sciences. Vol. 1: Theory.* New York: Seminar Press, 1972.

Shipstone, E. I. Some variables affecting pattern conception. *Psychological Monographs* (1960), 74, whole no. 504.

Shweder, R. A. Likeness and likelihood in everyday thought: Magical thinking and everyday judgments about personality. In P. N. Johnson-Laird and P. C. Wason, eds., *Thinking: Readings in Cognitive Science.* Cambridge: Cambridge University Press, 1977.

Sidman, M. *Tactics of Scientific Research: Evaluating Experimental Data in Psychology.* New York: Basic Books, 1960.

Siegler, R. S., ed. *Children's Thinking: What Develops?* Hillsdale, N. J.: Lawrence Erlbaum Associates, 1978.

Simon, H. A. Scientific discovery and the psychology of problem solving. In R. G. Colodny, ed., *Mind and Cosmos: Essays in Contemporary Science and Philosophy.* Pittsburgh: University of Pittsburgh Press, 1966.

—— *The Sciences of the Artificial.* Cambridge, Mass.: MIT Press, 1969.

—— Complexity and the representation of patterned sequences of symbols. *Psychological Review* (1972), 79:369–82.

—— Does scientific discovery have a logic? *Philosophy of Science* (1973), 40:471–80.

Simon, H. A., ed. *Models of Discovery.* Boston: D. Reidel, 1977.

Simon, H. A. and J. R. Hayes. The understanding process: Problem isomorphs. *Cognitive Psychology* (1976), 8:165–90.

Simon, H. A. and K. Kotovsky. Human acquisition of concepts for sequential patterns. *Psychological Review* (1963), 70:534–46.

Simon, H. A. and S. K. Reed. Modeling strategy shifts in a problem-solving task. *Cognitive Psychology* (1976) 8:86–97.

Simon, H. A. and R. K. Sumner. Pattern in music. In B. Kleinmuntz, ed., *Formal Representation of Human Judgment.* New York: Wiley, 1968.

Skinner, B. F. *Science and Human Behavior.* New York: Macmillan, 1953.

—— *Beyond Freedom and Dignity.* New York: Knopf, 1971.

Smedslund, J. The concept of correlation in adults. In C. P. Duncan, ed., *Thinking: Current Experimental Studies*. Philadelphia: Lippincott, 1967.

Smith, M. B. Criticisms of a social science. *Science* (1973), 180:610–12.

Snyder, M. and W. B. Swann, Jr. Behavioral confirmation in social interaction: From social perception to social reality. *Journal of Experimental Social Psychology* (1978), 14:148–62.

Spiegel-Rossing, I. and D. de Solla Price, eds. *Science, Technology, and Society: A Cross-disciplinary Perspective*. London: Sage Publications, 1977.

Spranger, E. *Types of Men: The Psychology and Ethics of Personality*. Translated from the fifth German edition by P. J. W. Pigors. Halle: Max Niemeyer, 1928.

Steinmann, D. O. and M. E. Doherty. A lens model analysis of a bookbag and poker chip experiment. A methodological note. *Organizational Behavior and Human Performance* (1972), 8:450–55.

Sternberg, R. J. Component processes in analogical reasoning. *Psychological Review* (1977), 84:353–78.

Sternberg, S. Memory scanning—mental processes revealed by reaction-time experiments. *American Scientist* (1969), 57:421–57.

Stevens, S. S. Psychology and the science of science. *Psychological Bulletin* (1939), 36:221–63.

Stewart, D. *Philosophical Essays*. 3d ed. Edinburgh: for Archibald Constable, 1818. First published 1810.

Storer, N. W. Introduction. In R. K. Merton, *The Sociology of Science: Theoretical and Empirical Investigations*. Chicago: University of Chicago Press, 1973.

Thackery, A. and R. K. Merton. On discipline-building: The paradoxes of George Sarton. *Isis* (1972), 63:473–95.

Titchener, E. B. Prolegomena to a study of introspection. *American Journal of Psychology* (1912), 23:427–48.

Toulmin, S. E. *Human Understanding. Vol. 1.: The Collective Use and Evolution of Concepts*. Princeton, N. J.: Princeton University Press, 1972.

Troutman, C. M. and J. Shanteau. Inferences based on nondiagnostic information. *Organizational Behavior and Human Performance* (1977), 19:43–55.

Tulving, E. Memory research: What kind of progress? In L. G. Nilsson, ed., *Perpectives on Memory Research: Essays in Honor of Uppsala University's 500th Anniversary*. Hillsdale, N. J.: Lawrence Erlbaum Associates, 1979.

Tune, G. S. Response preferences: A review of some relevant literature. *Psychological Bulletin* (1964), 61:286–302.

Tversky, A. Features of similarity. *Psychological Review* (1977), 84:327–52.

Tversky, A., and D. Kahneman. Belief in the law of small numbers. *Psychological Bulletin* (1971), 76:105–10.

—— Judgment under uncertainty: Heuristics and biases. *Science* (1974), 185:1124–31.

Tweney, R. D. Isaac Newton's two uses of hypothetical reasoning: Dual influences on the history of psychology. *Storia Critica della Psicologia* (1980), 1.

Tweney, R. D., M. E. Doherty, W. Worner, D. Pliske, C. R. Mynatt, K. Gross, and D. Arkkelin. Strategies of rule discovery in an inference task. *Quarterly Journal of Experimental Psychology* (1980), 32:109–123.

Tweney, R. D., G. W. Heiman, and H. W. Hoemann. Psychological processing of sign language: Effect of visual disruption on sign intelligibility. *Journal of Experimental Psychology: General* (1977), 106:255–68.

Tyndall, J. *Address Delivered before the British Association at Belfast. With Additions.* London: Longmans, Green, & Co., 1874.

Vygotsky, L. S *Thought and language.* E. Hanfmann & G. Vakar, eds. and trans. Cambridge: The MIT Press, 1962. Originally published 1934.

Wallsten, T. S. Complex sequential judgments given neutral and disconfirming information. Paper presented at the Annual Meeting of the Psychonomic Society, St. Louis, Missouri, November 1976.

Wason, P. C. On the failure to eliminate hypotheses in a conceptual task. *Quarterly Journal of Experimental Psychology* (1960), 12:129–40.

—— "On the failure to eliminate hypotheses. . ."—A second look. In P. C. Wason and P. Johnson-Laird, eds., *Thinking and Reasoning.* Baltimore: Penguin, 1968. (a)

—— Reasoning about a rule. *Quarterly Journal of Experimental Psychology* (1968), 23:273–81. (b)

—— Problem solving and reasoning. *British Medical Bulletin* (1971), 27:206–10.

Wason, P. C. and J. St. B. T. Evans. Dual processes in reasoning? *Cognition* (1975), 3:141–54.

Wason, P. C. and P. N. Johnson-Laird. *Psychology of Reasoning: Structure and Content.* Cambridge: Harvard University Press, 1972.

Wason, P. C. and D. Shapiro. Natural and contrived experience in a reasoning problem. *Quarterly Journal of Experimental Psychology* (1971), 23:63–71.

Watson, J. D. *The Double Helix.* New York: Atheneum, 1968.

Watson, J. D. and F. H. C. Crick. General implications of the structure of deoxyribonucleic acid. *Nature* (1953), 171:964–67.

Wechsler, J., ed. *On Aesthetics in Science.* Cambridge, Mass. MIT Press, 1978.

Weimer, W. B. The psychology of inference and expectation: Some preliminary remarks. In G. Maxwell and R. M. Anderson, eds., *Minnesota Studies in the Philosophy of Science, vol. 6: Induction, Probability, Confirmation.* Minneapolis: University of Minnesota Press, 1975.

—— *Notes on the Methodology of Scientific Research.* Hillsdale, N. J. Lawrence Erlbaum Associates, 1979.

—— *Psychology and the Conceptual Foundations of Science.* Hillsdale, N. J.: Lawrence Erlbaum Associates, in press.

Weitzenfeld, J. and G. A. Klein. *Analogical Reasoning as a Discovery Logic.* Air Force Office of Scientific Research, Research Report TR-SCR-79-5, 1979 (Available from Klein Associates, 740 Wright Street, Yellow Springs, Ohio 45387).

Werner, H. *Comparative Psychology of mental development.* New York: Harper & Row, 1940. E. B. Garside, trans. Originally published 1926.

Wertheimer, M. Experimentelle studien uber das seken von bewegung. *Zeitschift für Psychologie* (1912), 61:161–265.

—— *Productive Thinking.* New York: Harper, 1959.

West, S. S. The ideology of academic scientists. *IRE Transactions on Engineering Management,* June 1960, pp. 54–62.

Westman, R. S. The Melanchthon circle, Rheticus, and the Wittenberg interpretation of the Copernican theory. *Isis* (1975), 66:165–93.

Whewell, W. *History of the Inductive Sciences.* 3 vols. London: 1837.

—— *Novum organun renovatum.* London, 1858. Pt. II of the 3d edition of *Philosophy of the Inductive Sciences.* (Selection in R. E. Butts, ed., *William Whewell's Theory of Scientific Method.* Pittsburgh: University of Pittsburgh Press, 1968.)

—— *On the Philosophy of Discovery, Chapters Historical and Critical.* London: J. W. Parker & Son, 1860.

Whitfield, J. W. An experiment in problem solving. *Quarterly Journal of Experimental Psychology* (1951), 3:184–97.

Whitley, R., ed. *Social Processes of Scientific Development.* London: Routledge & Kegan Paul, 1974.

Wiggins, J. S. Despair and optimism in Minneapolis. *Contemporary Psychology* (1973), 18:605–6. (a)

—— *Personality and Prediction: Principles of Personality Assessment.* Reading, Mass. Addison-Wesley, 1973. (b)

Wilkes, J. M. Cognitive issues arising from study in the sociology of science. Paper presented at the annual meeting of the American Psychological Association, New York, September 1979.

Wilkins, J. *An Essay towards a Real Character, and a Philosophical Language.* London: for S. Gellibrand and J. Martyn, 1668.

Williams, D. S. Computer program organization induced from problem examples. In H. A. Simon and L. Siklossy, eds., *Representation and Meaning.* Englewood Cliffs, N. J.: Prentice-Hall, 1972.

Williams, L. P. *Michael Faraday: A Biography.* New York: Basic Books, 1965.

Woodward, W. R. Young Piaget revisited: From the grasp of consciousness to *Decalage. Genetic Psychology Monographs* (1979), 79:131–161.

Woodworth, R. S. *Dynamic Psychology*. New York: Columbia University Press, 1918.

—— *Experimental Psychology*. New York: Henry Holt and Company, 1938.

Wundt, W. *Outlines of Psychology*. Leipzig: Wilhelm Engelmann, 1907. C. H. Judd, trans. Third rev. English ed. from the 7th rev. German ed., published in 1905. Originally published in 1896.

—— *Elements of Folk Psychology. Outlines of a Psychological History of the Development of Mankind* London: George Allen & Unwin, 1916. E. L. Schaub trans., from the German edition of 1912.

Yachanin, S. A. An investigation of differential effects of thematic materials and order of rule presentation on a reasoning task. Master's thesis. Bowling Green (Ohio) State University, 1980.

Ziman, J. *Reliable Knowledge*. London: Cambridge University Press, 1978.

Zuckerman, H. *Scientific Elite: Nobel Laureates in the United States*. New York: Free Press, 1977.

Zuckerman, H. and R. K. Merton. Patterns of evolution in science: Institutionalisation, structure and functions of the referee system. *Minerva* (1971), 9:66–100.

# Index

Italicized page numbers in name entries refer to selections.